Lecture Notes in Computer Science   11784

More information about this series at http://www.springer.com/series/7408

Paul Fodor · Marco Montali ·
Diego Calvanese · Dumitru Roman (Eds.)

Rules and Reasoning

Third International Joint Conference, RuleML+RR 2019
Bolzano, Italy, September 16–19, 2019
Proceedings

Springer

Editors
Paul Fodor
Stony Brook University
Stony Brook, NY, USA

Marco Montali
Free University of Bozen-Bolzano
Bolzano, Italy

Diego Calvanese
Free University of Bozen-Bolzano
Bolzano, Italy

Dumitru Roman
SINTEF AS
Oslo, Norway

ISSN 0302-9743   ISSN 1611-3349   (electronic)
Lecture Notes in Computer Science
ISBN 978-3-030-31094-3   ISBN 978-3-030-31095-0   (eBook)
https://doi.org/10.1007/978-3-030-31095-0

LNCS Sublibrary: SL2 – Programming and Software Engineering

This Springer imprint is published by the registered company Springer Nature Switzerland AG
The registered company address is: Gewerbestrasse 11, 6330 Cham, Switzerland

Preface

These are the proceedings of the Third International Joint Conference on Rules and Reasoning (RuleML+RR). RuleML+RR joined the efforts of two well-established conference series: the International Web Rule Symposia (RuleML) and the Web Reasoning and Rule Systems (RR) conferences.

The RuleML symposia and RR conferences have been held since 2002 and 2007, respectively. The RR conferences have been a forum for discussion and dissemination of new results on Web reasoning and rule systems, with an emphasis on rule-based approaches and languages. The RuleML symposia were devoted to disseminating research, applications, languages, and standards for rule technologies, with attention to both theoretical and practical developments, to challenging new ideas, and to industrial applications. Building on the tradition of both, RuleML and RR, the joint conference series RuleML+RR aims at bridging academia and industry in the field of rules, and at fostering the cross-fertilization between the different communities focused on the research, development, and applications of rule-based systems. RuleML+RR aims at being the leading conference series for all subjects concerning theoretical advances, novel technologies, and innovative applications about knowledge representation and reasoning with rules.

To leverage these ambitions, RuleML+RR 2019 was organized as part of the Bolzano Rules and Artificial INtelligence Summit (BRAIN 2019). This summit was hosted by the Free University of Bozen-Bolzano at its main site in Bolzano, Italy. With its special focus theme on "beneficial AI," a core objective of BRAIN 2019 was to present and discuss the latest advancements in AI and rules, and their adoption in IT systems, towards improving key fields such as environment, health, and societies. To this end, BRAIN 2019 brought together a range of events with related interests. In addition to RuleML+RR, this included the 5th Global Conference on Artificial Intelligence (GCAI 2019), the DecisionCAMP 2019, and the Reasoning Web Summer School (RW 2019).

The RuleML+RR conference, moreover, included several subevents:

1. Doctoral Consortium, organized by Mantas Simkus (TU Wien, Austria) and Guido Governatori (Data61, Australia). The doctoral consortium was an initiative to attract and promote student research in rules and reasoning, with the opportunity for students to present and discuss their ideas, and benefit from close contact with leading experts in the field. This year, the DC was organized jointly with GCAI 2019, to favor interaction and exchange of ideas among students interested in rules and reasoning, and students interesting in various facets of artificial intelligence.
2. International Rule Challenge, organized by Sotiris Moschoyiannis (University of Surrey, UK) and Ahmet Soylu (Norwegian University of Science and Technology, SINTEF, Norway). The aim of this initiative was to provide competition among work in progress and new visionary ideas concerning innovative rule-oriented applications, aimed at both research and industry.

3. Posters and Interaction, organized by Petros Stefaneas (National Technical University of Athens, Greece) and Alexander Steen (University of Luxembourg, Luxembourg). The goal of this initiative was to showcase promising, preliminary research results and implemented systems, in the form of a poster or interactive demos.

The technical program of the main track of RuleML+RR 2019 included the presentation of ten full research papers and five short papers. These contributions were carefully selected by the Program Committee among 26 high-quality submissions to the event. Each paper was carefully reviewed and discussed by members of the PC. The technical program was then enriched with the additional contributions from the Doctoral Consortium and the Rule Challenge.

At RuleML+RR 2019 the following invited keynotes and tutorials were presented by experts in the field:

- Keynote by Marie-Laure Mugnier (University of Montpellier, France): "Existential Rules: a Study Through Chase Termination, FO-Rewritability and Boundedness"
- Keynote by Mike Gualtieri (VP and Principal Analyst, Forrester Research, USA): "The Future of Enterprise AI and Digital Decisions"
- Tutorial by Monica Palmirani (University of Bologna, Italy): "LegalRuleML and RAWE"

The keynotes were shared with GCAI 2019, consequently giving the opportunity to the RuleML+RR 2019 audience to also attend the two GCAI keynotes by Giuseppe De Giacomo (Sapienza Università di Roma, Italy) and Marlon Dumas (University of Tartu, Estonia). In addition, a shared session with DecisionCAMP 2019 provided insights on the most recent industrial trends in decision management.

The chairs sincerely thank the keynote and tutorial speakers for their contribution to the success of the event. The chairs also thank the Program Committee members and the additional reviewers for their hard work in the careful assessment of the submitted papers. Further thanks go to all authors of contributed papers, in particular, for their efforts in the preparation of their submissions and the camera-ready versions within the established schedule. Sincere thanks are due to the chairs of the additional tracks and subevents, namely the Doctoral Consortium, the Rule Challenge, and the Poster and Interaction Track, and to the chairs of all co-located BRAIN 2019 events. The chairs finally thank the entire organization team including the Publicity, Proceedings, Sponsorship, Speaker Support, and Social Program Chairs, who actively contributed to the organization and the success of the event.

A special thanks goes to all the sponsors of RuleML+RR 2019 and BRAIN 2019: the Free University of Bozen-Bolzano, the Transregional Collaborative Research Centre 248 "Foundations of Perspicuous Software Systems", Robert Bosch GmbH, the Artificial Intelligence Journal, oXygen, Hotel Greif, Ontopic S.r.l., and Springer. A special thanks also goes to the publisher, Springer, for their cooperation in editing this volume and publishing of the proceedings.

July 2019

Paul Fodor
Marco Montali
Diego Calvanese
Dumitru Roman

Organization

Summit Chair (BRAIN 2019)

Diego Calvanese — Free University of Bozen-Bolzano, Italy

General Chair (RuleML+RR 2019)

Diego Calvanese — Free University of Bozen-Bolzano, Italy

Program Chairs

Paul Fodor — Stony Brook University, USA
Marco Montali — Free University of Bozen-Bolzano, Italy

Proceedings Chairs

Dumitru Roman — SINTEF AS, University of Oslo, Norway
Adrian Giurca — Brandenburg University of Technology, Germany

Posters and Interactions Chairs

Petros Stefaneas — National Technical University of Athens, Greece
Alexander Steen — University of Luxembourg, Luxembourg

Doctoral Consortium (Joint with GCAI 2019)

Guido Governatori — NICTA, Australia
Mantas Simkus — TU Wien, Austria

Challenge Chairs

Sotiris Moschoyiannis — University of Surrey, UK
Ahmet Soylu Norwegian — University of Science and Technology, Norway

Publicity Chair

Nick Bassiliades Aristotle — University of Thessaloniki, Greece

Web Chair

Paolo Felli Free — University of Bozen-Bolzano, Italy

Program Committee

Leopoldo Bertossi	Relational AI Inc., Carleton University, Canada
Mehul Bhatt	Örebro University, Sweden
Meghyn Bienvenu	French National Center for Scientific Research, France
Andrea Billig	Fraunhofer FOKUS, Germany
Pedro Cabalar Fernández	University of La Corunna, Spain
Iliano Cervesato	Carnegie Mellon University, Qatar
Horatiu Cirstea	Inria, France
Stefania Costantini	University of Aquila, Italy
Giovanni De Gasperis	Università degli Studi dell'Aquila, Italy
Marc Denecker	KU Leuven, Belgium
Juergen Dix Clausthal	University of Technology, Germany
Wolfgang Faber	Alpen-Adria-Universität Klagenfurt, Austria
Thom Fruehwirth	University of Ulm, Germany
Tiantian Gao	Stony Brook University, USA
Giancarlo Guizzardi	University of Bozen-Bolzano, Italy
Tomás Kliegr	University of Economics in Prague, Czech Republic
Matthias Klusch	German Research Center for Artificial Intelligence, Germany
Michael Kohlhase	Erlangen FAU Jacobs University Bremen, Germany
Roman Kontchakov	University of London, UK
Manolis Koubarakis	University of Athens, Greece
Markus Krötzsch	TU Dresden, Germany
Domenico Lembo	Sapienza University of Rome, Italy
Maurizio Lenzerini	Sapienza University of Rome, Italy
Francesca Lisi	University of Bari, Italy
Thomas Lukasiewicz	University of Oxford, UK
Marco Manna	University of Calabria, Italy
Marco Maratea	University of Genoa, Italy
Angelo Montanari	University of Udine, Italy
Grzegorz J. Nalepa	AGH University of Science and Technology, Poland
Magdalena Ortiz	TU Vienna, Austria
Jeff Pan	University of Aberdeen, UK
Monica Palmirani	University of Bologna, Italy
Adrian Paschke	Free University of Berlin, Germany
Rafael Penaloza	Free University of Bozen-Bolzano, Italy
Andreas Pieris	University of Edinburgh, UK
Luca Pulina	University of Sassari, Italy
Jan Rauch	University of Economics in Prague, Czech Republic
Sebastian Rudolph	TU Dresden, Germany
Emanuel Sallinger	University of Oxford, UK
Konstantin Schekotihin	Alpen-Adria Universität, Austria
Stefan Schlobach	Free University of Amsterdam, The Netherlands
Umberto Straccia	ISTI-CNR, Italy
Rolf Schwitter	Macquarie University, Australia

Theresa Swift	Coherent Knowledge, USA
Giorgos Stamou	National Technical University of Athens, Greece
Giorgos Stoilos	Athens University of Economics and Business, Greece
Sergio Tessaris	Free University of Bolzen-Bolzano, Italy
Kia Teymourian	Boston University, USA
Nikos Triantafyllou	National Technical University of Athens, Greece
Anni-Yasmin Turhan	Dresden University of Technology, Germany
Riccardo Zese	University of Ferrara, Italy

RuleML+RR 2019 Sponsors

Free University of
Bozen-Bolzano

Center for
Perspicuous
Computing

Bosch

Artificial Intelligence
Journal

Oxygen XML Editor

Hotel Greif

Ontopic - A unibz
spinoff

Keynote Talks

The Future of Enterprise AI
and Digital Decisions

Mike Gualtieri

Forrester Research, 60 Acorn Park Drive, Cambridge, MA 02140
mgualtieri@forrester.com

Abstract. Machines are amazing learners. Humans are amazing learners. AI is best when powered by both. AI solutions that employ both machine learning and knowledge-engineered rules learn continuously from data whilst at the same time are informed by wisdom and commonsense expressed in rules. Forrester Research Vice President and Principal Analyst, Mike Gualtieri, will convey the key trends in enterprise AI and rules/decision management; and discuss how leading enterprises can use them in combination to build truly learning AI-infused applications at scale.

Existential Rules: A Study Through Chase Termination, FO-Rewritability and Boundedness

Marie-Laure Mugnier[1,2]

[1] University of Montpellier and LIRMM
mugnier@lirmm.fr
[2] Inria, Montpellier, France

Abstract. Existential rules, also known as Datalog+, are an expressive knowledge representation and reasoning language, which has been mainly investigated in the context of ontological query answering. This talk will first review the landscape of decidable classes of existential rules with respect to two fundamental problems, namely chase termination (does a given set of rules ensure that the chase terminates for any factbase?) and FO-rewritability (does a given set of rules ensure that any conjunctive query can be rewritten as a first-order query?). Regarding the chase, we will specifically focus on four well-known variants: the oblivious chase, the semi-oblivious (or skolem) chase, the restricted chase, and the core chase. We will then study the relationships between chase termination and FO-rewritability, which have been little investigated so far. This study leads us to another fundamental problem, namely boundedness (does a given set of rules ensure that the chase terminates for any factbase within a predefined depth?). The boundedness problem was deeply investigated in the context of datalog. It is known that boundedness and FO-rewritability are equivalent properties for datalog rules. Such an equivalence does not hold for general existential rules. We will provide a characterization of boundedness in terms of chase termination and FO-rewritability for the oblivious and semi-oblivious chase variants. Interesting questions remain open. This talk will rely on results from the literature and joint work published at ICDT 2019 and IJCAI 2019.

Contents

Technical Communication Papers

Full Papers

Finding New Diamonds: Temporal Minimal-World Query Answering over Sparse ABoxes

Stefan Borgwardt, Walter Forkel, and Alisa Kovtunova[✉]

Chair for Automata Theory, Technische Universität Dresden, Dresden, Germany
{stefan.borgwardt,walter.forkel,alisa.kovtunova}@tu-dresden.de

Abstract. Lightweight temporal ontology languages have become a very active field of research in recent years. Many real-world applications, like processing electronic health records (EHRs), inherently contain a temporal dimension, and require efficient reasoning algorithms. Moreover, since medical data is not recorded on a regular basis, reasoners must deal with sparse data with potentially large temporal gaps. In this paper, we introduce a temporal extension of the tractable language $\mathcal{ELH}_\bot$, which features a new class of *convex diamond* operators that can be used to bridge temporal gaps. We develop a completion algorithm for our logic, which shows that entailment remains tractable. Based on this, we develop a *minimal-world* semantics for answering metric temporal conjunctive queries with negation. We show that query answering is combined first-order rewritable, and hence in polynomial time in data complexity.

1 Introduction

Temporal description logics (DLs) combine terminological and temporal knowledge representation capabilities and have been investigated in detail in the last decades [3,28,32]. To obtain tractable reasoning procedures, lightweight temporal DLs have been developed [4,20]. The idea is to use temporal operators, often from the linear temporal logic LTL, inside DL axioms. For example, $\Diamond \exists$diagnosis.BrokenLeg $\sqsubseteq$ $\exists$treatment.LegCast states that after breaking a leg one has to wear a cast. However, this basic approach cannot represent the distance of events, e.g., that the cast only has to be worn for a fixed amount of time. Recently, metric temporal ontology languages have been investigated [7,14,21], which allow to replace $\Diamond$ in the above axiom with $\Diamond_{[-8,0]}$, i.e., wearing the cast is required only if the leg was broken ≤ 8 time points (e.g., weeks) ago.

Such knowledge representation capabilities are important for biomedical applications. For example, many clinical trials contain temporal eligibility criteria [16] such as: "type 1 diabetes with duration at least 12 months"[1]; "known

---

[1] https://clinicaltrials.gov/ct2/show/NCT02280564.

This work was partially supported by DFG grant 389792660 as part of TRR 248 and the DFG project BA 1122/19-1 (GOASQ).

© Springer Nature Switzerland AG 2019
P. Fodor et al. (Eds.): RuleML+RR 2019, LNCS 11784, pp. 3–18, 2019.
https://doi.org/10.1007/978-3-030-31095-0_1

history of heart disease or heart rhythm abnormalities"[2]; "CD4+ lymphocytes count > 250/mm3, for at least 6 months"[3]; or "symptomatic recurrent paroxysmal atrial fibrillation (PAF) (> 2 episodes in the last 6 months)"[4]. Moreover, measurements, diagnoses, and treatments in a patients' EHR are clearly valid only for a certain amount of time. To automatically screen patients according to the temporal criteria above, one needs a sufficiently powerful formalism that can reason about biomedical and temporal knowledge. This is an active area of current research [11,16,22]. For the atemporal part, one can use existing large biomedical ontologies that are based on lightweight (atemporal) DLs, e.g., SNOMED CT[5], which is formulated using the DL $\mathcal{ELH}$.

Since EHRs only contain information for specific points in time, it is especially important to be able to infer what happened to the patient in the meantime. For example, if a patient is diagnosed with a (currently) incurable disease like Diabetes, they will still have the disease at any future point in time. Similarly, if the EHR contains two entries of CD4Above250 four weeks apart, one may reasonably infer that this was true for the whole four weeks. Qualitative temporal DLs such as $\mathcal{TEL}_{\mathsf{infl}}^{\Diamond}$ [20] can express the former statement by declaring Diabetes as *expanding* via the axiom $\Diamond$Diabetes ⊑ Diabetes. We propose to extend this logic by adding a special kind of metric temporal operators to write $\Diamond_4$CD4Above250 ⊑ CD4Above250, making the measurement *convex* for a specified length of time n (e.g., 4 weeks). This means that information is interpolated between time points of distance less than n, thereby computing a convex closure of the available information. The threshold n allows us to distinguish the case where two mentions of CD4Above250 are years apart, and are therefore unrelated.

The distinguishing feature of $\mathcal{TEL}_{\mathsf{infl}}^{\Diamond}$ is that $\Diamond$-operators are only allowed on the left-hand side of concept inclusions [20], which is also common for temporal DLs based on *DL-Lite* [2,5]. Apart from adding convex metric temporal operators to this logic, we allow temporal roles like $\Diamond_2$hasTreatment ⊑ hasTreatment, and deal with the problem of having large gaps in the data, e.g., in patient records. We show that reasoning in the extended logic $\mathcal{TELH}_{\bot}^{\Diamond,\mathsf{lhs}}$ remains tractable.

Additionally, we consider the problem of answering temporal queries over $\mathcal{TELH}_{\bot}^{\Diamond,\mathsf{lhs}}$ knowledge bases. As argued in [6,12], evaluating clinical trial criteria over patient records requires both negated and temporal queries, but standard certain answer semantics is not suitable to deal with negation over patient records, which is why we adopt the *minimal-world* semantics from [12] for our purposes. Our query language extends the temporal conjunctive queries from [8] by metric temporal operators [7,21] and negation. For example, we can use queries like $\Box_{[-12,0]}(\exists y.\text{diagnosedWith}(x,y) \wedge \text{Diabetes}(y))$ to detect whether the first criterion from above is satisfied.

---

[2] https://clinicaltrials.gov/ct2/show/NCT02873052.
[3] https://clinicaltrials.gov/ct2/show/NCT02157311.
[4] https://clinicaltrials.gov/ct2/show/NCT00969735.
[5] https://www.snomed.org/.

Using a combined rewriting approach, we show that the data complexity of query answering is not higher than for positive atemporal queries in $\mathcal{ELH}_\perp$, and also provide a tight combined complexity result of EXPSPACE. Unlike most research on temporal query answering [2,8], we do not assume that input data is given for all time points in a certain interval, but rather at sporadic time points with arbitrarily large gaps. The main technical difficulty is to determine which additional time points are relevant for answering a query, and how to access these time points without having to fill all the gaps.

Full proofs can be found in the extended version at https://tu-dresden.de/inf/lat/papers.

2   The Lightweight Temporal Logic $\mathcal{TELH}_\perp^{\lozenge,\mathsf{lhs}}$

We first introduce the metric LTL operators that we will use and analyze their properties. LTL formulas are formulated over a finite set P of *propositional variables*. In this section, we consider only formulas built according to the syntax rule $\varphi ::= p \mid \varphi \wedge \varphi \mid \varphi \vee \varphi \mid \lozenge_I \varphi$, where $p \in P$ and I is an interval in $\mathbb{Z}$. The semantics is given by *LTL-structures* $\mathfrak{W} = (w_i)_{i \in \mathbb{Z}}$, where $w_i \subseteq P$. We write

$$\mathfrak{W}, i \vDash p \text{ iff } p \in w_i \text{ if } p \in P, \qquad \mathfrak{W}, i \vDash \varphi \wedge \psi \text{ iff } \mathfrak{W}, i \vDash \varphi \text{ and } \mathfrak{W}, i \vDash \psi,$$

$$\mathfrak{W}, i \vDash \lozenge_I \varphi \text{ iff } \exists k \in I: \mathfrak{W}, i+k \vDash \varphi, \qquad \mathfrak{W}, i \vDash \varphi \vee \psi \text{ iff } \mathfrak{W}, i \vDash \varphi \text{ or } \mathfrak{W}, i \vDash \psi.$$

More specifically, we only consider the following derived operators, where $n \geq 1$:

$$\begin{aligned}
\lozenge\varphi &:= \lozenge_{(-\infty,\infty)}\varphi & \overrightarrow{\lozenge}\varphi &:= \lozenge_{[0,\infty)}\varphi & \overleftarrow{\lozenge}\varphi &:= \lozenge_{(-\infty,0]}\varphi \\
\diamondsuit\varphi &:= \lozenge_{(-\infty,0]}\varphi \wedge \lozenge_{[0,\infty)}\varphi & \diamondsuit_n\varphi &:= \bigvee_{\substack{k,m\geq 0 \\ k+m=n-1}} (\lozenge_{[-k,0]}\varphi \wedge \lozenge_{[0,m]}\varphi) & & \quad(1)
\end{aligned}$$

The operator $\lozenge$ is the "eventually" operator of classical LTL, and $\overrightarrow{\lozenge}, \overleftarrow{\lozenge}$ are two variants that refer to the past, or to both past and future, respectively. The operator $\diamondsuit$ requires that φ holds *both* in the past and in the future, thereby distinguishing time points that lie within an interval enclosed by time points at which φ holds. This can be used to express the convex closure of time points, as described in the introduction. Finally, the operators $\diamondsuit_n$ represent a metric variant of $\diamondsuit$, requiring that different occurrences of φ are at most $n-1$ time points apart, i.e., enclose an interval of length n. To study the behavior of these operators, we consider their semantics in a more abstract way: given a set of time points where a certain information is available (e.g., a diagnosis), described by a propositional variable p, we consider the resulting set of time points at which $\lozenge p$ holds, where $\lozenge$ is a placeholder for one of the operators defined above (we will similarly use $\diamondsuit, \overrightarrow{\lozenge}, \overleftarrow{\lozenge}$ as placeholders for different $\lozenge$-operators in the following).

Definition 1. *We consider the sets* $\mathfrak{D}^c := \{\diamondsuit\} \cup \{\diamondsuit_i \mid i \geq 1\}$, $\mathfrak{D}^\pm = \{\lozenge, \overrightarrow{\lozenge}, \overleftarrow{\lozenge}\}$, *and* $\mathfrak{D} := \mathfrak{D}^\pm \cup \mathfrak{D}^c$ *of diamond operators. Each* $\lozenge \in \mathfrak{D}$ *induces a function* $\lozenge: 2^\mathbb{Z} \to 2^\mathbb{Z}$ *with* $\lozenge(M) := \{i \mid \mathfrak{W}_M, i \vDash \lozenge p\}$ *for all* $M \subseteq \mathbb{Z}$, *with the LTL-structure* $\mathfrak{W}_M := (w_i)_{i \in \mathbb{Z}}$ *such that* $w_i := \{p\}$ *if* $i \in M$, *and* $w_i := \varnothing$ *otherwise.*

We will omit the parentheses in $\Diamond(M)$ for a cleaner presentation. If M is empty, then $\Diamond M$ is also empty, for any $\Diamond \in \mathfrak{D}$. For any non-empty $M \subseteq \mathbb{Z}$, we obtain the following expressions, where $\max M$ may be ∞ and $\min M$ may be $-\infty$.

$$\Diamond M = \mathbb{Z} \quad \Diamond M = (-\infty, \max M] \quad \Diamond M = [\min M, \infty) \quad \Diamond M = [\min M, \max M]$$
$$\Diamond_1 M = M \quad \Diamond_n M = \{i \in \mathbb{Z} \mid \exists j, k \in M \text{ with } j \le i \le k \text{ and } k - j < n\}$$

In this representation, the convex closure operation behind $\Diamond$ becomes apparent. We now list several useful properties of these functions.

Lemma 2. *Using the pointwise inclusion order $\subseteq$ on the induced functions, we obtain the following ordered set $(\mathfrak{D}, \subseteq)$, where $\mathrm{id}_{2^{\mathbb{Z}}}$ is the identity function on $2^{\mathbb{Z}}$:*

$$\mathrm{id}_{2^{\mathbb{Z}}} = \Diamond_1 \subseteq \cdots \subseteq \Diamond_n \subseteq \Diamond_{n+1} \subseteq \cdots \subseteq \Diamond \begin{array}{c} \Diamond \\ \subseteq \quad \subseteq \\ \subseteq \quad \subseteq \\ \Diamond \end{array} \Diamond$$

The most important property is the following, which allows us to combine diamond operators without leaving the set $\mathfrak{D}$.

Lemma 3. *The set $\mathfrak{D}$ is closed under composition $\circ$, pointwise intersection $\cap$, and pointwise union $\cup$, and for any $\Diamond, \Diamond \in \mathfrak{D}$ these operators can be computed as:*

$$\Diamond \cap \Diamond = \inf_{(\mathfrak{D}, \subseteq)}\{\Diamond, \Diamond\} \quad \text{and} \quad \Diamond \circ \Diamond = \Diamond \cup \Diamond = \sup_{(\mathfrak{D}, \subseteq)}\{\Diamond, \Diamond\},$$

where $\inf_{(\mathfrak{D}, \subseteq)}$ denotes the infimum in $(\mathfrak{D}, \subseteq)$, and $\sup_{(\mathfrak{D}, \subseteq)}$ the supremum.

2.1 A New Temporal Description Logic

We define a new temporal description logic based on the operators in $\mathfrak{D}$. The main differences to $\mathcal{TEL}_{\mathrm{infl}}^{\Diamond}$ from [20] are that $\Diamond_n$-operators may occur in concept and role inclusions, and ABoxes may have gaps, which require special consideration during reasoning.

Syntax. Let $\mathbf{C}, \mathbf{R}, \mathbf{I}$ be disjoint sets of *concept, role,* and *individual names,* respectively. A *temporal role* is of the form $\Diamond r$ with $\Diamond \in \mathfrak{D}$ and $r \in \mathbf{R}$. A $\mathcal{TELH}_{\bot}^{\Diamond,\mathrm{lhs}}$ *concept* is built using the rule $C ::= A \mid \top \mid \bot \mid C \sqcap C \mid \exists r.C \mid \Diamond C$, where $A \in \mathbf{C}$, $\Diamond \in \mathfrak{D}$, and r is a temporal role. Such a C is an $\mathcal{ELH}_{\bot}$ concept (or *atemporal concept*) if it does not contain any diamond operators.

A $\mathcal{TELH}_{\bot}^{\Diamond,\mathrm{lhs}}$ *TBox* is a finite set of *concept inclusions (CIs)* $C \sqsubseteq D$ and *role inclusions (RIs)* $r \sqsubseteq s$, where C is a $\mathcal{TELH}_{\bot}^{\Diamond,\mathrm{lhs}}$ concept, D is an atemporal concept, r is a temporal role, and $s \in \mathbf{R}$. We write $C \equiv D$ to abbreviate the two inclusions $C \sqsubseteq D$, $D \sqsubseteq C$, and similarly for role inclusions. An *ABox* is a finite set of *concept assertions* $A(a, i)$ and *role assertions* $r(a, b, i)$, where $A \in \mathbf{C}$, $r \in \mathbf{R}$, $a, b \in \mathbf{I}$, and $i \in \mathbb{Z}$. We denote the set of time points $i \in \mathbb{Z}$ occurring in $\mathcal{A}$ by $\mathrm{tem}(\mathcal{A})$. Additionally, we assume that each time point is encoded in binary with at most n digits. A *knowledge base (KB)* (or *ontology*) $\mathcal{K} = \mathcal{T} \cup \mathcal{A}$ consists of a TBox $\mathcal{T}$ and an ABox $\mathcal{A}$. In the following, we always assume a KB $\mathcal{K} = \mathcal{T} \cup \mathcal{A}$ to be given.

Semantics. An *interpretation* $\mathcal{I} = (\Delta^{\mathcal{I}}, \cdot^{\mathcal{I}})$ has a *domain* $\Delta^{\mathcal{I}} \supseteq \mathbf{I}$ and assigns to each $A \in \mathbf{C}$ a set $A^{\mathcal{I}} \subseteq \Delta^{\mathcal{I}}$ and to each $r \in \mathbf{R}$ a binary relation $r^{\mathcal{I}} \subseteq \Delta^{\mathcal{I}} \times \Delta^{\mathcal{I}}$. A *temporal interpretation* $\mathfrak{J} = (\Delta^{\mathfrak{J}}, (\mathcal{I}_i)_{i \in \mathbb{Z}})$, is a collection of interpretations $\mathcal{I}_i = (\Delta^{\mathfrak{J}}, \cdot^{\mathcal{I}_i})$, $i \in \mathbb{Z}$, over $\Delta^{\mathfrak{J}}$. The functions $\cdot^{\mathcal{I}_i}$ are extended as follows.

$$(\Diamondblack r)^{\mathcal{I}_i} := \{(d, e) \in \Delta^{\mathfrak{J}} \times \Delta^{\mathfrak{J}} \mid i \in \Diamondblack\{j \mid (d, e) \in r^{\mathcal{I}_j}\}\} \qquad \top^{\mathcal{I}_i} := \Delta^{\mathfrak{J}} \qquad \bot^{\mathcal{I}_i} := \varnothing$$

$$(C \sqcap D)^{\mathcal{I}_i} := C^{\mathcal{I}_i} \cap D^{\mathcal{I}_i} \qquad\qquad (\exists r.C)^{\mathcal{I}_i} := \{d \in \Delta^{\mathfrak{J}} \mid \exists e \in C^{\mathcal{I}_i} : (d, e) \in r^{\mathcal{I}_i}\}$$

$$(\Diamondblack C)^{\mathcal{I}_i} := \{d \in \Delta^{\mathfrak{J}} \mid i \in \Diamondblack\{j \mid d \in C^{\mathcal{I}_j}\}\}$$

$\mathfrak{J}$ is a *model* of (or *satisfies*) a concept inclusion $C \sqsubseteq D$ if $C^{\mathcal{I}_i} \subseteq D^{\mathcal{I}_i}$ holds for all $i \in \mathbb{Z}$, a role inclusion $r \sqsubseteq s$ if $r^{\mathcal{I}_i} \subseteq s^{\mathcal{I}_i}$ holds for all $i \in \mathbb{Z}$, a concept assertion $A(a, i)$ if $a \in A^{\mathcal{I}_i}$, a role assertion $r(a, b, i)$ if $(a, b) \in r^{\mathcal{I}_i}$, and the KB $\mathcal{K}$ if it satisfies all axioms in $\mathcal{K}$. This fact is denoted by $\mathfrak{J} \vDash \alpha$, where α is an *axiom* (i.e., inclusion or assertion) or a KB. An ontology $\mathcal{K}$ is *consistent* if it has a model, and it *entails* α, written $\mathcal{K} \vDash \alpha$, if all models of $\mathcal{K}$ satisfy α. $\mathcal{K}$ is inconsistent iff $\mathcal{K} \vDash \top \sqsubseteq \bot$, and thus we focus on deciding entailment. In $\mathcal{ELH}_\bot$, this is possible in polynomial time [9].

We do not allow diamonds to occur on the right-hand side of CIs, because that would make the logic undecidable [4]. As usual, we can simulate CIs involving complex concepts by introducing fresh concept and role names as abbreviations. For example, $\exists \Diamondblack r.\Diamond A \sqsubseteq B$ can be split into $\Diamondblack r \sqsubseteq r'$, $\Diamond A \sqsubseteq A'$, and $\exists r'.A' \sqsubseteq B$. Hence, we can restrict ourselves w.l.o.g. to CIs in the following *normal form*:

$$\Diamond A \sqsubseteq B, \quad A_1 \sqcap A_2 \sqsubseteq B, \quad \Diamondblack r \sqsubseteq s, \quad \Diamond A \sqsubseteq \exists r.B, \quad \exists r.A \sqsubseteq B, \qquad (2)$$

where $\Diamond \in \mathfrak{D}$, $A, A_1, A_2, B \in \mathbf{C} \cup \{\bot, \top\}$, and $r, s \in \mathbf{R}$.

Convex Names. When considering axioms of the form $\Diamond A \sqsubseteq A$ for $A \in \mathbf{C}$, we can first observe that the converse direction $A \sqsubseteq \Diamond A$, although syntactically not allowed, trivially holds in all interpretations. Moreover, the following implications between such equivalences follow from Lemma 2:

Since $\{A \equiv \Diamonddown A, A \equiv \Diamondup A\}$ entails $A \equiv \Diamond A$, it thus makes sense to consider the *unique strongest* such axiom that is entailed by $\mathcal{K}$ (for a given A). We call A *rigid* if $A \equiv \Diamond A$ is the strongest such axiom, *shrinking* in case of $A \equiv \Diamonddown A$, *expanding* for $A \equiv \Diamondup A$, and *(n-)convex* for $A \equiv \Diamond_{(n)} A$, i.e., whenever A is satisfied at two time points (with distance $< n$), then it is also satisfied at all time points in between.

1-convex concept names do not satisfy any special property, and are also called *flexible*. We use the same terms for role names.

2.2   A Completion Algorithm

We use the completion rules in Fig. 1 to derive new axioms from $\mathcal{K}$. For simplicity, we treat $\top$ and $\bot$ like concept names, and thus allow assertions of the form $\top(a,i)$ and $\bot(a,i)$ here. It is clear that we cannot derive all possible entailments of the forms $\Diamond A \sqsubseteq B$ or $A(a,i)$, because (1) $\mathfrak{D}$ is infinite, and (2) $\mathbb{Z}$ is infinite. Moreover, there may be arbitrarily many time points between two assertions in $\mathcal{A}$ (exponentially many in the size of $\mathcal{A}$ if we assume time points to be encoded in binary). To deal with (1), we restrict the rule applications to the operators that occur in $\mathcal{K}$, in addition to $\Diamond$ and $\Diamondblack$, which are the only elements of $\mathfrak{D}$ that can be obtained via $\cup$, $\cap$, or $\circ$ from other $\Diamond$-operators, namely from $\Diamond$ and $\Diamondblack$. For (2), we consider the set of time points $\mathrm{tem}(\mathcal{A})$ (of linear size). Additionally, consider a maximal interval $[i,j]$ in $\mathbb{Z} \setminus \mathrm{tem}(\mathcal{A})$ (where i may be $-\infty$ and j may be ∞). To represent this interval, we choose a single representative time point $k \in [i,j]$, which is denoted by $|\ell| := k$ for all $\ell \in [i,j]$. For consistency, the representative $|i|$ for any $i \in \mathrm{tem}(\mathcal{A})$ is defined as i itself. Moreover, for any $k \in \mathbb{Z}$, we denote by $\lfloor k \rfloor := \max\{i \in \mathrm{tem}(\mathcal{A}) \mid i \le k\}$ the maximal element of $\mathrm{tem}(\mathcal{A})$ below (or equal to) k, which we consider to be $-\infty$ in the case that there is no such element, and similarly define $\lceil k \rceil$. Note that $\lfloor i \rfloor = i = \lceil i \rceil$ whenever $i \in \mathrm{tem}(\mathcal{A})$, and otherwise

$$\text{T1} \; \frac{}{\Diamond_1 A \sqsubseteq A} \qquad \text{T2} \; \frac{}{\Diamond A \sqsubseteq \top} \qquad \text{T3} \; \frac{}{\Diamond_1 r \sqsubseteq r} \qquad \text{T4} \; \frac{\Diamond A_1 \sqsubseteq A_2 \quad \Diamond A_2 \sqsubseteq A_3}{(\Diamond \circ \Diamond) A_1 \sqsubseteq A_3}$$

$$\text{T5} \; \frac{\Diamond r_1 \sqsubseteq r_2 \quad \Diamond r_2 \sqsubseteq r_3}{(\Diamond \circ \Diamond) r_1 \sqsubseteq r_3} \qquad \text{T6} \; \frac{\Diamond A \sqsubseteq A_1 \quad \Diamond A \sqsubseteq A_2 \quad A_1 \sqcap A_2 \sqsubseteq B}{(\Diamond \sqcap \Diamond) A \sqsubseteq B}$$

$$\text{T7} \; \frac{}{\exists r.\bot \sqsubseteq \bot} \qquad \text{T8} \; \frac{\Diamond A \sqsubseteq \exists r.A_1 \quad \Diamond r \sqsubseteq s \quad \Diamond A_1 \sqsubseteq B_1 \quad \exists s.B_1 \sqsubseteq B}{\Diamond A \sqsubseteq B}$$

$$\text{T8}' \; \frac{\Diamond A \sqsubseteq \exists r.A_1 \quad \Diamond r \sqsubseteq s \quad \Diamond A_1 \sqsubseteq B_1 \quad \exists s.B_1 \sqsubseteq B \quad (\Diamond \sqcap \Diamond) \in \mathfrak{D}^\pm}{((\Diamond \sqcap \Diamond) \circ \Diamond) A \sqsubseteq B}$$

$$\text{A1} \; \frac{}{\top(a,i)} \qquad \text{A2} \; \frac{i \in \Diamond A(a) \quad \Diamond A \sqsubseteq B}{B(a,i)} \qquad \text{A3} \; \frac{i \in \Diamond r(a,b) \quad \Diamond r \sqsubseteq s}{s(a,b,i)}$$

$$\text{A4} \; \frac{A_1(a,i) \quad A_2(a,i) \quad A_1 \sqcap A_2 \sqsubseteq B}{B(a,i)} \qquad \text{A5} \; \frac{r(a,b,i) \quad A(b,i) \quad \exists r.A \sqsubseteq B}{B(a,i)}$$

Fig. 1. Completion rules for $\mathcal{TELH}_\bot^{\Diamond,\mathrm{lhs}}$ knowledge bases

$\lfloor i \rfloor < i < \lceil i \rceil$. By restricting all assertions to the finite set of representative time points

$$\mathrm{rep}(\mathcal{A}) := \{\lfloor i \rfloor \mid i \in \mathbb{Z}\} \supset \mathrm{tem}(\mathcal{A}),$$

we can encode infinitely many entailments in a finite set. We also define the following abbreviations, for all $A \in \mathbf{C}$, $r \in \mathbf{R}$, and $a, b \in \mathbf{I}$ ($\mathcal{K}$ refers to the KB after possibly already applying some completion rules):

$$A(a) := \{i \in \mathrm{rep}(\mathcal{A}) \mid A(a, i) \in \mathcal{K}\}$$
$$r(a, b) := \{i \in \mathrm{rep}(\mathcal{A}) \mid r(a, b, i) \in \mathcal{K}\}$$

Hence, we can write $\lozenge A(a)$ in the completion rules to refer to the set of time points at which $\lozenge A$ is inferred to be satisfied by a, given only the assertions in $\mathcal{A}$.

In the rules of Fig. 1, we allow to instantiate A, B, A_1, A_2, A_3, B_1 by $\top, \bot$ or (normalized) $\mathcal{ELH}_\bot$ concepts from $\mathcal{K}$, r, s, r_1, r_2, r_3 by role names from $\mathcal{K}$, $\lozenge, \lozenge, \lozenge$ by $\lozenge, \lozenge$ or elements of $\mathfrak{D}$ occurring in $\mathcal{K}$, a, b by individual names from $\mathcal{K}$, and i by values from $\mathrm{rep}(\mathcal{A})$, such that the resulting axioms are in normal form. The side conditions $(\lozenge \cap \lozenge) \in \mathfrak{D}^\pm$, $i \in \lozenge A(a)$, $i \in \lozenge r(a, b)$ can be checked in polynomial time. All rules also apply to axioms without diamonds since we can treat A as $\lozenge_1 A$.

If $\mathcal{K}$ contains all axioms in the precondition of an instantiated rule, we consider the axiom in its conclusion. If it is a new assertion, we add it to $\mathcal{K}$. If it is a concept inclusion $\lozenge A \sqsubseteq B$, we check whether $\mathcal{K}$ already contains a CI of the form $\lozenge A \sqsubseteq B$. If not, then we simply add $\lozenge A \sqsubseteq B$ to $\mathcal{K}$; otherwise, and if $\lozenge \cup \lozenge \neq \lozenge$, we replace $\lozenge A \sqsubseteq B$ by the new axiom $(\lozenge \cup \lozenge)A \sqsubseteq B$, in order to reflect the validity of both axioms at once. RIs are handled in the same way. For example, if we know that $\lozenge A \sqsubseteq B$ holds, and have just inferred that $\lozenge A \sqsubseteq B$ holds as well, then $\lozenge A \sqsubseteq B$ is a valid entailment, because $\lozenge \sqsubseteq \lozenge \cup \lozenge$, and thus whenever an element satisfies $\lozenge A$, it must satisfy either $\lozenge A$ or $\lozenge A$. In this way, for any two concepts A, B, the KB always contains at most one axiom $\lozenge A \sqsubseteq B$, and similarly for roles.

Let $\mathcal{K}^*$ be the KB obtained by exhaustive application of the completion rules in Fig. 1 to $\mathcal{K}$, where we assume (for technical reasons explained in the extended version) that A2 and A3 are always applied at the same time for all $i \in \lozenge A(a)$ and $i \in \lozenge r(a, b)$, respectively. This process terminates since we only produce axioms of the form $\lozenge A \sqsubseteq B$, $\lozenge r \sqsubseteq s$, $A(a, i)$, or $r(a, b, i)$, where $\lozenge$ was already present in the initial $\mathcal{K}$ or it belongs to $\{\lozenge_1, \lozenge, \lozenge\}$, $i \in \mathrm{rep}(\mathcal{A})$, and A, B, r, s, a, b are from $\mathcal{K}$; there are only polynomially many such axioms.

To decide whether a concept assertion $D(a, i)$ follows from $\mathcal{K}$, we then simply look up whether $D(a, \lfloor i \rfloor)$ belongs to $\mathcal{K}^*$. For a concept inclusion $\lozenge C \sqsubseteq D$ with $C, D \in \mathbf{C}$, we check whether $\mathcal{K}^*$ contains an inclusion of the form $\lozenge C \sqsubseteq D$ with $\lozenge \sqsubseteq \lozenge$, which can be done in polynomial time (see Lemma 2). One can also check entailment of role axioms in a similar way, but we omit them here for brevity.

Lemma 4. $\mathcal{K}$ *is inconsistent iff* $\bot(a, i) \in \mathcal{K}^*$ *for some* $a \in \mathbf{I}$ *and* $i \in \mathrm{rep}(\mathcal{A})$.

Let now $\mathcal{K}$ be consistent, C be a $\mathcal{TELH}_{\perp}^{\Diamond,\text{lhs}}$ concept, D be an $\mathcal{ELH}_{\perp}$ concept, and $\Diamond \in \mathfrak{D}$. Then $\mathcal{K} \models \Diamond C \sqsubseteq D$ iff either there is $\Diamond \in \mathfrak{D}$ with $\Diamond C \sqsubseteq \perp \in \mathcal{K}^*$, or there is $\Diamond \supseteq \Diamond$ with $\Diamond C \sqsubseteq D \in \mathcal{K}^*$. Moreover, $\mathcal{K} \models D(a, i)$ iff $D(a, |i|) \in \mathcal{K}^*$.

We obtain the following result, where the lower bound follows from propositional Horn logic [23].

Theorem 5. *Entailment in $\mathcal{TELH}_{\perp}^{\Diamond,\text{lhs}}$ is P-complete.*

Example 6. Consider *rheumatoid arthritis*, an autoimmune disorder that cannot be healed. In irregular intervals, it produces so-called *flare ups*, that cause pain in the joints. We formalize this knowledge as follows:

$$\text{RheumatoidArthritisPatient} \equiv \exists \text{diagnosedWith.RheumatoidArthritis} \quad (3)$$

$$\text{FlareUpPatient} \sqsubseteq \text{RheumatoidArthritisPatient} \quad (4)$$

$$\Diamond\text{RheumatoidArthritisPatient} \sqsubseteq \text{RheumatoidArthritisPatient} \quad (5)$$

$$\Diamond_2\text{FlareUpPatient} \sqsubseteq \text{FlareUpPatient} \quad (6)$$

We make the assumption that a flare up is 2-month convex, hence if two flare ups are reported at most 2 months apart, we assume that they refer to the same flare up and hence the flare up also present in between the two reports. By applying Rule T4 from the completion algorithm to axioms (4) and (5), we can add

$$\Diamond\text{FlareUpPatient} \sqsubseteq \text{RheumatoidArthritisPatient}$$

to the KB. Suppose the ABox consists of the assertions $\text{FlareUpPatient}(p_1, i)$, $i \in \{0, 4, 5, 7\}$, for a patient p_1. The completed ABox, denoted by $\mathcal{A}^*$, is illustrated below, where for simplicity we omit the individual name p_1.

Here, RheumatoidArthritisPatient and FlareUpPatient are abbreviated by their first letters, respectively. Representatives $-1, 2, 6$ and 8 have been introduced and the intervals they represent are illustrated in gray.

3 Minimal-World Semantics for Metric Temporal Conjunctive Queries with Negation

We now consider the reasoning problem of query answering, which generalizes entailment of assertions. We develop a new temporal query language and follow an approach from [12] to find an appropriate closed-world semantics for negation.

Let **V** be a set of *variables*, and **T** := **I** ∪ **V** be the set of terms. An *atom* is either a *concept atom* of the form $A(\tau)$ or a *role atom* of the form $r(\tau, \rho)$, where $A \in \mathbf{C}$, $r \in \mathbf{R}$ and $\tau, \rho \subseteq \mathbf{T}$. A *conjunctive query (CQ)* $\phi(\mathbf{x})$ is a first-order formula of the form $\exists \mathbf{y}.\psi(\mathbf{x}, \mathbf{y})$, where ψ is a finite conjunction of atoms over the free variables $\mathbf{x}$ (also called the *answer variables*) and the quantified variables $\mathbf{y}$. *Conjunctive queries with (guarded) negation (NCQs)* are constructed by extending CQs with negated concept atoms $\neg A(\tau)$ and negated role atoms $\neg r(\tau, \rho)$ in such a way that, for any negated atom over terms τ (and ρ), the query contains at least one positive atom over τ (and ρ) containing all the variables of the negated atom. An NCQ is *rooted* if its variables are all connected via role atoms to an answer variable (from $\mathbf{x}$) or an individual name. An NCQ is *Boolean* if it does not have answer variables. To determine whether $\mathcal{I} \vDash \phi$ holds for an NCQ ϕ and an atemporal interpretation $\mathcal{I}$, we use standard first-order semantics.

We now extend the temporal CQs from [8] by metric operators [1,7,21] and negation.

Definition 7. Metric temporal conjunctive queries with negation (MTNCQs) *are built by the grammar rule*

$$\phi ::= \psi \mid \top \mid \bot \mid \neg\phi \mid \phi \wedge \phi \mid \phi \vee \phi \mid \phi \mathcal{U}_I \phi \mid \phi \mathcal{S}_I \phi, \tag{7}$$

where ψ is an NCQ, and I is an interval over $\mathbb{N}$. An MTNCQ ϕ is rooted/Boolean if all NCQs in it are rooted/Boolean.

ϕ	$\mathfrak{I}, i \vDash \phi$ iff
CQ ψ	$\mathcal{I}_i \vDash \psi$
$\top$	true
$\bot$	false
$\neg\phi$	$\mathfrak{I}, i \not\vDash \phi$
$\phi \wedge \psi$	$\mathfrak{I}, i \vDash \phi$ and $\mathfrak{I}, i \vDash \psi$
$\phi \vee \psi$	$\mathfrak{I}, i \vDash \phi$ or $\mathfrak{I}, i \vDash \psi$
$\phi \mathcal{U}_I \psi$	$\exists k \in I$ such that $\mathfrak{I}, i+k \vDash \psi$ and $\forall j : 0 \leq j < k : \mathfrak{I}, i+j \vDash \phi$
$\phi \mathcal{S}_I \psi$	$\exists k \in I$ such that $\mathfrak{I}, i-k \vDash \psi$ and $\forall j : 0 \leq j < k : \mathfrak{I}, i-j \vDash \phi$

Fig. 2. Semantics of (Boolean) MTNCQs for $\mathfrak{I} = (\Delta^{\mathfrak{I}}, (\mathcal{I}_i)_{i \in \mathbb{Z}})$ and $i \in \mathbb{Z}$.

We employ the standard semantics shown in Fig. 2. One can define the *next* operator as $\bigcirc\phi := \top\mathcal{U}_{[1,1]}\phi$, and similarly $\bigcirc^-\phi := \top\mathcal{S}_{[1,1]}\phi$. We can also express $\Diamond_I\phi := (\top\mathcal{S}_{-(I \cap (-\infty,0])}\phi) \vee (\top\mathcal{U}_{I \cap [0,\infty)}\phi)$ and $\Box_I\phi := \neg\Diamond_I\neg\phi$, and hence, by (1), the $\Diamond_n$-operators from Sect. 2. An *MTCQ* (or *positive MTNCQ*) is an MTNCQ without negation, where we assume that the operator $\Box_I$ is nevertheless included as part of the syntax of MTCQs.

Example 8. Consider the criterion "Diagnosis of Rheumatoid Arthritis (RA) of more than 6 months and less than 15 years."[6] This can be expressed as an MTNCQ as follows:

$$\phi(x) := \square_{[-6,0]} \left(\exists y.\text{diagnosedWith}(x,y) \wedge \text{RheumatoidArthritis}(y) \right)$$
$$\wedge \neg \, \square_{[-180,0]} \left(\exists y.\text{diagnosedWith}(x,y) \wedge \text{RheumatoidArthritis}(y) \right)$$

The semantics are defined model-theoretically as usual: Let $\mathcal{K} = (\mathcal{T}, \mathcal{A})$ be a $\mathcal{TELH}_\bot^{\Diamond,\text{lhs}}$ KB, $\phi(\mathbf{x})$ an MTNCQ, $\mathbf{a}$ a tuple of individual names from $\mathcal{A}$, $i \in \text{tem}(\mathcal{A})$, and $\mathfrak{I}$ a temporal interpretation. The pair $(\mathbf{a}, i)$ is an *answer* to $\phi(\mathbf{x})$ w.r.t. $\mathfrak{I}$ if $\mathfrak{I}, i \models \phi(\mathbf{a})$. The set of all answers for ϕ w.r.t. $\mathfrak{I}$ is denoted $\text{ans}(\phi, \mathfrak{I})$. The tuple $(\mathbf{a}, i)$ is a *certain answer* to ϕ w.r.t. $\mathcal{K}$ if it is an answer in every model of $\mathcal{K}$; all these tuples are collected in the set $\text{cert}(\phi, \mathcal{K})$.

Query answering is the decision problem of checking $(\mathbf{a}, i) \in \text{cert}(\phi, \mathcal{K})$ when given $\mathbf{a}$, i, ϕ, and $\mathcal{K} = (\mathcal{T}, \mathcal{A})$. CQ answering over $\mathcal{ELH}_\bot$ KBs is NP-complete in general, and P-complete in *data complexity*, where the query ϕ and the TBox $\mathcal{T}$ are not considered as part of the input [24,25,29]. However, certain answer semantics for NCQ answering over $\mathcal{ELH}_\bot$ is CONP-hard [19]. To achieve tractable reasoning in data-oriented applications, we extend the *minimal-world semantics* from [12], which allows for NCQ answering in polynomial time, and gives intuitive semantics to negated query atoms.

3.1   Minimal-World Semantics for MTNCQs

Our goal is to extend the approach from [12] to find a *minimal canonical model* of a $\mathcal{TELH}_\bot^{\Diamond,\text{lhs}}$ KB. Similarly to the *core* chase [17], the main idea is that this model should not contain redundant elements. Particularly, the minimum necessary number of anonymous objects together with the closed-world semantics adequately represents negative knowledge about the objects; for a detailed discussion, see [12]. We consider here the sublogic $\mathcal{TELH}_\bot^{\Diamond,\text{lhs},-}$ of $\mathcal{TELH}_\bot^{\Diamond,\text{lhs}}$ without temporal roles $\Diamond r$, because temporal roles interfere with the *minimality*: by propagating through time, a temporal role can easily violate the "local" minimality of interpretations at other time points, which could lead to unintuitive answers. In the definition of the model, we make use of entailment in $\mathcal{TELH}_\bot^{\Diamond,\text{lhs},-}$, which can be checked in polynomial time. Thus, we can exclude w.l.o.g. equivalent concept and role names. Also, for simplicity, in the following we assume w.l.o.g. that all CIs are in the following stronger normal form (cf. (2)):

$$\Diamond A \sqsubseteq B, \quad A_1 \sqcap A_2 \sqsubseteq B, \quad r \sqsubseteq s, \quad A \sqsubseteq \exists r.B, \quad \exists r.A \sqsubseteq B,$$

i.e., $\Diamond$-operators are allowed only in CIs of the form $\Diamond A \sqsubseteq B$. In particular, disallowing CIs of the form $\Diamond A \sqsubseteq \exists r.B$ allows us to draw a stronger connection to the original construction in [12]; see in particular Step 3(a) in Definition 9 below.

---

[6] https://clinicaltrials.gov/ct2/show/NCT01198002.

We need one more auxiliary definition from [12] to define the minimal temporal canonical model. Given a set V of existential restrictions, we say that $\exists r.A \in V$ is *minimal* in V if there is no other $\exists s.B \in V$ such that $\mathcal{K} \models s \sqsubseteq r$ and $\mathcal{K} \models B \sqsubseteq A$.

Definition 9. *The* minimal temporal canonical model $\mathfrak{I}_\mathcal{K} = (\Delta^{\mathfrak{I}_\mathcal{K}}, (\mathcal{I}_i)_{i \in \mathbb{Z}})$ *of a KB* $\mathcal{K} = (\mathcal{T}, \mathcal{A})$ *is obtained by the following steps.*

1. *Set* $\Delta^{\mathfrak{I}_\mathcal{K}} := \mathbf{I}$ *and* $a^{\mathcal{I}_i} := a$ *for all* $a \in N_I$ *and* $i \in \mathbb{Z}$.
2. *For each time point* $i \in \mathbb{Z}$, *define* $A^{\mathcal{I}_i} := \{a \mid \mathcal{K} \models A(a, i)\}$ *for all* $A \in \mathbf{C}$ *and* $r^{\mathcal{I}_i} := \{(a, b) \mid \mathcal{K} \models r(a, b, i)\}$ *for all* $r \in \mathbf{R}$.
3. *Repeat the following steps:*
   (a) *Select an element* $d \in \Delta^{\mathfrak{I}_\mathcal{K}}$ *that has not been selected before and, for each* $i \in \mathbb{Z}$, *let* $V_i := \{\exists r.B \mid d \in A^{\mathcal{I}_i},\ d \notin (\exists r.B)^{\mathcal{I}_i},\ \mathcal{K} \models A \sqsubseteq \exists r.B,\ A, B \in \mathbf{C}\}$.
   (b) *For each* $\exists r.B$ *that is minimal in some* V_i, *add a fresh element* e_{rB} *to* $\Delta^{\mathfrak{I}_\mathcal{K}}$. *For all* $i \in \mathbb{Z}$ *and* $\mathcal{K} \models B \sqsubseteq A$, *add* e_{rB} *to* $A^{\mathcal{I}_i}$.
   (c) *For all* $i \in \mathbb{Z}$, *minimal* $\exists r.B$ *in* V_i, *and* $\mathcal{K} \models r \sqsubseteq s$, *add* (d, e_{rB}) *to* $s^{\mathcal{I}_i}$.

We denote by $\mathfrak{I}_\mathcal{A}$ the result of executing only Steps 1 and 2 of this definition, i.e., restricting $\mathfrak{I}_\mathcal{K}$ to the named individuals. Since there are only finitely many elements of $\mathbf{I}$, $\mathbf{C}$, and $\mathbf{R}$ that are relevant for this definition (i.e., those that occur in $\mathcal{K}$), for simplicity we often treat $\mathfrak{I}_\mathcal{A}$ as if it had a finite object (but still infinite time) domain.

In $\mathfrak{I}_\mathcal{K}$, there may exist anonymous objects that are not connected to any named individuals in $\mathcal{I}_i$ and are not relevant for the satisfaction of the KB. For this reason, in the following we consider only rooted MTNCQs, which can be evaluated only over the parts of $\mathfrak{I}_\mathcal{K}$ that are connected to the named individuals. We show that $\mathfrak{I}_\mathcal{K}$ is actually a model of $\mathcal{K}$ and is canonical in the usual sense that it can be used to answer *positive* queries over $\mathcal{K}$ under certain answer semantics.

Lemma 10. *Let* $\mathcal{K}$ *be a consistent* $\mathcal{TELH}_\bot^{\diamondsuit,\mathsf{lhs},-}$ *KB. Then* $\mathfrak{I}_\mathcal{K}$ *is a model of* $\mathcal{K}$ *and, for every rooted MTCQ* ϕ, *we have* $\mathrm{cert}(\phi, \mathcal{K}) = \mathrm{ans}(\phi, \mathfrak{I}_\mathcal{K})$.

Thus, the following *minimal-world* semantics is compatible with certain answer semantics for positive (rooted) queries, while keeping a tractable data complexity.

Definition 11. *The set of* minimal-world answers *to an MTNCQ* q *over a consistent* $\mathcal{TELH}_\bot^{\diamondsuit,\mathsf{lhs},-}$ *KB* $\mathcal{K}$ *is* $\mathrm{mwa}(\phi, \mathcal{K}) := \mathrm{ans}(\phi, \mathfrak{I}_\mathcal{K})$.

3.2 A Combined Rewriting for MTNCQs

Since the minimal canonical model $\mathfrak{I}_\mathcal{K}$ may still be infinite, we now show that rooted MTNCQ answering under minimal-world semantics is combined first-order rewritable [27], i.e., to compute $\mathrm{mwa}(\phi, \mathcal{K})$ we can equivalently evaluate a rewritten query over a finite interpretation (of polynomial size). Since the rewriting depends only on the query and the TBox, its size is irrelevant for data complexity, and it can be evaluated in polynomial time. We proceed in two steps.

1. We rewrite ϕ into a *metric first-order temporal logic (MFOTL)* formula ϕ_T, which combines first-order formulas via metric temporal operators; for details, see [10]. This query can be evaluated over $\mathfrak{I}_{\mathcal{A}}$ instead of $\mathfrak{I}_{\mathcal{K}}$. Hence, we reduce the infinite object domain to the finite set $\mathbf{I}(\mathcal{K})$.
2. We then further rewrite ϕ_T into a three-sorted first-order formula (with explicit variables for time points), which is then evaluated over a restriction $\mathfrak{I}_{\mathcal{A}}^{\mathrm{fin}}$ of $\mathfrak{I}_{\mathcal{A}}$ that contains only finitely many time points (essentially those in $\mathrm{rep}(\mathcal{A})$, although we modify them slightly).

For the first step, we rewrite a rooted MTNCQ ϕ by replacing each (rooted) NCQ ψ with the first-order rewriting ψ_T from [12].[7] The result is a special kind of MFOTL formula ϕ_T [10], in which atemporal first-order formulas can be nested inside MTL-operators, similarly as in MTNCQs. The semantics is based on a satisfaction relation $\mathfrak{I}, i \models \phi_T$ that is defined in much the same way as in Fig. 2, the only exception being that $\mathfrak{I}, i \models \psi_T$ for a first-order formula ψ_T is defined by $\mathcal{I}_i \models \psi_T$, using the standard first-order semantics. We can lift the atemporal rewritability result from [12] in a straightforward way to our temporal setting.

Lemma 12. *Let* $\mathcal{K} = (\mathcal{T}, \mathcal{A})$ *be a consistent* $\mathcal{TELH}_{\bot}^{\diamondsuit,\mathsf{lhs},-}$ *KB and* ϕ *be a rooted MTNCQ. Then* $\mathrm{mwa}(\phi, \mathcal{K}) = \mathrm{ans}(\phi_T, \mathfrak{I}_{\mathcal{A}})$.

For the second rewriting step, we restrict ourselves to finitely many time points. More precisely, we consider the finite structure $\mathfrak{I}_{\mathcal{A}}^{\mathrm{fin}}$, which is obtained from $\mathfrak{I}_{\mathcal{A}}$ by restricting the set of time points to $\mathrm{rep}(\mathcal{A})$. By Lemma 4, the information contained in this structure is already sufficient to answer atomic queries. We extend this structure a little, by considering the *two* representatives i, j for each maximal interval $[i, j]$ in $\mathbb{Z} \setminus \mathrm{tem}(\mathcal{A})$. In this way, we ensure that the "border" elements are always representatives for their respective intervals. The size of the resulting structure $\mathfrak{I}_{\mathcal{A}}^{\mathrm{fin}}$ is polynomial in the size of $\mathcal{K}$.

Example 13. Let $\mathcal{A} = \{B(a,0), B(a,2), C(a,9)\}$ and $\mathcal{T} = \{\diamondsuit\diamondsuit_3 B \sqcap \diamondsuit C \sqsubseteq A\}$. Below one can see the finite structure $\mathfrak{I}_{\mathcal{A}}^{\mathrm{fin}}$ over the representative time points $\{-1, 0, 1, 2, 3, 8, 9, 10\}$, where for simplicity we omit the individual name.

The rewriting from Lemma 12 can refer to time instants outside of $\mathrm{rep}(\mathcal{A})$. However, when we want to evaluate a pure FO formula over the finite structure $\mathfrak{I}_{\mathcal{A}}^{\mathrm{fin}}$, this is not possible anymore, because the first-order quantifiers must quantify over the domain of $\mathfrak{I}_{\mathcal{A}}^{\mathrm{fin}}$. Moreover, since the query ϕ_T can contain metric temporal operators, we need to keep track of the distance between the time points in $\mathrm{tem}(\mathcal{A})$. Hence, in the following we assume that $\mathfrak{I}_{\mathcal{A}}^{\mathrm{fin}}$ is given as

---

[7] Strictly speaking, ψ_T in [12] is a *set* of first-order formulas, which is however equivalent to the disjunction of all these formulas.

a first-order structure with the domain $\mathbf{I} \cup \{b_1, \ldots, b_n\} \cup \mathrm{rep}(\mathcal{A})$ and additional predicates bit and sign such that $\mathrm{bit}(i, j)$, $1 \le j \le n$, is true iff the jth bit of the binary representation of the time stamp i is 1, and $\mathrm{sign}(i)$ is true iff i is non-negative.

Thus, we now consider three-sorted first-order formulas with the three sorts $\mathbf{I}$ (for objects), $\{b_1, \ldots, b_n\}$ (for bits) and $\mathrm{rep}(\mathcal{A})$ (for time stamps). We denote variables of sort $\mathrm{rep}(\mathcal{A})$ by t, t', t''. To simplify the presentation, we do not explicitly denote the sort of all variables, but this is always clear from the context. Every concept name is now accessed as a binary predicate of sort $\mathbf{I} \times \mathrm{rep}(\mathcal{A})$, e.g., $A(a, i)$ refers to the fact that individual a satisfies A at time point i. Similarly, role names correspond to ternary predicates of sort $\mathbf{I} \times \mathbf{I} \times \mathrm{rep}(\mathcal{A})$. It is clear that the expressions $t' \bowtie t$ and even $t' - t \bowtie m$ for some constant m and $\bowtie \in \{\ge, >, =, <, \le\}$ are definable as first-order formulas using the natural order $<$ on $\{1, \ldots, m\}$.

Lemma 14. *For ϕ_T there is a constant $N \in \mathbb{N}$ such that, for every subformula ψ of ϕ_T, every maximal interval J in $\mathbb{Z} \setminus \bigcup\{[i - N, i + N] \mid i \in \mathrm{tem}(\mathcal{A})\}$, all $k, \ell \in J$, and all relevant tuples $\mathbf{a}$ over $\mathbf{I}$, we have $\mathfrak{I}_\mathcal{A}, k \models \psi(\mathbf{a})$ iff $\mathfrak{I}_\mathcal{A}, \ell \models \psi(\mathbf{a})$.*

Hence, for evaluating subformulas of ϕ_T, it suffices to keep track of time points up to N steps away from the elements of $\mathrm{rep}(\mathcal{A})$; this includes at least one element from each of the intervals J mentioned in Lemma 14, since every element of $\mathrm{tem}(\mathcal{A})$ is immediately surrounded by two elements of $\mathrm{rep}(\mathcal{A})$.

We exploit Lemma 14 in the following definition of the three-sorted first-order formula $[\psi]^n(\mathbf{x}, t)$ that simulates the behavior of $\psi(\mathbf{x})$ at the "virtual" time point $t + n$, where $n \in [-N, N]$. Whenever we use a formula $[\psi]^n(\mathbf{x}, t)$, we require that t denotes a representative for $t + n$. Due to our assumption that each maximal interval from $\mathbb{Z} \setminus \mathrm{tem}(\mathcal{A})$ is represented by its endpoints (see Example 13), we know that t is a representative for $t + n$ iff there is no element of $\mathrm{rep}(\mathcal{A})$ between t and $t + n$. We can encode this check in an auxiliary formula:

$$\mathbf{rep}^n(t) := \neg \exists t'. (t + n \le t' < t) \vee (t < t' \le t + n).$$

Example 15. In Example 13, 3 and 8 are representatives for the missing time points 4–7, and we have $\mathfrak{I}_\mathcal{A}^{\mathrm{fin}} \models \mathbf{rep}^1(3)$ (with $N = 1$). However, for $\phi_T = \bigcirc \neg C(x)$, we have $\mathfrak{I}_\mathcal{A}, 3 \models \phi_T(a)$, but $\mathfrak{I}_\mathcal{A}, 8 \not\models \phi_T(a)$, i.e., the behavior at 3 and 8 differs. To distinguish this, we need to refer to the "virtual" time point 4 (gray circled "v") that is not included in $\mathfrak{I}_\mathcal{A}^{\mathrm{fin}}$, via the formula $[\neg C(x)]^1(a, 3)$. By Lemma 14, it is sufficient to consider 4, because this determines the behavior at 5–7.

We now define $[\psi]^n(\mathbf{x}, t)$ recursively, for each subformula ψ of ϕ_T. If ψ is a single rewritten NCQ, then $[\psi]^n(\mathbf{x}, t)$ is obtained by replacing each atemporal atom $A(x)$ by $A(x, t)$, and similarly for role atoms. The parameter n can be ignored here, because we assumed that t is a representative for $t + n$, and hence the time points t and $t + n$ are interpreted in $\mathfrak{I}_\mathcal{A}$ equally. For conjunctions, we set $[\psi_1 \wedge \psi_2]^n(\mathbf{x}, t) := [\psi_1]^n(\mathbf{x}, t) \wedge [\psi_2]^n(\mathbf{x}, t)$ and similarly for the other Boolean constructors. Finally, we demonstrate the translation for $\mathcal{U}$-formulas (the case of $\mathcal{S}$-formulas is analogous). We define $[\psi_1 \,\mathcal{U}_{[c_1,c_2]}\psi_2]^n(\mathbf{x},t)$ as

$$
\exists t'. \bigvee_{n'\in[-N,N]} \Big((t+n+c_1 \le t'+n' \le t+n+c_2) \wedge \mathtt{rep}^{n'}(t') \wedge [\psi_2]^{n'}(\mathbf{x},t') \wedge
$$

$$
\forall t''. \bigwedge_{n''\in[-N,N]} \Big(\big((t+n \le t''+n'' < t'+n') \wedge \mathtt{rep}^{n''}(t'')\big) \rightarrow [\psi_1]^{n''}(\mathbf{x},t'')\Big)\Big),
$$

where c_2 may be ∞, in which case the upper bound of $t+n+c_2$ can be removed.

Lemma 16. *Let* $\mathcal{K} = (\mathcal{T}, \mathcal{A})$ *be a consistent* $\mathcal{TELH}_{\bot}^{\diamondsuit,\mathsf{lhs},-}$ *KB and* ϕ *be an MTNCQ. Then* $\mathrm{ans}([\phi_{\mathcal{T}}]^0(\mathbf{x},t), \mathfrak{J}_{\mathcal{A}}^{\mathrm{fin}}) = \mathrm{ans}(\phi_{\mathcal{T}}, \mathfrak{J}_{\mathcal{A}})$.

This lemma allows us to compute in polynomial time that patient p_1 from Example 6 is an answer to $\phi(x)$ from Example 8 exactly at time point 7. Below we summarize our tight complexity results, which by Lemma 10 also hold for rooted MTCQs under certain answer semantics.

Theorem 17. *Answering rooted MTNCQs under minimal-world semantics over* $\mathcal{TELH}_{\bot}^{\diamondsuit,\mathsf{lhs},-}$ *KBs is* EXPSPACE-*complete, and* P-*complete in data complexity.*

Proof. EXPSPACE-hardness is inherited from propositional MTL [1,18]. Moreover, first-order formulas over finite structures can be evaluated in PSPACE [31]. Finally, the size of $[\phi_{\mathcal{T}}]^0(\mathbf{x},t)$ is bounded exponentially in the size of ϕ and $\mathcal{T}$: each rewritten NCQ $\psi_{\mathcal{T}}$ may be exponentially larger than ψ, and each $[\psi_1 \,\mathcal{U}_I\psi_2]^n(\mathbf{x},t)$ introduces exponentially many disjuncts and conjuncts (but the nesting depth of constructors in this formula is linear in the nesting depth of $\psi_1 \,\mathcal{U}_I\psi_2$).

For data complexity, hardness is inherited from atemporal $\mathcal{EL}$ [15]. Evaluating FO($<$, $\mathtt{bit}$)-formulas is in DLogTime-uniform AC^0 in data complexity [26], and the size of our rewriting only depends on the query and the TBox. By Lemmas 12 and 16 and since $\mathfrak{J}_{\mathcal{A}}^{\mathrm{fin}}$ is of size polynomial in the size of $\mathcal{A}$, deciding whether a tuple $\mathbf{a}$ is a minimal-world answer of an MTNCQ w.r.t. a $\mathcal{TELH}_{\bot}^{\diamondsuit,\mathsf{lhs},-}$ KB is possible in P. □

4   Related Work and Discussion

For a general overview of temporal ontology and query languages, see [3,28]. In the presence of a single rigid role, allowing the operator $\diamondsuit$ on both sides of $\mathcal{EL}$ CIs makes subsumption undecidable [4]. In [20], a variety of restrictions are investigated to regain decidability. In particular, allowing the qualitative operators $\diamondsuit, \diamondsuit, \diamondsuit, \diamondsuit$ only on the left-hand side of CIs makes the logic tractable. Adding LTL operators to concepts was also investigated in other DLs, like $\mathcal{ALC}$ (without temporal roles) [28,32] and *DL-Lite* [4]. Only recently, also metric variants of such logics were considered [7,21,30]. There is a multitude of proposals for (non-metric) temporal query answering for lightweight DLs [2,5,8,13,14].

We extend previous results by introducing a tractable temporal extension of $\mathcal{ELH}_\perp$ that allows metric temporal operators, and a metric temporal query language. For MTNCQs under minimal-world semantics, we show that the complexity of query answering does not increase from the classical case. Future work includes representing numeric information, such as measurements and dosages of medications, which are important for evaluating eligibility criteria of clinical trials [11,16] and extending the set $\mathfrak{D}$. It seems possible to allow other diamond operators in $\mathcal{TELH}_\perp^{\lozenge,\mathsf{lhs}}$ axioms if they satisfy the relevant properties (see Lemmas 2 and 3). Currently, we are working on an optimized implementation of this method for temporal queries over large medical ontologies such as SNOMED CT.

References

1. Alur, R., Henzinger, T.A.: A really temporal logic. J. ACM **41**(1), 181–204 (1994)
2. Artale, A., Kontchakov, R., Kovtunova, A., Ryzhikov, V., Wolter, F., Zakharyaschev, M.: First-order rewritability of ontology-mediated temporal queries. In: Proceedings IJCAI, pp. 2706–2712. AAAI Press (2015)
3. Artale, A., Kontchakov, R., Kovtunova, A., Ryzhikov, V., Wolter, F., Zakharyaschev, M.: Ontology-mediated query answering over temporal data: a survey (invited talk). In: Proceedings TIME, pp. 1:1–1:37. Schloss Dagstuhl (2017)
4. Artale, A., Kontchakov, R., Lutz, C., Wolter, F., Zakharyaschev, M.: Temporalising tractable description logics. In: Proceedings TIME, pp. 11–22. IEEE Press (2007)
5. Artale, A., Kontchakov, R., Wolter, F., Zakharyaschev, M.: Temporal description logic for ontology-based data access. In: Proceedings IJCAI, pp. 711–717. AAAI Press (2013)
6. Baader, F., Borgwardt, S., Forkel, W.: Patient selection for clinical trials using temporalized ontology-mediated query answering. In: Proceedings HQA, pp. 1069–1074. ACM (2018)
7. Baader, F., Borgwardt, S., Koopmann, P., Ozaki, A., Thost, V.: Metric temporal description logics with interval-rigid names. In: Dixon, C., Finger, M. (eds.) FroCoS 2017. LNCS (LNAI), vol. 10483, pp. 60–76. Springer, Cham (2017). https://doi.org/10.1007/978-3-319-66167-4_4
8. Baader, F., Borgwardt, S., Lippmann, M.: Temporal query entailment in the description logic $\mathcal{SHQ}$. J. Web Sem. **33**, 71–93 (2015)
9. Baader, F., Brandt, S., Lutz, C.: Pushing the $\mathcal{EL}$ envelope. In: Proceedings IJCAI, pp. 364–369. Professional Book Center (2005)
10. Basin, D., Klaedtke, F., Müller, S., Zălinescu, E.: Monitoring metric first-order temporal properties. J. ACM **62**(2), 1–45 (2015)
11. Bonomi, L., Jiang, X.: Patient ranking with temporally annotated data. J. Biomed. Inf. **78**, 43–53 (2018)
12. Borgwardt, S., Forkel, W.: Closed-world semantics for conjunctive queries with negation over $\mathcal{ELH}_\perp$ ontologies. In: Calimeri, F., Leone, N., Manna, M. (eds.) JELIA 2019. LNCS (LNAI), vol. 11468, pp. 371–386. Springer, Cham (2019). https://doi.org/10.1007/978-3-030-19570-0_24
13. Borgwardt, S., Lippmann, M., Thost, V.: Temporalizing rewritable query languages over knowledge bases. J. Web Sem. **33**, 50–70 (2015)

14. Brandt, S., Kalaycı, E.G., Ryzhikov, V., Xiao, G., Zakharyaschev, M.: Querying log data with metric temporal logic. J. Artif. Intell. Res. **62**, 829–877 (2018)
15. Calvanese, D., De Giacomo, G., Lembo, D., Lenzerini, M., Rosati, R.: Data complexity of query answering in description logics. Art. Intell. **195**, 335–360 (2013)
16. Crowe, C.L., Tao, C.: Designing ontology-based patterns for the representation of the time-relevant eligibility criteria of clinical protocols. AMIA Jt. Summits Transl. Sci. Proc. **2015**, 173–177 (2015)
17. Deutsch, A., Nash, A., Remmel, J.B.: The chase revisited. In: Proceedings of the Twenty-Seventh ACM SIGMOD-SIGACT-SIGART Symposium on Principles of Database Systems, PODS 2008, 9–11 June 2008, Vancouver, BC, Canada, pp. 149–158 (2008)
18. Furia, C.A., Spoletini, P.: Tomorrow and all our yesterdays: MTL satisfiability over the integers. In: Fitzgerald, J.S., Haxthausen, A.E., Yenigun, H. (eds.) ICTAC 2008. LNCS, vol. 5160, pp. 126–140. Springer, Heidelberg (2008). https://doi.org/10.1007/978-3-540-85762-4_9
19. Gutiérrez-Basulto, V., Ibáñez-García, Y., Kontchakov, R., Kostylev, E.V.: Queries with negation and inequalities over lightweight ontologies. J. Web Sem. **35**, 184–202 (2015)
20. Gutiérrez-Basulto, V., Jung, J.C., Kontchakov, R.: Temporalized $\mathcal{EL}$ ontologies for accessing temporal data: complexity of atomic queries. In: Proceedings IJCAI, pp. 1102–1108. AAAI Press (2016)
21. Gutiérrez-Basulto, V., Jung, J.C., Ozaki, A.: On metric temporal description logics. In: Proceedings ECAI, pp. 837–845. IOS Press (2016)
22. Hripcsak, G., Zhou, L., Parsons, S., Das, A.K., Johnson, S.B.: Modeling electronic discharge summaries as a simple temporal constraint satisfaction problem. J. Am. Med. Inform. Assn. **12**(1), 55–63 (2005)
23. Jones, N.D., Laaser, W.T.: Complete problems for deterministic polynomial time. Theore. Comput. Sci. **3**(1), 105–117 (1976)
24. Krisnadhi, A., Lutz, C.: Data complexity in the $\mathcal{EL}$ family of description logics. In: Dershowitz, N., Voronkov, A. (eds.) LPAR 2007. LNCS (LNAI), vol. 4790, pp. 333–347. Springer, Heidelberg (2007). https://doi.org/10.1007/978-3-540-75560-9_25
25. Krötzsch, M., Rudolph, S.: Conjunctive queries for $\mathcal{EL}$ with role composition. In: Proceedings DL, pp. 355–362 (2007)
26. Lindell, S.: A purely logical characterization of circuit uniformity. In: Proceedings of the 7th Annual Structure in Complexity Theory Conference, pp. 185–192 (1992)
27. Lutz, C., Toman, D., Wolter, F.: Conjunctive query answering in the description logic $\mathcal{EL}$ using a relational database system. In: Proceedings IJCAI, pp. 2070–2075. AAAI Press (2009)
28. Lutz, C., Wolter, F., Zakharyaschev, M.: Temporal description logics: a survey. In: Proceedings of the 15th International Symposium on Temporal Representation and Reasoning (TIME 2008), pp. 3–14. IEEE Press (2008)
29. Rosati, R.: On conjunctive query answering in $\mathcal{EL}$. In: Proceedings DL, pp. 451–458 (2007)
30. Thost, V.: Metric temporal extensions of DL-Lite and interval-rigid names. In: Proceedings KR, pp. 665–666. AAAI Press (2018)
31. Vardi, M.Y.: The complexity of relational query languages (extended abstract). In: Proceedings STOC, pp. 137–146. ACM (1982)
32. Wolter, F., Zakharyaschev, M.: Temporalizing description logics. In: Frontiers of Combining Systems 2, pp. 379–402. Research Studies Press/Wiley (2000)

Reasoning on *DL-Lite$_\mathcal{R}$* with Defeasibility in ASP

Loris Bozzato[1]([⊠])(iD), Thomas Eiter[2](iD), and Luciano Serafini[1](iD)

[1] Fondazione Bruno Kessler, Via Sommarive 18, 38123 Trento, Italy
`{bozzato,serafini}@fbk.eu`
[2] Institute of Logic and Computation, Technische Universität Wien,
Favoritenstraße 9-11, 1040 Vienna, Austria
`eiter@kr.tuwien.ac.at`

Abstract. Reasoning on defeasible knowledge is a topic of interest in the area of description logics, as it is related to the need of representing exceptional instances in knowledge bases. In this direction, in our previous works we presented a framework for representing (contextualized) OWL RL knowledge bases with a notion of justified exceptions on defeasible axioms: reasoning in such framework is realized by a translation into ASP programs. The resulting reasoning process for OWL RL, however, introduces a complex encoding in order to capture reasoning on the negative information needed for reasoning on exceptions. In this paper, we apply the justified exception approach to knowledge bases in *DL-Lite$_\mathcal{R}$*, i.e. the language underlying OWL QL. We provide a definition for *DL-Lite$_\mathcal{R}$* knowledge bases with defeasible axioms and study their semantic and computational properties. The limited form of *DL-Lite$_\mathcal{R}$* axioms allows us to formulate a simpler encoding into ASP programs, where reasoning on negative information is managed by direct rules. The resulting materialization method gives rise to a complete reasoning procedure for instance checking in *DL-Lite$_\mathcal{R}$* with defeasible axioms.

1 Introduction

Representing defeasible information is a topic of interest in the area of description logics (DLs), as it is related to the need of accommodating the presence of exceptional instances in knowledge bases. This interest led to different proposals for non-monotonic features in DLs based on different notions of defeasibility, e.g. [2,4,10,14,18]. In this direction, we presented in [6] an approach to represent defeasible information in contextualized DL knowledge bases by introducing a notion of *justifiable exceptions*: general *defeasible axioms* can be overridden by more specific exceptional instances if their application would provably lead to inconsistency. Reasoning in $\mathcal{SROIQ}$-RL (i.e. OWL RL) knowledge bases is realized by a translation to datalog, which provides a complete *materialization calculus* [19] for instance checking and conjunctive query (CQ) answering. While the translation covers the full $\mathcal{SROIQ}$-RL language, it needs a complex encoding to represent reasoning on exceptions. In particular, it relies on proofs by contradiction to ensure completeness in presence of negative disjunctive information.

© Springer Nature Switzerland AG 2019
P. Fodor et al. (Eds.): RuleML+RR 2019, LNCS 11784, pp. 19–35, 2019.
https://doi.org/10.1007/978-3-030-31095-0_2

In this paper, we consider the case of knowledge bases with defeasible axioms in *DL-Lite*$_\mathcal{R}$ [13], which corresponds to the language underlying the OWL QL fragment [21]. It is indeed interesting to show the applicability of our defeasible reasoning approach to the well-known *DL-Lite* family: in particular, by adopting *DL-Lite*$_\mathcal{R}$ as the base logic we need to take unnamed individuals introduced by existential formulas into account, especially for the justifications of exceptions. Moreover, we show that due to the restricted form of its axioms, the *DL-Lite*$_\mathcal{R}$ language allows us to give a less involved datalog encoding in which reasoning on negative information is directly encoded in datalog rules (cf. discussion on "justification safeness" in [6]).

The contributions of this paper can be summarized as follows:

- In Sect. 3 we provide a definition of defeasible DL knowledge base (DKB) with justified models that draws from the definition of *Contextualized Knowledge Repositories (CKR)* [7,8,23] with defeasible axioms provided in [6]. This allows us to concentrate on the defeasible reasoning aspects without considering the aspects related to context representation. In the case of *DL-Lite*$_\mathcal{R}$, we focus on models in which exceptions only occur on individuals named in the DKB (called *exception-safe*), and we present a condition which ensures this property for the justified models of the DKB.
- For such exception-safe DKBs based on *DL-Lite*$_\mathcal{R}$,we provide in Sect. 4 a translation to datalog (under answer set semantics [16]) that alters the translation in [5,6] and prove its correctness for instance checking. In particular, the fact that reasoning on negative disjunctive information is not needed allows us to provide a simpler translation (without the use of the involving "test" environments mechanism of [6]).
- In Sect. 5 we provide complexity results for reasoning problems on exception-safe DKBs based on *DL-Lite*$_\mathcal{R}$. Deciding satisfiability of such a DKB with respect to justified models is tractable, while inference of an axiom under cautious (i.e., certainty) semantics is co-NP-complete in general.

2   Preliminaries

Description Logics and *DL-Lite*$_\mathcal{R}$ Language. We assume the common definitions of description logics [1] and the definition of the logic *DL-Lite*$_\mathcal{R}$ [13]: we summarize in the following the basic definitions used in this work.

A *DL vocabulary* Σ consists of the mutually disjoint countably infinite sets NC of *atomic concepts*, NR of *atomic roles*, and NI of *individual constants*. Complex *concepts* are then recursively defined as the smallest sets containing all concepts that can be inductively constructed using the constructors of the considered DL language. A *DL-Lite*$_\mathcal{R}$ *knowledge base* $\mathcal{K} = \langle \mathcal{T}, \mathcal{R}, \mathcal{A} \rangle$ consists of: a TBox $\mathcal{T}$ containing *general concept inclusion (GCI)* axioms $C \sqsubseteq D$ where C, D are concepts, of the form:

$$C := A \mid \exists R \qquad\qquad D := A \mid \neg C \mid \exists R$$

where $A \in$ NC and $R \in$ NR;[1] an RBox $\mathcal{R}$ containing *role inclusion (RIA)* axioms $S \sqsubseteq R$, reflexivity, irreflexivity, inverse and role disjointness axioms, where S, R are roles; and an ABox $\mathcal{A}$ composed of assertions of the forms $D(a)$, $R(a,b)$, with $R \in$ NR and $a, b \in$ NI.

A *DL interpretation* is a pair $\mathcal{I} = \langle \Delta^\mathcal{I}, \cdot^\mathcal{I} \rangle$ where $\Delta^\mathcal{I}$ is a non-empty set called *domain* and $\cdot^\mathcal{I}$ is the *interpretation function* which assigns denotations for language elements: $a^\mathcal{I} \in \Delta^\mathcal{I}$, for $a \in$ NI; $A^\mathcal{I} \subseteq \Delta^\mathcal{I}$, for $A \in$ NC; $R^\mathcal{I} \subseteq \Delta^\mathcal{I} \times \Delta^\mathcal{I}$, for $R \in$ NR. The interpretation of non-atomic concepts and roles is defined by the evaluation of their description logic operators (see [13] for *DL-Lite$_\mathcal{R}$*). An interpretation $\mathcal{I}$ *satisfies* an axiom ϕ, denoted $\mathcal{I} \models_{\mathrm{DL}} \phi$, if it verifies the respective semantic condition, in particular: for $\phi = D(a)$, $a^\mathcal{I} \in D^\mathcal{I}$; for $\phi = R(a,b)$, $\langle a^\mathcal{I}, b^\mathcal{I} \rangle \in R^\mathcal{I}$; for $\phi = C \sqsubseteq D$, $C^\mathcal{I} \subseteq D^\mathcal{I}$ (resp. for RIAs). $\mathcal{I}$ is a *model* of $\mathcal{K}$, denoted $\mathcal{I} \models_{\mathrm{DL}} \mathcal{K}$, if it satisfies all axioms of $\mathcal{K}$.

Without loss of generality, we adopt the *standard name assumption (SNA)* in the DL context (see [11,15] for more details). That is, we assume an infinite subset $\mathrm{NI}_S \subseteq$ NI of individual constants, called *standard names* s.t. in every interpretation $\mathcal{I}$ we have (i) $\Delta^\mathcal{I} = \mathrm{NI}_S^\mathcal{I} = \{c^\mathcal{I} \mid c \in \mathrm{NI}_S\}$; (ii) $c^\mathcal{I} \neq d^\mathcal{I}$, for every distinct $c, d \in \mathrm{NI}_S$. Thus, we may assume that $\Delta^\mathcal{I} = \mathrm{NI}_S$ and $c^\mathcal{I} = c$ for each $c \in \mathrm{NI}_S$. The *unique name assumption (UNA)* corresponds to assuming $c \neq d$ for all constants in $\mathrm{NI} \setminus \mathrm{NI}_S$ resp. occurring in the knowledge base.

Datalog Programs and Answer Sets. We express our rules in *datalog with negation* under answer sets semantics. In fact, we use here two kinds of negation[2]: strong ("classical") negation $\neg$ and weak *(default) negation* **not** under the interpretation of answer sets semantics [16]; the latter is in particular needed for representing defeasibility.

A *signature* is a tuple $\langle \mathbf{C}, \mathbf{P} \rangle$ of a finite set $\mathbf{C}$ of *constants* and a finite set $\mathbf{P}$ of *predicates*. We assume a set $\mathbf{V}$ of *variables*; the elements of $\mathbf{C} \cup \mathbf{V}$ are *terms*. An *atom* is of the form $p(t_1, \ldots, t_n)$ where $p \in \mathbf{P}$ and $t_1, \ldots, t_n$, are terms. A *literal* l is either a *positive literal* p or a *negative literal* $\neg p$, where p is an atom and $\neg$ is strong negation. Literals of the form p, $\neg p$ are *complementary*. We denote with $\neg.l$ the opposite of literal l, i.e., $\neg.p = \neg p$ and $\neg.\neg p = p$ for an atom p. A (datalog) rule r is an expression:

$$a \leftarrow b_1, \ldots, b_k, \mathbf{not}\ b_{k+1}, \ldots, \mathbf{not}\ b_m. \tag{1}$$

where $a, b_1, \ldots, b_m$ are literals and **not** is negation as failure (NAF). We denote with $Head(r)$ the head a of rule r and with $Body(r) = \{b_1, \ldots, b_k, \mathbf{not}\ b_{k+1}, \ldots, \mathbf{not}\ b_m\}$ the body of r, respectively. A (datalog) *program* P is a finite set of rules. An atom (rule etc.) is *ground*, if no variables occur in it. A *ground substitution* σ for $\langle \mathbf{C}, \mathbf{P} \rangle$ is any function $\sigma : \mathbf{V} \rightarrow \mathbf{C}$; the *ground instance* of an atom (rule, etc.)

---

[1] In the following, we will use C to denote a left-side concept and D as a right-side concept.

[2] Strong negation can be easily emulated using fresh atoms and weak negation resp. constraints. While it does not yield higher expressiveness, it is more convenient for presentation.

χ from σ, denoted $\chi\sigma$, is obtained by replacing in χ each occurrence of variable $v \in \mathbf{V}$ with $\sigma(v)$. A *fact* H is a ground rule r with empty body. The *grounding* of a rule r, $grnd(r)$, is the set of all ground instances of r, and the *grounding* of a program P is $grnd(P) = \bigcup_{r \in P} grnd(r)$.

Given a program P, the *(Herbrand) universe* U_P of P is the set of all constants occurring in P and the *(Herbrand) base* B_P of P is the set of all the ground literals constructable from the predicates in P and the constants in U_P. An *interpretation* $I \subseteq B_P$ is any satisfiable subset of B_P (i.e., not containing complementary literals); a literal l is *true* in I, denoted $I \models l$, if $l \in I$, and l is *false* in I if $\neg.l$ is true. Given a rule $r \in grnd(P)$, we say that $Body(r)$ is true in I, denoted $I \models Body(r)$, if (i) $I \models b$ for each literal $b \in Body(r)$ and (ii) $I \not\models b$ for each literal $\mathtt{not}\, b \in Body(r)$. A rule r is *satisfied* in I, denoted $I \models r$, if either $I \models Head(r)$ or $I \not\models Body(r)$. An interpretation I is a *model* of P, denoted $I \models P$, if $I \models r$ for each $r \in grnd(P)$; moreover, I is *minimal*, if $I' \not\models P$ for each subset $I' \subset I$.

Given an interpretation I for P, the (Gelfond-Lifschitz) *reduct* of P w.r.t. I, denoted by $G_I(P)$, is the set of rules obtained from $grnd(P)$ by (i) removing every rule r such that $I \models l$ for some $\mathtt{not}\, l \in Body(r)$; and (ii) removing the NAF part from the bodies of the remaining rules. Then I is an *answer set* of P, if I is a minimal model of $G_I(P)$; the minimal model is unique and exists iff $G_I(P)$ has some model. Moreover, if M is an answer set for P, then M is a minimal model of P. We say that a literal $a \in B_P$ is a *consequence* of P and write $P \models a$ if every answer set M of P fulfills $M \models a$.

3   DL Knowledge Base with Justifiable Exceptions

In this paper we concentrate on reasoning on a DL knowledge base enriched with *defeasible axioms*, whose syntax and interpretation are analogous to [6]. With respect to the contextual framework presented in [6], this corresponds to reasoning inside a single local context: while this simplifies presentation of the defeasibility aspects and the resulting reasoning method for the case of *DL-Lite$_{\mathcal{R}}$*, it can be generalized to the original case of multiple local contexts.

Syntax. Given a DL language $\mathcal{L}_\Sigma$ based on a DL vocabulary $\Sigma = \mathrm{NC}_\Sigma \cup \mathrm{NR}_\Sigma \cup \mathrm{NI}_\Sigma$, a *defeasible axiom* is any expression of the form $\mathrm{D}(\alpha)$, where $\alpha \in \mathcal{L}_\Sigma$.

We denote with $\mathcal{L}_\Sigma^{\mathrm{D}}$ the DL language extending $\mathcal{L}_\Sigma$ with the set of defeasible axioms in $\mathcal{L}_\Sigma$. On the base of such language, we provide our definition of knowledge base with defeasible axioms.

Definition 1 (defeasible knowledge base, DKB). *A defeasible knowledge base (DKB) $\mathcal{K}$ on a vocabulary Σ is a DL knowledge base over $\mathcal{L}_\Sigma^{\mathrm{D}}$.*

In the following, we tacitly consider DKBs based on *DL-Lite$_{\mathcal{R}}$*.

Example 1. We introduce a simple example showing the definition and interpretation of a defeasible existential axiom. In the organization of a university research department, we want to specify that "in general" department members need also to teach at least a course. On the other hand, PhD students, while recognized as department members, are not allowed to hold a course. We can represent this scenario as a DKB $\mathcal{K}_{dept}$ where:

$$\mathcal{K}_{dept} : \left\{ \begin{array}{l} \mathrm{D}(DeptMember \sqsubseteq \exists hasCourse),\, Professor \sqsubseteq DeptMember, \\ PhDStudent \sqsubseteq DeptMember,\, PhDStudent \sqsubseteq \neg\exists hasCourse, \\ Professor(alice),\, PhDStudent(bob) \end{array} \right\}$$

Intuitively, we want to override the fact that there exists some course assigned to the PhD student *bob*. On the other hand, for the individual *alice* no overriding should happen and the defeasible axiom can be applied.   ◇

Semantics. We can now define a model based interpretation of DKBs, in particular by providing a semantic characterization to defeasible axioms.

Similarly to the case of $\mathcal{SROIQ}$-RL in [6], we can express *DL-Lite$_\mathcal{R}$* knowledge bases in first-order (FO) logic, where every axiom $\alpha \in \mathcal{L}_\Sigma$ is translated into an equivalent FO-sentence $\forall \boldsymbol{x}.\phi_\alpha(\boldsymbol{x})$ where $\boldsymbol{x}$ contains all free variables of ϕ_α depending on the type of the axiom. The translation, depending on the axiom types, can be defined analogously to the FO-translation presented in [6]. In the case of existential axioms of the kind $\alpha = A \sqsubseteq \exists R$, the FO-translation $\phi_\alpha(\boldsymbol{x})$ is defined as:

$$A(x_1) \rightarrow R(x_1, f_\alpha(x_1));$$

that is, we introduce a Skolem function $f_\alpha(x_1)$ which represents new "existential" individuals. Formally, for every right existential axiom $\alpha \in \mathcal{L}_\Sigma$, we define a Skolem function $f_\alpha : \mathrm{NI} \mapsto \mathcal{E}$ where $\mathcal{E}$ is a set of new individual constants not appearing in NI. In particular, for a set of individual names $N \subseteq \mathrm{NI}$, we will write $sk(N)$ to denote the extension of N with the set of Skolem constants for elements in N.

After this transformation the resulting formulas $\phi_\alpha(\boldsymbol{x})$ amount semantically to Horn formulas, since left-side concepts C can be expressed by an existential positive FO-formula, and right-side concepts D by a conjunction of Horn clauses. The following property from [6, Section 3.2] is then preserved for *DL-Lite$_\mathcal{R}$* knowledge bases.

Lemma 1. *For a DL knowledge base $\mathcal{K}$ on $\mathcal{L}_\Sigma$, its FO-translation $\phi_\mathcal{K} := \bigwedge_{\alpha \in \mathcal{K}} \forall \boldsymbol{x} \phi_\alpha(\boldsymbol{x})$ is semantically equivalent to a conjunction of universal Horn clauses.*

With these considerations on the definition of FO-translation, we can now provide our definition of axiom instantiation:

Definition 2 (axiom instantiation). *Given an axiom $\alpha \in \mathcal{L}_\Sigma$ with FO-translation $\forall \boldsymbol{x}.\phi_\alpha(\boldsymbol{x})$, the instantiation of α with a tuple $\mathbf{e}$ of individuals in NI_Σ, written $\alpha(\mathbf{e})$, is the specialization of α to $\mathbf{e}$, i.e., $\phi_\alpha(\mathbf{e})$, depending on the type of α.*

Note that, since we are assuming standard names, this basically means that we can express instantiations (and exceptions) to any element of the domain (identified by a standard name in NI_Σ). We next introduce clashing assumptions and clashing sets.

Definition 3 (clashing assumptions and sets). *A* clashing assumption *is a pair* $\langle \alpha, \mathbf{e} \rangle$ *s.t.* $\alpha(\mathbf{e})$ *is an instantiation for an axiom* $\alpha \in \mathcal{L}_\Sigma$. *A* clashing set *for a clashing assumption* $\langle \alpha, \mathbf{e} \rangle$ *is a satisfiable set S that consists of ABox assertions over* $\mathcal{L}_\Sigma$ *and negated ABox assertions of the forms* $\neg C(a)$ *and* $\neg R(a, b)$ *such that* $S \cup \{\alpha(\mathbf{e})\}$ *is unsatisfiable.*

A clashing assumption $\langle \alpha, \mathbf{e} \rangle$ represents that $\alpha(\mathbf{e})$ is not satisfiable, while a clashing set S provides an assertional "justification" for the assumption of local overriding of α on $\mathbf{e}$. We can then extend the notion of DL interpretation with a set of clashing assumptions.

Definition 4 (CAS-interpretation). *A* CAS-interpretation *is a structure* $\mathcal{I}_{CAS} = \langle \mathcal{I}, \chi \rangle$ *where* $\mathcal{I} = \langle \Delta^\mathcal{I}, \cdot^\mathcal{I} \rangle$ *is a DL interpretation for* Σ *and* χ *is a set of clashing assumptions.*

By extending the notion of satisfaction with respect to CAS-interpretations, we can disregard the application of defeasible axioms to the exceptional elements in the sets of clashing assumptions. For convenience, we call two DL interpretations $\mathcal{I}_1$ and $\mathcal{I}_2$ NI-*congruent*, if $c^{\mathcal{I}_1} = c^{\mathcal{I}_2}$ holds for every $c \in NI$.

Definition 5 (CAS-model). *Given a DKB* $\mathcal{K}$, *a CAS-interpretation* $\mathcal{I}_{CAS} = \langle \mathcal{I}, \chi \rangle$ *is a CAS-model for* $\mathcal{K}$ *(denoted* $\mathcal{I}_{CAS} \models \mathcal{K}$), *if the following holds:*

(i) for every $\alpha \in \mathcal{L}_\Sigma$ *in* $\mathcal{K}$, $\mathcal{I} \models \alpha$;
(ii) for every $D(\alpha) \in \mathcal{K}$ *(where* $\alpha \in \mathcal{L}_\Sigma$), *with* $|\mathbf{x}|$-*tuple* $\mathbf{d}$ *of elements in* NI_Σ *such that* $\mathbf{d} \notin \{\mathbf{e} \mid \langle \alpha, \mathbf{e} \rangle \in \chi\}$, *we have* $\mathcal{I} \models \phi_\alpha(\mathbf{d})$.

We say that a clashing assumption $\langle \alpha, \mathbf{e} \rangle \in \chi$ is *justified* for a *CAS* model $\mathcal{I}_{CAS} = \langle \mathcal{I}, \chi \rangle$, if some clashing set $S = S_{\langle \alpha, \mathbf{e} \rangle}$ exists such that, for every CAS-model $\mathcal{I}'_{CAS} = \langle \mathcal{I}', \chi \rangle$ of $\mathcal{K}$ that is NI-congruent with $\mathcal{I}_{CAS}$, it holds that $\mathcal{I}' \models S_{\langle \alpha, \mathbf{e} \rangle}$. We then consider as DKB models only the CAS-models where all clashing assumptions are justified.

Definition 6 (justified CAS model and DKB model). *A* CAS *model* $\mathcal{I}_{CAS} = \langle \mathcal{I}, \chi \rangle$ *of a DKB* $\mathcal{K}$ *is* justified, *if every* $\langle \alpha, \mathbf{e} \rangle \in \chi$ *is justified. An interpretation* $\mathcal{I}$ *is a DKB model of* $\mathcal{K}$ *(in symbols,* $\mathcal{I} \models \mathcal{K}$), *if* $\mathcal{K}$ *has some justified CAS model* $\mathcal{I}_{CAS} = \langle \mathcal{I}, \chi \rangle$.

Example 2. Reconsidering $\mathcal{K}_{dept}$ in Example 1, a CAS-model providing the intended interpretation of defeasible axioms is $\mathcal{I}_{CAS_{dept}} = \langle \mathcal{I}, \chi_{dept} \rangle$ where $bob^\mathcal{I} \neq alice^\mathcal{I}$ and $\chi_{dept} = \{\langle \alpha, bob \rangle\}$ with $\alpha = DeptMember \sqsubseteq \exists hasCourse$. The fact that this model is justified is verifiable considering that for the clashing set $S = \{DeptMember(bob), \neg \exists hasCourse(bob)\}$ we have $\mathcal{I} \models S$. On the other hand, note that a similar clashing assumption for *alice* is not justifiable: it is not possible from the contents of $\mathcal{K}_{dept}$ to derive a clashing set S' such that $S' \cup \{\alpha(alice)\}$ is unsatisfiable. By Definition 5, this allows us to apply α to this individual as expected and thus $\mathcal{I} \models \exists hasCourse(alice)$. ◇

DKB-models have interesting properties similar as CKR-models in [6]. An example is non-redundancy, cf. [6, Prop. 6, minimality of justification] (the proof is similar).

Proposition 1. *Suppose $\mathcal{I}_{CAS} = \langle \mathcal{I}, \chi \rangle$ and $\mathcal{I}'_{CAS} = \langle \mathcal{I}', \chi' \rangle$ are NI-congruent justified CAS-models of a DKB $\mathcal{K}$, then $\chi \not\subset \chi'$ holds.*

We are interested here in DKB-models $\mathcal{I}_{CAS} = \langle \mathcal{I}, \chi \rangle$ in which clashing assumptions $\langle \alpha, \mathbf{e} \rangle \in \chi$ are only over the individuals of the knowledge base; that is, exceptions can not be expressed on unnamed individuals introduced by existential axioms. A condition ensuring this is that no clashing set $S_{\langle \alpha, \mathbf{e} \rangle}$ for a defeasible axiom α where $\mathbf{e}$ contains some unnamed individual (i.e., some skolem term) can be derived from $\mathcal{K}$ if all defeasible axioms are turned into strict axioms (denote this knowledge base by $\mathcal{K}_s$). Formally, denote by $N_{\mathcal{K}}$ the individuals occurring in $\mathcal{K}$. We say that $\mathcal{K}$ is *exception-safe*, if no clashing set $S_{\langle \alpha, \mathbf{e} \rangle}$ can be derived by unfolding the axioms from $\mathcal{K}_s$ that contains an assertion $D(e)$ (or $R(e_1, e_2)$) on individuals not appearing in $N_{\mathcal{K}}$ (i.e. not named in $\mathcal{K}$).

Proposition 2. *Let $\mathcal{I}_{CAS} = \langle \mathcal{I}, \chi \rangle$ be a CAS-model of DKB $\mathcal{K}$ and let $\mathcal{K}'$ result from $\mathcal{K}$ by pushing equality w.r.t. $\mathcal{I}$, i.e., replace all $a, b \in N_{\mathcal{K}}$ s.t. $a^{\mathcal{I}} = b^{\mathcal{I}}$ by one representative. If $\mathcal{K}'$ is exception-safe, then $\mathcal{I}_{CAS}$ is justified only if every $\langle \alpha, \mathbf{e} \rangle \in \chi$ is over $N_{\mathcal{K}}$.*

Proof (Sketch). Suppose $\mathcal{I}_{CAS}$ is justified and some $\langle \alpha, \mathbf{e} \rangle \in \chi$ is not over $N_{\mathcal{K}}$. Then, by definition of justification, some clashing set S for $\langle \alpha, \mathbf{e} \rangle$ with $\mathbf{e}$ outside $N_{\mathcal{K}}$ is satisfied in all CAS-models $\mathcal{I}'_{CAS}$ of $\mathcal{K}$ that are NI-congruent with $\mathcal{I}_{CAS}$. This means that S can be derived with axiom unfolding restricted by the clashing assumptions in χ. But then S can also be derived without restrictions, and thus from the knowledge base $\mathcal{K}'_s$. However, this means that $\mathcal{K}'$ is not exception-safe, which is a contradiction. □

Table 1. Normal form for $\mathcal{K}$ axioms from $\mathcal{L}_{\Sigma}$

for $A, B, C \in NC_{\Sigma}, R \in NR_{\Sigma}, a, b \in NI_{\Sigma}$:					
$A(a)$	$R(a,b)$	$\neg A(a)$	$\neg R(a,b)$	$A \sqsubseteq B$	$A \sqsubseteq \neg C$
$A_{\exists R} \sqsubseteq B$	$A \sqsubseteq A_{\exists R}$	$R \sqsubseteq T$	$Dis(R,S)$	$Inv(R,S)$	$Irr(R)$

Example 3 (Ex. 2 cont'd). Reconsider the CAS-model $\mathcal{I}_{CAS_{dept}} = \langle \mathcal{I}, \chi_{dept} \rangle$ where $\chi_{dept} = \{\langle \alpha, bob \rangle\}$ with $bob^{\mathcal{I}} \neq alice^{\mathcal{I}}$, $\chi_{dept} = \{\langle \alpha, bob \rangle\}$ and $\alpha = DeptMember \sqsubseteq \exists hasCourse$. If we make α strict, we cannot derive a clashing set $S = \{DeptMember(e), \neg \exists hasCourse(e)\}$ where e is an unnamed individual; to derive $DeptMember(e)$, it would require some axiom $\exists R^- \sqsubseteq DeptMember$ where some unnamed individual is introduced by some axiom $A \sqsubseteq \exists R$; however, no such former axioms can be derived, and thus $\mathcal{K}_{dept}$ is exception-safe. $\Diamond$

We remark that the conditions for exception-safety can be tested in polynomial time, by non-deterministically unfolding the axioms (resolution-style, or in a chase) to derive clashing sets in logarithmic workspace. Syntactic classes ensuring this property can be singled out, which we omit here. In the sequel, we tacitly assume exception safe DKBs under UNA, unless stated otherwise.

4 Datalog Translation for *DL-Lite$_\mathcal{R}$* DKB

We present a datalog translation for reasoning on *DL-Lite$_\mathcal{R}$* DKBs which refines the translation provided in [6]. The translation provides a reasoning method for positive instance queries w.r.t. entailment (on preferred models). An important aspect of this translation is that, due to the form of *DL-Lite$_\mathcal{R}$* axioms, no inference on disjunctive negative information is needed for the reasoning on derivations of clashing sets. Thus, reasoning by contradiction using "test environments" is not needed and we can directly encode negative reasoning as rules on negative literals: with respect to the discussion in [6], we can say that *DL-Lite$_\mathcal{R}$* thus represents an inherently "justification safe" fragment which then allows us to formulate such a direct datalog encoding. With respect to the interpretation of right-hand side existential axioms, we follow the approach of [19]: for every axiom of the kind $\alpha = A \sqsubseteq \exists R$, an auxiliary abstract individual aux^α is added in the translation to represent the class of all R-successors introduced by α.

We introduce a *normal form* for axioms of *DL-Lite$_\mathcal{R}$* (in Table 1) which allows us to simplify the formulation of reasoning rules. We can provide rules to transform any *DL-Lite$_\mathcal{R}$* DKB into normal form and show that the rewritten DKB is equivalent to the original. In Table 1, we introduce new symbols $A_{\exists R}$ to simplify the management of existential formulas in rules for defeasible axioms: in the normalization, we assume that, for every new symbol $A_{\exists R}$, axioms $A_{\exists R} \sqsubseteq \exists R$, $\exists R \sqsubseteq A_{\exists R}$ are added to the KB.

Translation Rules Overview. We can now present the components of our datalog translation for *DL-Lite$_\mathcal{R}$* based DKBs. As in the original formulation in [5,6], which extended the encoding without defeasibility proposed in [8] (inspired by the materialization calculus in [19]), the translation includes sets of *input rules* (which encode DL axioms and signature in datalog), *deduction rules* (datalog rules providing instance level inference) and *output rules* (that encode, in terms of a datalog fact, the ABox assertion to be proved). The translation is composed by the following sets of rules:

DL-Lite$_R$ Input and Output Rules: rules in I_{dlr} encode as datalog facts the *DL-Lite$_R$* axioms and signature of the input DKB. For example, in the case of existential axioms,[3] these are translated as $A \sqsubseteq \exists R \mapsto \{\text{supEx}(A, R, aux^\alpha)\}$: note that this rule, in the spirit of [19], introduces an auxiliary element aux^α, which intuitively represents the class of all new R-successors generated by the axiom α. Similarly, output rules in O encode in datalog the ABox assertions to be proved. These rules are provided in Table 2.

Table 2. *DL-Lite$_R$* input, deduction and output rules

DL-Lite$_R$ input translation $I_{dlr}(S)$

(idlr-nom)	$a \in \text{NI} \mapsto \{\text{nom}(a)\}$		(idlr-supnot)	$A \sqsubseteq \neg B \mapsto \{\text{supNot}(A, B)\}$
(idlr-cls)	$A \in \text{NC} \mapsto \{\text{cls}(A)\}$		(idlr-subex)	$\exists R \sqsubseteq B \mapsto \{\text{subEx}(R, B)\}$
(idlr-rol)	$R \in \text{NR} \mapsto \{\text{rol}(R)\}$		(idlr-supex)	$A \sqsubseteq \exists R \mapsto \{\text{supEx}(A, R, aux^\alpha)\}$

(idlr-inst)	$A(a) \mapsto \{\text{insta}(a, A)\}$		(idlr-subr)	$R \sqsubseteq S \mapsto \{\text{subRole}(R, S)\}$
(idlr-inst2)	$\neg A(a) \mapsto \{\neg\text{insta}(a, A)\}$		(idlr-dis)	$\text{Dis}(R, S) \mapsto \{\text{dis}(R, S)\}$
(idlr-triple)	$R(a, b) \mapsto \{\text{triplea}(a, R, b)\}$		(idlr-inv)	$\text{Inv}(R, S) \mapsto \{\text{inv}(R, S)\}$
(irl-ntriple)	$\neg R(a, b) \mapsto \{\neg\text{triplea}(a, R, b)\}$		(idlr-irr)	$\text{Irr}(R) \mapsto \{\text{irr}(R)\}$
(idlr-subc)	$A \sqsubseteq B \mapsto \{\text{subClass}(A, B)\}$		(idlr-ref)	$\text{Ref}(R) \mapsto \{\text{ref}(R)\}$

DL-Lite$_R$ deduction rules P_{dlr}

(pdlr-instd)	$\text{instd}(x, z) \leftarrow \text{insta}(x, z).$
(pdlr-tripled)	$\text{tripled}(x, r, y) \leftarrow \text{triplea}(x, r, y).$
(pdlr-subc)	$\text{instd}(x, z) \leftarrow \text{subClass}(y, z), \text{instd}(x, y).$
(pdlr-supnot)	$\neg\text{instd}(x, z) \leftarrow \text{supNot}(y, z), \text{instd}(x, y).$
(pdlr-subex)	$\text{instd}(x, z) \leftarrow \text{subEx}(v, z), \text{tripled}(x, v, x').$
(pdlr-supex)	$\text{tripled}(x, r, x') \leftarrow \text{supEx}(y, r, x'), \text{instd}(x, y).$
(pdlr-subr)	$\text{tripled}(x, w, x') \leftarrow \text{subRole}(v, w), \text{tripled}(x, v, x').$
(pdlr-dis1)	$\neg\text{tripled}(x, u, y) \leftarrow \text{dis}(u, v), \text{tripled}(x, v, y).$
(pdlr-dis2)	$\neg\text{tripled}(x, v, y) \leftarrow \text{dis}(u, v), \text{tripled}(x, u, y).$
(pdlr-inv1)	$\text{tripled}(y, v, x) \leftarrow \text{inv}(u, v), \text{tripled}(x, u, y).$
(pdlr-inv2)	$\text{tripled}(y, u, x) \leftarrow \text{inv}(u, v), \text{tripled}(x, v, y).$
(pdlr-irr)	$\neg\text{tripled}(x, u, x) \leftarrow \text{irr}(u), \text{const}(x).$
(pdlr-ref)	$\text{tripled}(x, u, x) \leftarrow \text{ref}(u), \text{const}(x).$
(pdlr-ninstd)	$\neg\text{instd}(x, z) \leftarrow \neg\text{insta}(x, z).$
(pdlr-ntripled)	$\neg\text{tripled}(x, r, y) \leftarrow \neg\text{triplea}(x, r, y).$
(pdlr-nsubc)	$\neg\text{instd}(x, y) \leftarrow \text{subClass}(y, z), \neg\text{instd}(x, z).$
(pdlr-nsupnot)	$\neg\text{instd}(x, y) \leftarrow \text{supNot}(y, z), \neg\text{instd}(x, z).$
(pdlr-nsubex)	$\neg\text{tripled}(x, v, x') \leftarrow \text{subEx}(v, z), \text{const}(x'), \neg\text{instd}(x, z).$
(pdlr-nsupex)	$\neg\text{instd}(x, y) \leftarrow \text{supEx}(y, r, w), \text{const}(x), \text{all\_nrel}(x, r).$
(pdlr-nsubr)	$\neg\text{tripled}(x, v, x') \leftarrow \text{subRole}(v, w), \neg\text{tripled}(x, w, x').$
(pdlr-ninv1)	$\neg\text{tripled}(y, v, x) \leftarrow \text{inv}(u, v), \neg\text{tripled}(x, u, y).$
(pdlr-ninv2)	$\neg\text{tripled}(y, u, x) \leftarrow \text{inv}(u, v), \neg\text{tripled}(x, v, y).$
(pdlr-allnrel1)	$\text{all\_nrel\_step}(x, r, y) \leftarrow \text{first}(y), \neg\text{tripled}(x, r, y).$
(pdlr-allnrel2)	$\text{all\_nrel\_step}(x, r, y) \leftarrow \text{all\_nrel\_step}(x, r, y'), \text{next}(y', y), \neg\text{tripled}(x, r, y).$
(pdlr-allnrel3)	$\text{all\_nrel}(x, r) \leftarrow \text{last}(y), \text{all\_nrel\_step}(x, r, y).$

Output translation $O(\alpha)$

(o-concept)	$A(a) \mapsto \{A(a)\}$
(o-role)	$R(a, b) \mapsto \{R(a, b)\}$

---

[3] Note that, by the normal form above, this kind of axioms is in the form $A_{\exists R} \sqsubseteq \exists R$.

DL-Lite$_\mathcal{R}$ Deduction Rules: rules in P_{dlr} (in Table 2) add deduction rules for ABox reasoning. In the case of existential axioms, the rule (pdlr-supex) introduces a new relation to the auxiliary individual as follows:

$$\texttt{tripled}(x, r, x') \leftarrow \texttt{supEx}(y, r, x'), \texttt{instd}(x, y).$$

In this translation the reasoning on negative information is directly encoded by "contrapositive" versions of the rules. For example, with respect to previous rule, we have:

$$\neg\texttt{instd}(x, y) \leftarrow \texttt{supEx}(y, r, w), \texttt{const}(x), \texttt{all\_nrel}(x, r).$$

where $\texttt{all\_nrel}(x, r)$ verifies that $\neg\texttt{triple}(x, r, y)$ holds for all $\texttt{const}(y)$ by an iteration over all constants.

Defeasible Axioms Input Translations: the set of input rules I_D (shown in Table 3) provides the translation of defeasible axioms $D(\alpha)$ in the DKB: in other words, they are used to specify that the axiom α needs to be considered as defeasible. For example, $D(A \sqsubseteq B)$ is translated to $\texttt{def\_subclass}(A, B)$. Note that, by the definition of the normal form, the existential axioms are "compiled out" from defeasible axioms (i.e. defeasible existential axioms can be expressed by using the newly added $A_{\exists R}$ concepts).

Overriding Rules: rules for defeasible axioms provide the different conditions for the correct interpretation of defeasibility: the overriding rules define conditions (corresponding to clashing sets) for recognizing an exceptional instance. For example, for axioms of the form $D(A \sqsubseteq B)$, the translation introduces the rule:

$$\texttt{ovr}(\texttt{subClass}, x, y, z) \leftarrow \texttt{def\_subclass}(y, z), \texttt{instd}(x, y), \neg\texttt{instd}(x, z).$$

Note that in this version of the calculus, the reasoning on negative information (of the clashing sets) is directly encoded in the deduction rules. Overriding rules in P_D are shown in Table 3.

Defeasible Application Rules: another set of rules in P_D defines the defeasible application of such axioms: intuitively, defeasible axioms are applied only to instances that have not been recognized as exceptional. For example, the rule (app-subc) applies a defeasible concept inclusion $D(A \sqsubseteq B)$:

$$\texttt{instd}(x, z) \leftarrow \texttt{def\_subclass}(y, z), \texttt{instd}(x, y), \texttt{not ovr}(\texttt{subClass}, x, y, z).$$

Defeasible application rules are provided in Table 3.

Translation Process. Given a DKB $\mathcal{K}$ in *DL-Lite$_\mathcal{R}$* normal form, a program $PK(\mathcal{K})$ that encodes query answering for $\mathcal{K}$ is obtained as:

$$PK(\mathcal{K}) = P_{dlr} \cup P_D \cup I_{dlr}(\mathcal{K}) \cup I_D(\mathcal{K})$$

Moreover, $PK(\mathcal{K})$ is completed with a set of supporting facts about constants: for every literal $\texttt{nom}(c)$ or $\texttt{supEx}(a, r, c)$ in $PK(\mathcal{K})$, $\texttt{const}(c)$ is added to $PK(\mathcal{K})$.

Table 3. Input and deduction rules for defeasible axioms

Input rules for defeasible axioms $I_D(S)$

(id-inst) $D(A(a)) \mapsto \{\, \texttt{def\_insta}(A, a). \,\}$

(id-triple) $D(R(a, b)) \mapsto \{\, \texttt{def\_triplea}(R, a, b). \,\}$ (id-subr) $D(R \sqsubseteq S) \mapsto \{\, \texttt{def\_subr}(R, S). \,\}$

(id-ninst) $D(\neg A(a)) \mapsto \{\, \texttt{def\_ninsta}(A, a). \,\}$ (id-dis) $D(\mathrm{Dis}(R, S)) \mapsto \{\, \texttt{def\_dis}(R, S). \,\}$

(id-ntriple) $D(\neg R(a, b)) \mapsto \{\, \texttt{def\_ntriplea}(R, a, b). \,\}$ (id-inv) $D(\mathrm{Inv}(R, S)) \mapsto \{\, \texttt{def\_inv}(R, S). \,\}$

(id-subc) $D(A \sqsubseteq B) \mapsto \{\, \texttt{def\_subclass}(A, B). \,\}$ (id-irr) $D(\mathrm{Irr}(R)) \mapsto \{\, \texttt{def\_irr}(R). \,\}$

(id-supnot) $D(A \sqsubseteq \neg B) \mapsto \{\, \texttt{def\_supnot}(A, B). \,\}$ (id-ref) $D(\mathrm{Ref}(R)) \mapsto \{\, \texttt{def\_ref}(R). \,\}$

Deduction rules for defeasible axioms P_D: overriding rules

(ovr-inst) $\texttt{ovr}(\texttt{insta}, x, y) \leftarrow \texttt{def\_insta}(x, y), \neg\texttt{instd}(x, y).$

(ovr-triple) $\texttt{ovr}(\texttt{triplea}, x, r, y) \leftarrow \texttt{def\_triplea}(x, r, y), \neg\texttt{tripled}(x, r, y).$

(ovr-ninst) $\texttt{ovr}(\texttt{ninsta}, x, y) \leftarrow \texttt{def\_ninsta}(x, y), \texttt{instd}(x, y).$

(ovr-ntriple) $\texttt{ovr}(\texttt{ntriplea}, x, r, y) \leftarrow \texttt{def\_ntriplea}(x, r, y), \texttt{tripled}(x, r, y).$

(ovr-subc) $\texttt{ovr}(\texttt{subClass}, x, y, z) \leftarrow \texttt{def\_subclass}(y, z), \texttt{instd}(x, y), \neg\texttt{instd}(x, z).$

(ovr-supnot) $\texttt{ovr}(\texttt{supNot}, x, y, z) \leftarrow \texttt{def\_supnot}(y, z), \texttt{instd}(x, y), \texttt{instd}(x, z).$

(ovr-subr) $\texttt{ovr}(\texttt{subRole}, x, y, r, s) \leftarrow \texttt{def\_subr}(r, s), \texttt{tripled}(x, r, y), \neg\texttt{tripled}(x, s, y).$

(ovr-dis) $\texttt{ovr}(\texttt{dis}, x, y, r, s) \leftarrow \texttt{def\_dis}(r, s), \texttt{tripled}(x, r, y), \texttt{tripled}(x, s, y).$

(ovr-inv1) $\texttt{ovr}(\texttt{inv}, x, y, r, s) \leftarrow \texttt{def\_inv}(r, s), \texttt{tripled}(x, r, y), \neg\texttt{tripled}(y, s, x).$

(ovr-inv2) $\texttt{ovr}(\texttt{inv}, x, y, r, s) \leftarrow \texttt{def\_inv}(r, s), \texttt{tripled}(y, s, x), \neg\texttt{tripled}(x, r, y).$

(ovr-irr) $\texttt{ovr}(\texttt{irr}, x, r) \leftarrow \texttt{def\_irr}(r), \texttt{tripled}(x, r, x).$

(ovr-ref) $\texttt{ovr}(\texttt{ref}, x, r) \leftarrow \texttt{def\_ref}(r), \neg\texttt{tripled}(x, r, x).$

Deduction rules for defeasible axioms P_D: application rules

(app-inst) $\texttt{instd}(x, z) \leftarrow \texttt{def\_insta}(x, z), \texttt{not}\ \texttt{ovr}(\texttt{insta}, x, z).$

(app-triple) $\texttt{tripled}(x, r, y) \leftarrow \texttt{def\_triplea}(x, r, y), \texttt{not}\ \texttt{ovr}(\texttt{triplea}, x, r, y).$

(app-subc) $\texttt{instd}(x, z) \leftarrow \texttt{def\_subclass}(y, z), \texttt{instd}(x, y), \texttt{not}\ \texttt{ovr}(\texttt{subClass}, x, y, z).$

(app-supnot) $\neg\texttt{instd}(x, z) \leftarrow \texttt{def\_supnot}(y, z), \texttt{instd}(x, y), \texttt{not}\ \texttt{ovr}(\texttt{supNot}, x, y, z).$

(app-subr) $\texttt{tripled}(x, w, x') \leftarrow \texttt{def\_subr}(v, w), \texttt{tripled}(x, v, x'), \texttt{not}\ \texttt{ovr}(\texttt{subRole}, x, y, v, w).$

(app-dis1) $\neg\texttt{tripled}(x, v, y) \leftarrow \texttt{def\_dis}(u, v), \texttt{tripled}(x, u, y), \texttt{not}\ \texttt{ovr}(\texttt{dis}, x, y, u, v).$

(app-dis2) $\neg\texttt{tripled}(x, u, y) \leftarrow \texttt{def\_dis}(u, v), \texttt{tripled}(x, v, y), \texttt{not}\ \texttt{ovr}(\texttt{dis}, x, y, u, v).$

(app-inv1) $\texttt{tripled}(y, v, x) \leftarrow \texttt{def\_inv}(u, v), \texttt{tripled}(x, u, y), \texttt{not}\ \texttt{ovr}(\texttt{inv}, x, y, u, v).$

(app-inv2) $\texttt{tripled}(x, u, y) \leftarrow \texttt{def\_inv}(u, v), \texttt{tripled}(y, v, x), \texttt{not}\ \texttt{ovr}(\texttt{inv}, x, y, u, v).$

(app-irr) $\neg\texttt{tripled}(x, u, x) \leftarrow \texttt{def\_irr}(u), \texttt{const}(x)\ \texttt{not}\ \texttt{ovr}(\texttt{irr}, x, u).$

(app-ref) $\texttt{tripled}(x, u, x) \leftarrow \texttt{def\_ref}(u), \texttt{const}(x)\ \texttt{not}\ \texttt{ovr}(\texttt{ref}, x, u).$

(app-ninst) $\neg\texttt{instd}(x, z) \leftarrow \texttt{def\_ninsta}(x, z), \texttt{not}\ \texttt{ovr}(\texttt{ninsta}, x, z).$

(app-ntriple) $\neg\texttt{tripled}(x, r, y) \leftarrow \texttt{def\_ntriplea}(x, r, y), \texttt{not}\ \texttt{ovr}(\texttt{ntriplea}, x, r, y).$

(app-nsubc) $\neg\texttt{instd}(x, y) \leftarrow \texttt{def\_subclass}(y, z), \neg\texttt{instd}(x, z), \texttt{not}\ \texttt{ovr}(\texttt{subClass}, x, y, z).$

(app-nsupnot) $\neg\texttt{instd}(x, y) \leftarrow \texttt{def\_supnot}(y, z), \texttt{instd}(x, z), \texttt{not}\ \texttt{ovr}(\texttt{supNot}, x, y, z).$

(app-nsubr) $\neg\texttt{tripled}(x, v, y) \leftarrow \texttt{def\_subr}(v, w), \neg\texttt{tripled}(x, w, y), \texttt{not}\ \texttt{ovr}(\texttt{subRole}, x, y, v, w).$

(app-ninv1) $\neg\texttt{tripled}(y, v, x) \leftarrow \texttt{def\_inv}(u, v), \neg\texttt{tripled}(x, u, y), \texttt{not}\ \texttt{ovr}(\texttt{inv}, x, y, u, v).$

(app-ninv2) $\neg\texttt{tripled}(x, u, y) \leftarrow \texttt{def\_inv}(u, v), \neg\texttt{tripled}(y, v, x), \texttt{not}\ \texttt{ovr}(\texttt{inv}, x, y, u, v).$

Then, given an arbitrary enumeration $c_0, \ldots, c_n$ s.t. each $\texttt{const}(c_i) \in PK(\mathcal{K})$, the facts $\texttt{first}(c_0)$, $\texttt{last}(c_n)$ and $\texttt{next}(c_i, c_{i+1})$ with $0 \leq i < n$ are added to $PK(\mathcal{K})$. Query answering $\mathcal{K} \models \alpha$ is then obtained by testing whether the (instance) query, translated to datalog by $O(\alpha)$, is a consequence of $PK(\mathcal{K})$, i.e., whether $PK(\mathcal{K}) \models O(\alpha)$ holds.

Correctness. The presented translation procedure provides a sound and complete materialization calculus for instance checking on *DL-Lite$_R$* DKBs in normal form.

As in [6], the proof for this result can be verified by establishing a correspondence between minimal justified models of $\mathcal{K}$ and answer sets of $PK(\mathcal{K})$. Besides the simpler structure of the final program, the proof is simplified by the direct formulation of rules for negative reasoning. Another new aspect of the proof in the case of *DL-Lite$_R$* resides in the management of existential axioms, since there is the need to define a correspondence between the auxiliary individuals in the translation and the interpretation of existential axioms in the semantics: we follow the approach of Krötzsch in [19], where, in building the correspondence with justified models, auxiliary constants aux^α are mapped to the class of Skolem individuals for existential axiom α.

As in [6], in our translation we consider UNA and *named models*, i.e. interpretations restricted to $sk(N_\mathcal{K})$. Thus, we can show the correctness result on Herbrand models, that will be denoted $\hat{\mathcal{I}}(\chi)$. Let $\mathcal{I}_{CAS} = \langle \mathcal{I}, \chi \rangle$ be a justified named CAS-model. We define the set of overriding assumptions $OVR(\mathcal{I}_{CAS}) = \{\, \texttt{ovr}(p(\mathbf{e})) \mid \langle \alpha, \mathbf{e} \rangle \in \chi,\ I_{dlr}(\alpha) = p \,\}$. Given a CAS-interpretation $\mathcal{I}_{CAS}$, we can define a corresponding Herbrand interpretation $I(\mathcal{I}_{CAS})$ for $PK(\mathcal{K})$ by including the following atoms in it:

(1). all facts of $PK(\mathcal{K})$;
(2). $\texttt{instd}(a, A)$, if $\mathcal{I} \models A(a)$ and $\neg\texttt{instd}(a, A)$, if $\mathcal{I} \models \neg A(a)$;
(3). $\texttt{tripled}(a, R, b)$, if $\mathcal{I} \models R(a, b)$ and $\neg\texttt{tripled}(a, R, b)$, if $\mathcal{I} \models \neg R(a, b)$;
(4). $\texttt{tripled}(a, R, aux^\alpha)$, if $\mathcal{I} \models \exists R(a)$ for $\alpha = A \sqsubseteq \exists R$;
(5). $\texttt{all\_nrel}(a, R)$ if $\mathcal{I} \models \neg\exists R(a)$;
(6). each $\texttt{ovr}$-literal from $OVR(\mathcal{I}_{CAS})$.

The next proposition shows that the least Herbrand model of $\mathcal{K}$ can be represented by the answer sets of the program $PK(\mathcal{K})$.

Proposition 3. *Let $\mathcal{K}$ be a DKB in DL-Lite$_R$ normal form. Then:*

(i). for every (named) justified clashing assumption χ, the interpretation $S = I(\hat{\mathcal{I}}(\chi))$ is an answer set of $PK(\mathcal{K})$;
(ii). every answer set S of $PK(\mathcal{K})$ is of the form $S = I(\hat{\mathcal{I}}(\chi))$ where χ is a (named) justified clashing assumption for $\mathcal{K}$.

Proof (Sketch). We consider $S = I(\hat{\mathcal{I}}(\chi))$ built as above and reason over the reduct $G_S(PK(\mathcal{K}))$ of $PK(\mathcal{K})$ with respect S. Basically, $G_S(PK(\mathcal{K}))$ contains all ground rules from $PK(\mathcal{K})$ that are not falsified by some NAF literal in S: in particular, this excludes application rules for the axiom instances that are recognized as overridden.

Item (i) can be proved by showing that given a justified χ, S is an answer set for $G_S(PK(\mathcal{K}))$ (and thus $PK(\mathcal{K})$): the proof follows the same reasoning of the one in [6], where the fact that $I(\hat{\mathcal{I}}(\chi))$ satisfies rules of the form (pdlr-supex) in $PK(\mathcal{K})$ is verified by the condition (4) on existential formulas in the construction of the model above.

For item (ii), we can show that from any answer set S we can build a justified model $\mathcal{I}_S$ for $\mathcal{K}$ such that $S = I(\hat{\mathcal{I}}(\chi))$ holds. The model can be defined similarly to the original proof, but we need to consider auxiliary individuals in the domain of $\mathcal{I}_S$, that is thus defined as: $\Delta^{\mathcal{I}_S} = \{c \mid c \in \mathrm{NI}_\Sigma\} \cup \{aux^\alpha \mid \alpha = A \sqsubseteq \exists R \in \mathcal{K}\}$. The result can then be proved by considering the effect of deduction rules for existential axioms in $G_S(PK(\mathcal{K}))$: auxiliary individuals provide the domain elements in $\mathcal{I}_S$ needed to verify this kind of axioms. The justification of the model follows by verifying that the new formulation of overriding rules correctly encodes the possible clashing sets for the input defeasible axioms. $\square$

The correctness result directly follows from Proposition 3.

Theorem 1. *Let $\mathcal{K}$ be a DKB in DL-Lite$_R$ normal form, and let $\alpha \in \mathcal{L}_\Sigma$ such that the output translation $O(\alpha)$ is defined. Then, $\mathcal{K} \models \alpha$ iff $PK(\mathcal{K}) \models O(\alpha)$.*

5   Complexity of Reasoning Problems

We first consider the satisfiability problem, i.e., deciding whether a given *DL-Lite$_R$* DKB has some DKB-model. As it turns out, defeasible axioms do not increase the complexity with respect to satisfiability of *DL-Lite$_R$*, due to the following property.

Proposition 4. *Let $\mathcal{K}$ be a DL-Lite$_R$ DKB, and let $\chi_0 = \{\langle \alpha, \mathbf{e} \rangle \mid D(\alpha) \in \mathcal{K},$ $\mathbf{e}$ is over $N_\mathcal{K}\}$ be the clashing assumption that makes an exception to every defeasible axiom over the individuals occurring in $\mathcal{K}$. Then, $\mathcal{K}$ has some DKB-model iff $\mathcal{K}$ has some CAS-model $\mathcal{I}_{CAS} = \langle \mathcal{I}, \chi_0 \rangle$.*

Informally, the only if direction holds because any DKB-model of $\mathcal{K}$ is also a CAS-model of $\mathcal{K}$; as justified exceptions are only on $N_\mathcal{K}$ and making more exceptions does not destroy CAS-modelhood, some CAS-model with clashing assumptions χ_0 exists. Conversely, if $\mathcal{K}$ has some CAS-model of the form $\mathcal{I}_{CAS} = \langle \mathcal{I}, \chi_0 \rangle$, a justified CAS-model can be obtained by setting $\chi = \chi_0$ and trying to remove, one by one, each clashing assumption $\langle \alpha, \mathbf{e} \rangle$ from χ; this is possible, if $\mathcal{K}$ has some NI-congruent model $\langle \mathcal{I}', \chi \setminus \{\langle \alpha, \mathbf{e} \rangle\} \rangle$. After looping through all clashing assumptions in χ_0, we have that some NI-congruent model $\langle \mathcal{I}', \chi \rangle$ exists that is justified.

Thus, DKB-satisfiability testing boils down to CAS-satisfiability checking, which can be done using the datalog encoding described in the previous section. From the particular form of that encoding, we obtain the following result.

Theorem 2. *Deciding whether a given DL-Lite$_R$ DKB $\mathcal{K}$ has some DKB-model is NLogSpace-complete in combined complexity and FO-rewritable in data complexity.*

To see this, the program $PK(\mathcal{K})$ for $\mathcal{K}$ has in each rule at most one literal with an intensional predicate in the body, i.e., a predicate that is defined by proper rules. Thus, we have a linear datalog program with bounded predicate arity, for which derivability of an atom is feasible in nondeterministic logspace, as this can be reduced to a graph reachability problem in logarithmic space. The NLogSpace-hardness is inherited from the combined complexity of KB satisfiability in $DL\text{-}Lite_\mathcal{R}$, which is NLogSpace-complete.

As regards data-complexity, it is well-known that instance checking and similarly satisfiability testing for $DL\text{-}Lite_\mathcal{R}$ are FO-rewritable [13]; this has been shown by a reformulation algorithm, which informally unfolds the axioms $\alpha(\boldsymbol{x})$ (i.e., performs resolution viewing axioms as clauses), such that deriving an instance $A(a)$ reduces to presence of certain assertions in the ABox. This unfolding can be adorned by typing each argument $x \in \boldsymbol{x}$ of an axiom to whether it is an individual from the DKB (type i), or an unnamed individual (type u); for example, $\alpha(x) = A \sqsubseteq B$ yields $\alpha_i(x)$ and $\alpha_u(x)$. The typing carries over to unfolded axioms. In unfolding, one omits typed versions of defeasible axioms $D(\alpha(\boldsymbol{x}))$, which w.l.o.g. have no existential restrictions; e.g., for $D(\alpha(x)) = D(B \sqsubseteq C)$, one omits $\alpha_i(x)$. In this way, instance derivation (and similarly satisfiability testing) is reduced to presence of certain ABox assertions again.

On the other hand, entailment checking is intractable: while some justified model is constructible in polynomial time, there can be exponentially many clashing assumptions for such models, even under UNA; finding a DKB model that violates an axiom turns out to be difficult.

Theorem 3. *Given a DKB $\mathcal{K}$ and an axiom α, deciding $\mathcal{K} \models \alpha$ is co-NP-complete; this holds also for data complexity and instance checking, i.e., α is of the form $A(a)$ for some assertion $A(a)$.*

Proof (Sketch). To refute $\mathcal{K} \models \alpha$, we can exhibit that a justified CAS-model $\mathfrak{I}_{CAS} = \langle \mathcal{I}, \chi \rangle$ of $\mathcal{K}$ named relative to $sk(N)$ exists such that $\mathcal{I} \not\models \alpha$, with $N_\mathcal{K} \subseteq N \subseteq \mathrm{NI} \setminus \mathrm{NI}_S$ and where N includes few fresh individual names such that $\mathcal{I}$ violates the instance of α for some elements $\mathbf{e}$ over $sk(N)$. We can guess clashing assumptions χ over $N_\mathcal{K}$, where each $\langle \alpha, \mathbf{e} \rangle \in \chi$ has a unique clashing set $S_{\alpha(\mathbf{e})}$, and a partial interpretation over N, and check derivability of all $S_{\alpha(\mathbf{e})}$ and that the interpretation extends to a model of $\mathcal{K}$ in polynomial time. Thus, we overall obtain membership of entailment in co-NP.

The co-NP-hardness can be shown by a reduction from inconsistency-tolerant reasoning from $DL\text{-}Lite_\mathcal{R}$ KBs under AR-semantics [20]. Given a $DL\text{-}Lite_\mathcal{R}$ KB $\mathcal{K} = \mathcal{A} \cup \mathcal{T}$ with ABox $\mathcal{A}$ and TBox $\mathcal{T}$, a repair is a maximal subset $\mathcal{A}' \subseteq \mathcal{A}$ such that $\mathcal{K}' = \mathcal{A}' \cup \mathcal{T}$ is satisfiable; an assertion α is AR-entailed by $\mathcal{K}$, if $\mathcal{K}' \models \alpha$ for every repair $\mathcal{K}'$ of $\mathcal{K}$. As shown by Lembo et al., deciding AR-entailment is co-NP-hard; this continues to hold under UNA and if all assertions involve only concept resp. role names.

Let $\hat{\mathcal{K}} = \mathcal{T} \cup \{D(\alpha) \mid \alpha \in \mathcal{A}\}$, i.e., all assertions from $\mathcal{K}$ are defeasible. As easily seen, the maximal repairs $\mathcal{A}'$ correspond to the justified clashing assumptions by $\chi = \{\langle \alpha, \mathbf{e} \rangle \mid \alpha(\mathbf{e}) \in \mathcal{A} \setminus \mathcal{A}'\}$. Thus, $\mathcal{K}$ AR-entails α iff $\hat{\mathcal{K}} \models \alpha$, proving co-NP-hardness.

For data complexity (without defeasible facts), we can adjust the transformation and emulate $D(A(a))$ by an axiom $D(A' \sqsubseteq A)$: we make the assertion

$A'(a)$, where A' is a fresh concept name; similarly $D(R(a,b))$ is emulated by $D(R' \sqsubseteq R)$ plus $R'(a,b)$, where R' is a fresh role name. As Lembo et al. proved co-NP-hardness under data-complexity, the claimed result follows.   □

We observe that the co-NP-hardness proof in [20] used many role restrictions and inverse roles; for combined complexity, co-NP-hardness of entailment in absence of any role names can be derived from results about propositional circumscription in [12]. In particular, [12, Theorem 16] showed that deciding whether an atom z is a circumscriptive consequence of a positive propositional 2CNF F if all variables except z are minimized (i.e., in circumscription notation $CIRC(F; P, \emptyset; \{z\}) \models z$), is co-NP-hard;[4] such an inference can be easily emulated by entailment from a DKB constructed from F and z, where propositional variables are used as concept names.

Indeed, for each clause $c = x \vee y$ in F, we add to $\mathcal{K}$ an axiom $x \sqsubseteq \neg y$ if $z \neq x, y$ and an axiom $x \sqsubseteq z$ (resp. $y \sqsubseteq z$) if $z = y$ (resp. $x = z$). Furthermore, for each variable $x \neq z$, we add $D(x(a))$, where a is a fixed individual. This effects that justified DKB-models of $\mathcal{K}$ correspond to the models of $CIRC(F; P, \emptyset; \{z\})$, where the minimality of exceptions in justified DKB-models emulates the minimality of circumscription models; thus, $\mathcal{K} \models z(a)$ iff $CIRC(F; P, \emptyset; \{z\}) \models z$. Similarly as above, defeasible assertions could be moved to defeasible axioms $D(c \sqsubseteq v)$ with a single assertion $c(a)$.

While this establishes co-NP-hardness of entailment for combined complexity under UNA when roles are absent, the data complexity is tractable; this is because we can consider the axioms for individuals a separately, and if the GCI axioms are fixed only few axioms per individual exist. This is similar if role axioms but no existential restrictions are permitted, as we can concentrate on the pairs a, b and b, a of individuals. The questions remains how much of the latter is possible while staying tractable.

6   Discussion and Conclusion

Related Works. The relation of the justified exception approach to nonmonotonic description logics was discussed in [6], where in particular an in-depth comparison w.r.t. typicality in DLs [18], normality [3] and overriding [2] was given. A distinctive feature of our approach, linked to the interpretation of exception candidates as different clashing assumptions, is the possibility to "reason by cases" inside the alternative justified models (cf. the discussion of the classic Nixon Diamond example [6, Section 7.4]).

The introduction of non-monotonic features in the *DL-Lite* family and, more in general, to low complexity DLs has been the subject of many works, mostly with the goal of preserving the low complexity properties of the base logic in the extension. For example, in [3] an in-depth study of the complexity of reasoning with circumscription in *DL-Lite$_\mathcal{R}$* and $\mathcal{EL}$ was presented: the work considers

---

[4] The models of $CIRC(F; P, \emptyset; \{z\})$ are all models M of F such that no model M' of F exists with $M' \setminus \{z\} \subset M \setminus \{z\}$.

defeasibility on inclusion axioms $C \sqsubseteq_n D$; in the case of circumscription in *DL-Lite$_\mathcal{R}$*, it is shown that by suitable syntactic restrictions instance checking can be limited to Π_2^p. Similarly, in [17] the authors studied the complexity of the application of their typicality approach to *DL-Lite$_c$* and $\mathcal{EL}^\perp$ in order to search for low complexity fragments of DLs with typicality: in the case of *DL-Lite$_c$* enriched with typical concept inclusions, entailment matches the Π_2^p complexity bound from [3]. A recent work in this direction is [22], where a defeasible version of $\mathcal{EL}^\perp$ was obtained, extending works on rational closure on DLs [14]: higher typicality is modelled by extending classical canonical models of $\mathcal{EL}^\perp$ with multiple representatives of concepts and individuals; inference is then defined on a canonical model of the extended domain.

Summary and Future Directions. In this paper, we considered the justified exception approach in [6] for reasoning on *DL-Lite$_\mathcal{R}$* KBs with defeasible axioms. We have shown that the limited language of *DL-Lite$_\mathcal{R}$* allows us to formulate a direct datalog translation to reason on derivations for negative information in instance checking.

Beyond instance queries, (Boolean) conjunctive queries (CQs) can be defined similar as in [6] and the ASP encoding can be extended in order to model CQ evaluation under cautious (certainty) semantics; the latter problem can be shown to be Π_2^p-complete. An interesting issue is how to manage settings that allow for exceptions on unnamed individuals (generated by existential axioms), and to provide them with suitable semantic characterizations. Multiple auxiliary elements aux^α may be necessary to enable different exceptions for unnamed individuals reached from different individuals; this remains for further investigation. Moreover, we plan to apply and evaluate the current results on *DL-Lite$_\mathcal{R}$* in the framework of Contextualized Knowledge Repositories with hierarchies as in [9].

References

1. Baader, F., Calvanese, D., McGuinness, D., Nardi, D., Patel-Schneider, P. (eds.): The Description Logic Handbook. Cambridge University Press, Cambridge (2003)
2. Bonatti, P.A., Faella, M., Petrova, I., Sauro, L.: A new semantics for overriding in description logics. Artif. Intell. **222**, 1–48 (2015)
3. Bonatti, P.A., Faella, M., Sauro, L.: Defeasible inclusions in low-complexity DLs. J. Artif. Intell. Research **42**, 719–764 (2011)
4. Bonatti, P.A., Lutz, C., Wolter, F.: Description logics with circumscription. In: KR 2006, pp. 400–410. AAAI Press (2006)
5. Bozzato, L., Eiter, T., Serafini, L.: Contextualized knowledge repositories with justifiable exceptions. In: DL2014. CEUR-WP, vol. 1193, pp. 112–123 (2014). CEUR-WS.org
6. Bozzato, L., Eiter, T., Serafini, L.: Enhancing context knowledge repositories with justifiable exceptions. Artif. Intell. **257**, 72–126 (2018)
7. Bozzato, L., Homola, M., Serafini, L.: Towards more effective tableaux reasoning for CKR. In: DL2012, CEUR-WP, vol. 846, pp. 114–124 (2012). CEUR-WS.org
8. Bozzato, L., Serafini, L.: Materialization calculus for contexts in the semantic web. In: DL2013, CEUR-WP, vol. 1014, pp. 552–572 (2013). CEUR-WS.org

9. Bozzato, L., Serafini, L., Eiter, T.: Reasoning with justifiable exceptions in contextual hierarchies. In: KR 2018, pp. 329–338. AAAI Press (2018)
10. Britz, K., Varzinczak, I.: Introducing role defeasibility in description logics. In: Michael, L., Kakas, A. (eds.) JELIA 2016. LNCS (LNAI), vol. 10021, pp. 174–189. Springer, Cham (2016). https://doi.org/10.1007/978-3-319-48758-8_12
11. de Bruijn, J., Eiter, T., Tompits, H.: Embedding approaches to combining rules and ontologies into autoepistemic logic. In: KR 2008, pp. 485–495. AAAI Press (2008)
12. Cadoli, M., Lenzerini, M.: The complexity of propositional closed world reasoning and circumscription. J. Comput. Syst. Sci. **48**(2), 255–310 (1994)
13. Calvanese, D., De Giacomo, G., Lembo, D., Lenzerini, M., Rosati, R.: Tractable reasoning and efficient query answering in description logics: the DL-Lite family. J. Autom. Reasoning **39**(3), 385–429 (2007)
14. Casini, G., Straccia, U.: Rational closure for defeasible description logics. In: Janhunen, T., Niemelä, I. (eds.) JELIA 2010. LNCS (LNAI), vol. 6341, pp. 77–90. Springer, Heidelberg (2010). https://doi.org/10.1007/978-3-642-15675-5_9
15. Eiter, T., Ianni, G., Lukasiewicz, T., Schindlauer, R., Tompits, H.: Combining answer set programming with description logics for the semantic web. Artif. Intell. **172**(12–13), 1495–1539 (2008)
16. Gelfond, M., Lifschitz, V.: Classical negation in logic programs and disjunctive databases. New Gener. Comput. **9**, 365–385 (1991)
17. Giordano, L., Gliozzi, V., Olivetti, N., Pozzato, G.L.: Reasoning about typicality in low complexity DLs: the logics $\mathcal{EL}^\perp T_{min}$ and $DL\text{-}Lite_c T_{min}$. In: Walsh, T. (ed.) IJCAI 2011, pp. 894–899. IJCAI/AAAI (2011)
18. Giordano, L., Gliozzi, V., Olivetti, N., Pozzato, G.L.: A non-monotonic description logic for reasoning about typicality. Artif. Intell. **195**, 165–202 (2013)
19. Krötzsch, M.: Efficient inferencing for OWL EL. In: Janhunen, T., Niemelä, I. (eds.) JELIA 2010. LNCS (LNAI), vol. 6341, pp. 234–246. Springer, Heidelberg (2010). https://doi.org/10.1007/978-3-642-15675-5_21
20. Lembo, D., Lenzerini, M., Rosati, R., Ruzzi, M., Savo, D.F.: Inconsistency-tolerant semantics for description logics. In: Hitzler, P., Lukasiewicz, T. (eds.) RR 2010. LNCS, vol. 6333, pp. 103–117. Springer, Heidelberg (2010). https://doi.org/10.1007/978-3-642-15918-3_9
21. Motik, B., Fokoue, A., Horrocks, I., Wu, Z., Lutz, C., Grau, B.C.: OWL 2 web ontology language profiles. In: W3C Recommendation, W3C, October 2009
22. Pensel, M., Turhan, A.: Reasoning in the defeasible description logic $\mathcal{EL}_\perp$ - computing standard inferences under rational and relevant semantics. Int. J. Approx. Reasoning **103**, 28–70 (2018)
23. Serafini, L., Homola, M.: Contextualized knowledge repositories for the semantic web. J. Web Semant. **12**, 64–87 (2012)

ODRL Policy Modelling and Compliance Checking

Marina De Vos[1], Sabrina Kirrane[2(✉)], Julian Padget[1], and Ken Satoh[3]

[1] University of Bath, Bath, UK
{m.d.vos,j.a.padget}@bath.ac.uk
[2] Vienna University of Economics and Business, Vienna, Austria
sabrina.kirrane@wu.ac.at
[3] National Institute of Informatics and Sokendai, Tokyo, Japan
ksatoh@nii.ac.jp

Abstract. This paper addresses the problem of constructing a policy pipeline that enables compliance checking of business processes against regulatory obligations. Towards this end, we propose an Open Digital Rights Language (ODRL) profile that can be used to capture the semantics of both business policies in the form of sets of required permissions and regulatory requirements in the form of deontic concepts, and present their translation into Answer Set Programming (via the Institutional Action Language (InstAL)) for compliance checking purposes. The result of the compliance checking is either a positive compliance result or an explanation pertaining to the aspects of the policy that are causing the non-compliance. The pipeline is illustrated using two (key) fragments of the General Data Protect Regulation, namely Articles 6 (Lawfulness of processing) and Articles 46 (Transfers subject to appropriate safeguards) and industrially-relevant use cases that involve the specification of sets of permissions that are needed to execute business processes. The core contributions of this paper are the ODRL profile, which is capable of modelling regulatory obligations and business policies, the exercise of modelling elements of GDPR in this semantic formalism, and the operationalisation of the model to demonstrate its capability to support personal data processing compliance checking, and a basis for explaining why the request is deemed compliant or not.

1 Introduction

The General Data Protection Regulation (GDPR), which came into effect in May 2018, provides data controllers and processors with legal requirements and guidelines concerning the processing and sharing of personal data. Although there are a number of self assessment tools that can be used by companies to manually assess GDPR compliance (cf., Information Commissioner's Office (ICO) UK [19], Microsoft Trust Center [24], Nymity [26]), automated compliance checking of business processes with respect to legal obligations is highly desirable, especially when such systems are regularly updated (e.g., maintenance and feature updates).

In order to support automated compliance checking it is necessary to encode both GDPR requirements and business processes in a machine understandable format.

© Springer Nature Switzerland AG 2019
P. Fodor et al. (Eds.): RuleML+RR 2019, LNCS 11784, pp. 36–51, 2019.
https://doi.org/10.1007/978-3-030-31095-0_3

Existing work in the area focuses on modelling legal requirements using semantic technologies [4–6,31]; reasoning over legal rules [2,11,29]; and reasoning over business policies [11,18,22]. At the same time there are several semantic technology based general policy languages, such as KAoS [9], Rei [21] and Protune [8], which could potentially be used to model and reason over legal requirements and business policies. However, there are no standardised mechanisms for representing policies, such that compliance can be checked automatically.

When it comes to Web standardisation activities relating to semantic based policy languages, the closest match is the Open Digital Rights Language (ODRL) information model and associated vocabularies, which enables various parties to specify permissions, prohibitions, and duties relating to actions performed on assets. Although ODRL is primarily used to specify licenses, it could be adapted/extended, via the ODRL profile mechanism, such that it is also possible to specify a broader set of policies.

Thus, in this paper, we introduce our ODRL Regulatory Compliance Profile (ORCP), which can be used to model both regulatory requirements and sets of permissions required to execute a business process, and discuss how these policies can be translated into Answer Set Programming (ASP) [17] rules such that it is possible to automatically check compliance, identify conflicts and propose resolution strategies. ASP is a declarative programming language with a mathematical foundation guaranteeing sound and completeness of the results. The translation to ASP is facilitated through InstAL [10,23,27], a domain specific action language for modelling normative systems and legal frameworks.

Summarising our contributions, we: (i) propose an ODRL profile, which is capable of modelling regulatory permissions, prohibitions, obligations, and dispensations, and permissions needed to execute business processes within a company; (ii) show how elements of the GDPR can be modelled using this semantic formalism; (iii) demonstrate how ASP rules based on InstAL can be used for automatic compliance checking; and (iv) propose a mechanism to provide evidence in support of the compliance decision and identify what is missing.

The remainder of the paper is structured as follows: Sect. 2 discusses related work on modelling and reasoning over legal requirements. Section 3 introduces our ODRL profile that can be used to encode both regulatory requirements and permissions required by business processes. Section 4 demonstrates how InstAL can be used for ODRL policy compliance checking and explanation generation. Section 5 offers some evaluation of and reflection on our proposal. Finally, Sect. 6 concludes the paper and identifies several open research questions.

2   Related Work

Over the years there have been several prominent works that focus specifically on modelling legal requirements using semantic technologies [4–6,31]. Boer et al. [6] build upon the MetaLex [5] eXtensible Markup Language legislation encoding mechanism, in order to define a language and vocabularies that cater for the interchange of legal knowledge, known as the Legal Knowledge Interchange Format (LKIF). Bartolini et al. [4] propose an ontology that can be used to model data protection requirements, and demonstrates how it can be integrated into existing business workflows. Like Bartolini et al. [4], Pandit et al. [31] focus specifically on data protection, however they demonstrate how the European Legislation Identifier (ELI) ontology can be used to model the GDPR as linked data. Outputs include a DCAT[1] catalog containing the official text of the GDPR and a SKOS[2] ontology defining concepts related to GDPR.

In addition, there is a body of work relating to legal reasoning [2, 11, 18, 18, 22, 29]. Palmirani et al. [29] and Athan et al. [2] demonstrate how LegalRuleML, an extension of RuleML [7], can be used to specify legal norms, guidelines, and policies. Dimyadi et al. [11] compares LegalRuleML and LKIF formalisms and highlights that one of the primary benefits of Semantic technology based approaches is the availability of mature reasoning engines. While, Lam and Hashmi [22] demonstrate how LegalRuleML can be translated into defeasible logic, which allows for modelling and reasoning over business policies. Governatori et al. [18] in turn shows how LegalRuleML together with Semantic technologies is used for business process regulatory compliance checking.

In the early days of Semantic Web research, researchers developed general policy languages with formal semantics (such as KAoS [9], Rei [21] and Protune [8]), that could potentially be used to model and reason over legal permissions, prohibitions, obligations, and dispensations. More recently, the Open Digital Rights Language, which was primarily intended to define rights to or to limit access to digital resources (cf. [30]), has demonstrated its potential as a general policy language. For instance, researchers have hinted as to how it could be used to express: access policies [33]; requests, data offers and agreements [32]; and basic regulatory policies [1]. While, Fornara and Colombetti [12] consider how to add obligations to (an earlier version) of ODRL and subsequently in Fornara et al. [13] how to reason over such ODRL extensions, using additional ontologies and semantic rules.

In this paper, we propose an ODRL regulatory compliance profile that can be used to model both regulatory requirements in terms of deontic concepts (permissions, prohibitions, obligations and dispensations), and business policies in the forms of sets of permissions required to execute the policy. The ODRL policies are subsequently translated into ASP rules, which not only cater for automatic compliance checking, but also for non compliance detection and explanation.

3   Modelling Legislative Requirements and Business Policies Using ODRL

We start by presenting our generalised ODRL information model and subsequently demonstrate how it can be used to encode legal requirements and permissions required by business processes. At this stage we do not aim to be exhaustive in terms of modelling, but rather our objective is to demonstrate that ODRL can be extended to cater for the modelling of regulatory policies (in the form of nested permissions, prohibitions, obligations, and dispensations) and business policies (in the form of discrete permissions that are needed to execute a business process). The full ODRL Regulatory

---

[1] DCAT, https://www.w3.org/TR/vocab-dcat/.

[2] SKOS, https://www.w3.org/TR/skos-reference/.

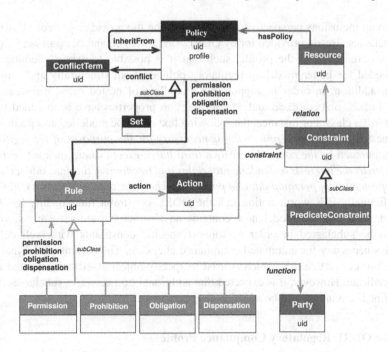

Fig. 1. A generalised ODRL information model

Compliance Profile, including examples in both Turtle and JSON-LD serialisations are available in the form of a draft specification[3] and an ontology[4]. Our expectation is that the ODRL Regulatory Compliance profile ontology will be extended by domain experts with additional classes and properties to support not only the modelling of relevant articles from the GDPR but also other legislation.

3.1 Generalising the ODRL Information Model

Figure 1 provides a high level overview of a generalised ODRL information model. Like ODRL, a `Policy` is composed of a `Set` of `Rules` each of which govern an `Action` that is performed by `Party`. Given that `Asset` is too specific, it is replaced by a more general `Resource` class, which is a superclass of `Asset`. As per ODRL, the `ConflictTerm` class is used to specify the conflict resolution strategy.

In terms of exclusions, the `Agreement` and `Offer` policy subclasses, the `Duty` rule subclass, and the `duty`, `failure`, `remedy`, and `consequence` properties are removed in the new model, as these classes and properties were strongly motivated by use cases relating to licensing.

---

[3] ODRL Regulatory Compliance Profile, https://ai.wu.ac.at/policies/orcp/regulatory-model.html.

[4] ODRL Regulatory Compliance Profile Ontology, https://ai.wu.ac.at/policies/orcp/odrl_regulatory_profile.ttl.

From an inclusions perspective, in addition to the `Permissions`, `Prohibitions` rule subclasses already provided for by ODRL, `Obligation` and `Dispensation` rule subclasses are added to the profile, such that it is possible to express deontic concepts needed for better modeling regulatory policies, both structurally and semantically. In addition, in order to support the modelling of nested rules, `permission`, `prohibition`, `obligation` and `dispensation` properties have been added to the abstract `Rule` class. For instance, the following text could be modelled as a permission with a nested prohibition: *"processing is necessary for the purposes of the legitimate interests pursued by the controller or by a third party, except where such interests are overridden by the interests or fundamental rights and freedoms of the data subject which require protection of personal data, in particular where the data subject is a child"*.

Additionally it is worth noting that the ODRL constraint functionality has been curtailed, such that the model now contains an abstract `Constraint` class, which needs to be subclassed in order to support specific constraints with well defined semantics necessary for automated compliance checking. The current model includes a `PredicateConstraint`, which is used to specify object assertions expected for a given predicate. However, it is expected that additional `Constraint` subclasses (with well defined semantics) will be added as the need arises.

3.2 The ODRL Regulatory Compliance Profile

In addition to the core classes and properties outlined in the previous section, based on our analysis of Article 6 and Article 46 of the GDPR, the profile defines several additional classes (e.g., `LegalBasis`, `Purpose`, and `Location`) and properties (e.g., `legalBasis`, `purpose`, `processingLocation`, `recipientLocation`, `organisationType`, `appropriateSafeguards`, and `dataSubjectProvisions`),

Art. 46 GDPR – Transfers subject to appropriate safeguards

1. In the absence of a decision pursuant to Article 45(3), a controller or processor may transfer personal data to a third country or an international organisation only if the controller or processor has provided appropriate safeguards, and on condition that enforceable data subject rights and effective legal remedies for data subjects are available.
2. The appropriate safeguards referred to in paragraph 1 may be provided for, without requiring any specific authorisation from a supervisory authority, by:
   (a) a legally binding and enforceable instrument between public authorities or bodies;
   (b) binding corporate rules in accordance with Article 47;
   (c) standard data protection clauses adopted by the Commission in accordance with the examination procedure referred to in Article 93(2);
   (d) standard data protection clauses adopted by a supervisory authority and approved by the Commission pursuant to the examination procedure referred to in Article 93(2);
   (e) an approved code of conduct pursuant to Article 40 together with binding and enforceable commitments of the controller or processor in the third country to apply the appropriate safeguards, including as regards data subjects' rights; or
   (f) an approved certification mechanism pursuant to Article 42 together with binding and enforceable commitments of the controller or processor in the third country to apply the appropriate safeguards, including as regards data subjects' rights.

Annotation key: party, resource, action, constraint

Fig. 2. Paragraphs 1 and 2 excerpted from GDPR Article 46

which are needed in order to check the compliance of business processes with said articles. The example ODRL Regulatory Compliance Profile policies (based on paragraphs 1 and 2 of Article 46 of the GDPR) presented in this paper are encoded using the Turtle serialisation syntax, with the `odrl` prefix used to denote the ODRL ontology[5] and the `orcp` prefix used to denote the proposed regulatory compliance ontology[6]. Figure 2 depicts an extract from Article 46 of the GDPR, where colour coding is used to highlight parties, resources, actions, and constraints that need to be modelled using our regulatory profile. Given that the objective is to enable companies to assert that various data subject provisions and safeguards exist, we treat each point under a paragraph as a single resource and do not model the parties, actions and constraints relating to these resources.

```
1   <http://example.com/policy:gdpr-article46> a orcp:Set ;
2       odrl:profile <http://example.com/odrl:profile:regulatory-compliance> ;
3       orcp:permission
4           [ odrl:action orcp:Transfer ;
5             orcp:data orcp:PersonalData ;
6             odrl:predicateConstraint
7               [ odrl:or (
8                   [ odrl:leftOperand orcp:organisationType ;
9                     odrl:operator odrl:isA ;
10                    odrl:rightOperand orcp:InternationalOrganisation
11                  ]
12                  [ odrl:leftOperand orcp:recipientLocation ;
13                    odrl:operator odrl:isA ;
14                    odrl:rightOperand orcp:ThirdCountry
15                ] )
16             ] ;
17         orcp:obligation
18             [ odrl:predicateConstraint
19                 [ odrl:leftOperand orcp:dataSubjectProvisions ;
20                   odrl:operator odrl:isA ;
21                   odrl:rightOperand orcp:EnforceableDataSubjectRights
22                 ]
23             ],
24             [ odrl:predicateConstraint
25                 [ odrl:leftOperand orcp:dataSubjectProvisions ;
26                   odrl:operator odrl:isA ;
27                   odrl:rightOperand orcp:LegalRemediesForDataSubjects
28                 ]
29             ],
30             [ odrl:predicateConstraint
31                 [ odrl:leftOperand orcp:appropriateSafeguards ;
32                   odrl:operator odrl:isAnyOf ;
33                   odrl:rightOperand ( orcp:LegallyBindingEnforceableInstrument
34                                       orcp:BindingCorporateRules
35                                       orcp:StandardDataProtectionClauses
36                                       orcp:ApprovedCodeOfConduct
37                                       orcp:ApprovedCertificateMechanism )
38                 ]
39             ]
40        ] .
```

Listing 1. ODRL/TTL representation of paragraphs 1 and 2 of GDPR Article 46

---

[5] <http://www.w3.org/ns/odrl/2/>.

[6] <http://example.com/odrl:profile:regulatory-compliance/>.

Listing 2. ODRL/TTL request for permission to transfer personal data

```
1   <http://example.com/policy:bp-transfer> a orcp:Set ;
2       odrl:profile <http://example.com/odrl:profile:regulatory-compliance> ;
3       orcp:permission
4           [ odrl:action orcp:Transfer ;
5             orcp:data orcp:PersonalData ;
6             orcp:responsibleParty orcp:Controller ;
7             orcp:organisationType orcp:InternationalOrganisation ;
8             orcp:sender <http://example.com/CompanyA_Ireland> ;
9             orcp:recipient <http://example.com/CompanyA_US> ;
10            orcp:recipientLocation orcp:ThirdCountry ;
11            orcp:purpose orcp:PersonalRecommendations ;
12            orcp:legalBasis orcp:Consent ;
13            odrl:dataSubjectProvisions orcp:EnforceableDataSubjectRights ;
14            odrl:dataSubjectProvisions orcp:LegalRemediesForDataSubjects
15           ] .
```

The ODRL Regulatory Model representation of Article 46 is presented in Listing 1 and the permission needed to execute the business process is presented in Listing 2. More specifically, the regulatory policy presented in Listing 1 states that the `Transfer` (action) of `PersonalData` (resource) to an `InternationalOrganisation` or `ThirdCountry` is permitted if `EnforceableDataSubject`, `LegalRemediesFor DataSubjects`, and `appropriateSafeguards` of type `LegallyBindingEn forceableInstrument`, `BindingCorporateRules`, `StandardDataProtection Clauses`, `ApprovedCodeOfConduct`, or `ApprovedCertificateMechanism` are asserted in the company policy. While, the permission needed to execute a business process, presented in Listing 2, states that a `Controller` (party) who is an `InternationalOrganisation` wishes to perform a `Transfer` (action) of `PersonalData` (resource), from `CompanyA_Ireland` to `CompanyA_USA` in a `ThirdCountry`, for generating `PersonalRecommendations`, where the lawfulness for processing is `Consent`. In addition the company policy asserts that the company has `EnforceableDataSubjectRights` and `LegalRemediesForDataSubjects` in place within the company.

4   Compliance Checking

The previous section describes and justifies the design of the ODRL policy compliance profile, which provides us with a means to represent a regulatory policy – such as fragments of GDPR – and to represent a company's particular business process, but it does not give the means to determine whether an implementation is compliant with the regulatory obligation. ODRL is not written using OWL-DL or even full OWL, it is purely RDF, and as such poses technical problems for some existing tools, for example the standard reasoners Pellet and Hermit cannot handle ODRL out of the box, while open-world reasoning over RDF has computational tractability issues. A practical, and common solution, is to switch from open-world to closed-world reasoning [25], which is the approach we take here, while also translating the ODRL policy representations into InstAL, which is subsequently compiled into an Answer Set Program [3], so that we may benefit from the computational capabilities of an Answer Set solver, in our case

CLINGO [15]. We choose to use InstAL partly since it is a familiar tool for us, and partly because InstAL has been designed as a domain specific language for representation and reasoning about regulations, and thus offers modelling elements suitable for the task.

In the rest of this section, we provide a brief primer on InstAL, before describing how compliance check data is encoded, and how ODRL policies are translated to InstAL. We conclude the section with an example of a data transfer compliance check, based on the content of Listing 2, which illustrates how the model detects conflicts between the process description and the regulatory policy.

4.1 Institutional Action Language

The Institutional Action Language (InstAL) is a domain-specific language (DSL) for writing models in terms of events and states, which translates to Answer Set Programming (ASP) for model evaluation under closed-world (non-monotonic) reasoning. Model state is expressed in terms of fluents – facts that are true if present and false if absent – which can be either inertial – true once initiated, until explicitly terminated – or non-inertial – whose presence is the result of a condition expressed over the model state. Inertial fluents model so-called institutional or normative facts, following the concepts of deontic logic [34], namely permission/prohibition and obligation, institutional power [20] and domain-specific facts. An InstAL specification has five kinds of rules: (i) x generates y: x is an event (action) and y is one or more events, whose generation can be conditional on the model state; (ii) x initiates y: x is an event and y is one or more fluents to add to the model state, subject to a condition as above; (iii) x terminates y: as initiates, but the fluents are deleted; (iv) x when y: x is a non-inertial fluent and y is a condition over the model state; and (v) initially x: x is one or more fluents that shall be part of the initial model state, again possibly subject to a condition.

4.2 Data Representation

The approach taken for the purposes of this paper is to map the ODRL representation into three-element term fluents, reflecting the underlying RDF triples:

```
type Subject;
type Predicate;
type Object;
fluent triple(Subject,Predicate,Object);
```

which while simplistic, offers a uniform representation to which it is straightforward to translate. Consequently, we may represent the data for a given compliance-check, such as the transfer process description in Listing 2, as a set of triples:

```
triple(bp_transfer,action,transfer)
triple(bp_transfer,resource,personalData)
triple(bp_transfer,responsibleParty,controller)
triple(bp_transfer,organisationType,internationalOrganisation)
triple(bp_transfer,sender,companyA_Ireland)
triple(bp_transfer,recipient,companyA_US)
triple(bp_transfer,recipientLocation,thirdCountry)
triple(bp_transfer,purpose,personalRecommendations)
triple(bp_transfer,legalBasis,consent)
triple(bp_transfer,dataSubjectProvisions,enforceableDataSubjectRights)
triple(bp_transfer,dataSubjectProvisions,effectiveLegalRemedies)
```

4.3 Policy Representation

The GDPR article fragments are represented as rules which determine the compliance of the process description depending on the data supplied (such as that given above), by means of non-inertial fluent (when) rules, whose left-hand side is true, if the expression on the right-hand side is true, to determine whether permission is given or not. The translation is driven off the tree structure of the ODRL specification, so that where an article (such as the fragment of Article 46 in Listing 1) has several sub-terms, checking is broken into one condition for each term:

```
article46_body(Process) when
  article46_body_term1(Process),
  article46_body_term2(Process),
  article46_body_term3(Process),
  article46_body_term4(Process);
```

and the (non-inertial) fluent `article46` is true when the corresponding body is true:

```
noninertial fluent article46(Subject);
article46(Process) when
  triple(Process,resource,personalData),
  article46_body(Process);
```

while the process starts through the _doCheck event, if the action is a transfer:

```
_doCheck(Process) initiates
   permission(Process,article46)
   if article46(Process), triple(Process,action,transfer);
_doCheck(Process) initiates
   prohibition(Process,article46)
   if not article46(Process), triple(Process,action,transfer);
```

which initiates one of the (inertial) fluents `permission` or `prohibited`, corresponding to the permission on line 3 of Listing 1.

Predicates in an ODRL policy are either implied – a sequence of terms is a conjunction, so all the elements must be true (e.g. for the permission to hold as in Listing 1) – or explicit, introduced by a `predicateConstraint`, whose body may be a disjunction of terms, or a binary operator, being either `isA`, which tests a subclass relationship, or `isAnyOf`, which tests that at least one of the right operands holds.

or: The `or` operator is, as expected, defined to be true when either or both its sub-terms are true, which is achieved by defining two rules, one for each sub-term:

```
article46_body_term1(Process) when
  article46_term1_or(Process);
```

```
article46_term1_or(Process) when
   article46_organisationType(Process);
article46_term1_or(Process) when
   article46_recipientLocation(Process);
```

In addition there are the corresponding `supports` and `lacks` fluent rules for the explanation process (see below).

isA: There is no defined class hierarchy yet for the classes of the policy profile, so the InstAL model currently uses equality rather than a proper implementation of `isA`. Since the hierarchy would be defined in its entirety at the time of translation, the subclass relationship can be grounded and subsequently queried according to the needs of a given compliance request.

isAnyOf: The `isAnyOf` operator is defined to be true when at least one of the right hand sides is true. The encoding is verbose and predictable, reflecting the (tree) structure of the source ODRL. The usage in Article 46 specifies five alternatives, but for the purposes of illustration we here show two cases. The first two fragments introduce the `isAnyOf` part of the article 46 encoding:

```
article46_body_term4(Process) when
  article46_term4_isAnyOf(Process);

article46_term4_isAnyOf(Process) when
  article46_appropriateSafeguards(Process);
```

This is followed by positive tests for the presence of each of the specified appropriate safeguards (of which we list two):

```
article46_appropriateSafeguards(Process) when
  triple(Process,appropriateSafeguards,bindingCorporateRules);
article46_appropriateSafeguards(Process) when
  triple(Process,appropriateSafeguards,legallyBindingEnforceableInstrument);
```

4.4   Explanation Representation

The purpose of the model is firstly to establish whether the process description is GDPR-compliant (noting the relevant article), and secondly, if not, the cause of non-compliance. Consequently, we define the following fluents to capture such justifications:

```
type Article;
fluent permission(Subject,Article);
fluent prohibition(Subject,Article);
noninertial fluent supports(Subject,Article,Predicate,Object);
noninertial fluent lacks(Subject,Article,Predicate,Object);
```

Then, we address the matter of how the model reports the absence of data in the description. As can be seen below, the second term checks for the presence of enforceable data subject rights. If this term is true, then the description `supports` the term:

```
article46_body_term2(Process) when
  triple(Process,dataSubjectProvisions,enforceableDataSubjectRights);
supports(Process,article46,dataSubjectProvisions,
        enforceableDataSubjectRights) when
  article46_body_term2(Process),
  triple(Process,dataSubjectProvisions,enforceableDataSubjectRights);
```

if not, the description `lacks` such data, and the respective non-inertial fluents become true:

```
lacks(Process,article46,dataSubjectProvisions,
        enforceableDataSubjectRights) when
  applies(Process,article46),
  not supports(Process,article46,dataSubjectProvisions,
                enforceableDataSubjectRights);
```

Note that we check which article applies to the description in order not to report lacks that are not relevant to the description. Similarly, we can determine whether appropriate safeguards are supported and which are lacking (corresponding to the two cases listed under the discussion of isAnyOf above):

```
supports(Process,article46,appropriateSafeguards,X) when
   article46_appropriateSafeguards(Process),
   triple(Process,appropriateSafeguards,X);
lacks(Process,article46,appropriateSafeguards,bindingCorporateRules) when
   applies(Process,article46),
   not article46_appropriateSafeguards(Process);
lacks(Process,article46,appropriateSafeguards,
      legallyBindingEnforceableInstruments) when
   applies(Process,article46),
   not article46_appropriateSafeguards(Process);
```

The complete implementation is published through the InstAL repository[7].

Operationalization of the Encoding. Putting the above together, we arrive at a specification for a partial model of Articles 6 and 46 of the GDPR. This model, along with its grounding data can now be fed into the answer set solver, outputting either a confirmation of permission for the process description, along with the facts that support the permission, or a prohibition, along with the facts that provide partial support and the facts that are lacking.

4.5   Data Transfer Example

The transfer process described in ODRL in Listing 2 is represented in InstAL as listed in the previous section (called `bp_transfer`), identifying action, resource, responsible party, organization type, sender, recipient, recipient location, purpose, legal basis (for the transfer) and two kinds of data subject provisions. We have taken some syntactic liberties to accommodate InstAL's constraints on the naming of terms and literals, while still conveying the intention. As is conventional in logic programming languages, variables start with a capital letter, e.g. Party, while lower case are literals, e.g. `bp_transfer`.

Solving for this data, results in an answer set that includes the facts given below, in which we can see that the transfer is prohibited, since it lacks any of the specified appropriate safeguards:

```
prohibition(bp_transfer,article46)

supports(bp_transfer,article6,responsibleParty,controller)
supports(bp_transfer,article6,legalBasis,consent)
supports(bp_transfer,article46,organisationType,internationalOrganisation)
supports(bp_transfer,article46,dataSubjectProvisions,enforceableDataSubjectRights)
supports(bp_transfer,article46,dataSubjectProvisions,effectiveLegalRemedies)

lacks(bp_transfer,article46,appropriateSafeguards,standardProtectionClauses)
lacks(bp_transfer,article46,appropriateSafeguards,
      legallyBindingEnforceableInstruments)
lacks(bp_transfer,article46,appropriateSafeguards,bindingCorporateRules)
lacks(bp_transfer,article46,appropriateSafeguards,approvedCodeOfConduct)
lacks(bp_transfer,article46,appropriateSafeguards,approvedCertificateMechanism)
```

---

[7] https://github.com/instsuite/instsuite.github.io/blob/master/gdpr.ial.

If we add the following data to the description:

```
triple(bp_transfer,appropriateSafeguards,bindingCorporateRules)
```

and solve again, the description is compliant according to the conditions of Article 46, thanks to the support of binding corporate rules (in addition to the same supports noted above):

```
supports(bp_transfer,article46,appropriateSafeguards,bindingCorporateRules)
permission(bp_transfer,article46)
```

5   Evaluation

In the evaluation of what we have presented in this paper, we assess the following aspects: (i) The adequacy of the modelling of the sample articles in ODRL; (ii) The adequacy of the mapping from ODRL to InstAL; and (iii) The performance of the InstAL model and corresponding ASP.

We start by assessing the suitability of ODRL for modelling legal requirements. The generalised ODRL Information Model appears to be quite similar to the ODRL Information Model, with the main changes relating to the replacement of Asset class with the more general Resource class, the conversion of the Constraint class to an abstract class and the inclusion of a single PredicateConstraint subclass, the replacement of the Duty class with the Obligation class, and the inclusion of a new Dispensation class. We deliberately excluded legal concepts from this generalised ODRL Information Model as we believe it can serve as the foundations for other ODRL profiles, for instance to express usage policies, social norms, and privacy policies. The profile itself and the corresponding ontology are quite different from the original ontology, due to the need to change the range of several classes to consider the Resource class as opposed to the Asset class. Additionally, we were often forced to define new vocabulary rather than reuse the existing ODRL vocabulary as the skos:definition, where skos denotes the Simple Knowledge Organization System ontology, was too specific/limited for our use. As for the modelling of the text of the GDPR using the proposed ODRL Regulatory Compliance Profile, rather than opting for a one to one modelling of the text as RDF, we chose instead to only model things that can be checked automatically. For instance, to enable companies to attest that certain provisions and safeguards exist, rather than actually carrying out, as part of the compliance checking process, the verification that such things exist. Here we assume we are dealing with companies that want to demonstrate compliance, and are using the compliance checking as a form of guidance with respect to their legal obligations, or as a means to verify that changes to business processes are still legally compliant. In this paper we assessed the suitability of the proposed ODRL Regulatory Compliance Profile using two (key) fragments of the General Data Protect Regulation, namely Articles 6 (Lawfulness of processing) and Articles 46 (Transfers subject to appropriate safeguards). Considering the extensible nature of RDF, the profile can easily be extended with additional vocabulary and constraints in order to extend this work to not only include other articles of the GDPR but also to model other legislation.

The second issue is the adequacy of the mapping from ODRL to InstAL. As illustrated in Sect. 4, we took a direct approach that effectively replicates the data in the

ODRL model as three element terms, while turning most of the internal nodes of the document tree into non-inertial fluents whose value is determined by the corresponding child nodes. This strategy has the benefits of (i) being able to associate supports and lacks rules with internal nodes where it is desirable to gather data about the justification or otherwise of the compliance result (ii) making the code generation highly localized in terms of dependence on data in the source document. The model contains just the one action, which is used to invoke a compliance check on a given Subject, such as bp_transfer in the example. As we note in the next aspect of the discussion on performance, the translation could be differently structured to reduce grounding costs at the expense of verbosity and legibility. There are four other elements of the ODRL information model for which to account, namely the different types of rules (permission, prohibition, obligation and dispensation). The working example of Article 46 contains only permission and obligation, so we discuss those first. The permission translates to the inertial fluent that is initiated as a result of compliance with the article46 rule, while the prohibition generation is an artefact of the operationalization of the compliance check to indicate the reporting of compliance failure. The obligation element of the ODRL information model provides syntactic structure and semantic annotation, to indicate that its subterms need to be checked, aligning with the legal interpretation of the notion of obligation, but has no actionable semantics of itself in respect of compliance checking, hence the translation skips over obligation to process its children. The same is effectively true for prohibition and dispensation, except that the former introduces a negation and the latter a side-condition on the child terms of the respective nodes.

The third issue we consider is the performance of the policy model. Answer Set Programming operates in two stages: grounding i.e. replacing variables with grounded terms, and solving, i.e. computing the answer sets. Both have high complexity in general that can often be tamed in practice. The solve step in this case is essentially polynomial because there is only one answer set and the program is stratified (see [16] for details). The grounding cost is, generally speaking, dependent on the number of distinct variables appearing in each rule and the number of alternative values a given variable might take. The encoding strategy used here is, as noted earlier, simplistic, and in consequence the state space size is a function of the number of combinations arising from the number of subject, predicate and object values, but since there is only one subject in each case (the example identifier, e.g. bp_transfer), this reduces to the product of the number of predicate and object values. The typing used in InstAL and the smart grounding processes used by the *gringo* grounder [14], reduces the search space significantly. This could be reduced further if the predicate could carry type information and hence define the range, restricting the values that the object might take. For example, instead of writing triple(Subject,Predicate,Object), the predicate can be encoded in the term: predicate(Subject,X), where X is the type denoting the set of object values in the range of predicate, which would be quite manageable in the context of a more sophisticated translation, although the resulting code might not be so human legible. As it stands, the grounding (and solving) costs are negligible, but could benefit from reconsideration given larger state spaces. Alternatively, we could consider enhancing InstAL's type system to support subsumption, which would allow the pre-grounding of most rules with the abstract types. The use of ASP allows us to guarantee the soundness and completeness of our approach (under the assumption that the modelling was correct). In other words, the system is always able to say if the business process is compliant or not and the answer is correct with respect to the modelling.

6 Conclusion

In this paper, we introduced our ODRL Regulatory Compliance Profile, which can be used to model both regulatory requirements via nested permissions, prohibitions, obligations, and dispensations, and business policies via discrete permissions that are needed to execute a compliant business process. We subsequently demonstrated how such policies can be translated into Answer Set Programming such that it is possible to automatically check compliance, and provide, if necessary, a basic explanation why compliance is not achieved.

We are currently working together with our industry partners to extend the proposed ODRL Regulatory Compliance Profile to cater for the representation of more detailed business policies, that specify which data are processed, for what purpose, where the processing takes places, for how long the data will be stored, and with whom the data is shared. Such information is needed in order to check compliance with a broader set of Articles from the GDPR. One of the primary challenges involves bridging the gap between very abstract legal requirements and the very detailed business policies. Here we plan to exploit class and property hierarchies by expanding our modelling and compliance checking to support subsumption based reasoning.

Further future work includes: (i) demonstrating how a broader set of articles can be modelled using our ODRL Regulatory Compliance Profile and automatically translated from ODRL policies into InstAL rules and Vice versa. (ii) providing a formal semantics for our ODRL Regulatory Compliance Profile, and adapting both our ontology and compliance checking to cater for legislative opening clauses that require reasoning over multiple pieces of legislation; and (iii) exploring how the ODRL policy description might form part of a description for a policy reasoning service [28] and thereby facilitate (semantic) discovery and use of such services.

Acknowledgements. This work was supported in part by the European Union's Horizon 2020 research and innovation programme under grant 731601 and by JSPS Grant-in-Aid for Scientific Research(S), Grant Number 17H06103. We would like to thank the SPECIAL project consortium for their feedback on the proposed profile.

References

1. Agarwal, S., Steyskal, S., Antunovic, F., Kirrane, S.: Legislative compliance assessment: framework, model and GDPR instantiation. In: Medina, M., Mitrakas, A., Rannenberg, K., Schweighofer, E., Tsorououlas, N. (eds.) APF 2018. LNCS, vol. 11079, pp. 131–149. Springer, Cham (2018). https://doi.org/10.1007/978-3-030-02547-2_8
2. Athan, T., Boley, H., Governatori, G., Palmirani, M., Paschke, A., Wyner, A.Z.: Oasis Legal-RuleML. In: ICAIL, vol. 13, pp. 3–12 (2013)
3. Baral, C.: Knowledge Representation, Reasoning and Declarative Problem Solving. Cambridge University Press, Cambridge (2003)

4. Bartolini, C., Muthuri, R., Santos, C.: Using ontologies to model data protection requirements in workflows. In: JSAI International Symposium on Artificial Intelligence (2015)
5. Boer, A., Hoekstra, R., Winkels, R., Van Engers, T., Willaert, F.: Metalex: legislation in XML. In: Legal Knowledge and Information Systems (Jurix 2002), pp. 1–10 (2002)
6. Boer, A., Winkels, R., Vitali, F.: MetaLex XML and the legal knowledge interchange format. In: Casanovas, P., Sartor, G., Casellas, N., Rubino, R. (eds.) Computable Models of the Law. LNCS (LNAI), vol. 4884, pp. 21–41. Springer, Heidelberg (2008). https://doi.org/10.1007/978-3-540-85569-9_2
7. Boley, H., Paschke, A., Shafiq, O.: RuleML 1.0: the overarching specification of web rules. In: Dean, M., Hall, J., Rotolo, A., Tabet, S. (eds.) RuleML 2010. LNCS, vol. 6403, pp. 162–178. Springer, Heidelberg (2010). https://doi.org/10.1007/978-3-642-16289-3_15
8. Bonatti, P.A. Olmedilla, D.: Rule-based policy representation and reasoning for the semantic web. In: Proceedings of the Third International Summer School Conference on Reasoning Web (2007)
9. Bradshaw, J.M.: Software Agents. MIT Press, Cambridge (1997)
10. Cliffe, O., De Vos, M., Padget, J.: Answer set programming for representing and reasoning about virtual institutions. In: Inoue, K., Satoh, K., Toni, F. (eds.) CLIMA 2006. LNCS (LNAI), vol. 4371, pp. 60–79. Springer, Heidelberg (2007). https://doi.org/10.1007/978-3-540-69619-3_4
11. Dimyadi, J., Pauwels, P., Amor, R.: Modelling and accessing regulatory knowledge for computer-assisted compliance audit. J. Inf. Technol. Constr. 21, 317–336 (2016)
12. Fornara, N., Colombetti, M.: Operational semantics of an extension of ODRL able to express obligations. In: Belardinelli, F., Argente, E. (eds.) EUMAS/AT -2017. LNCS (LNAI), vol. 10767, pp. 172–186. Springer, Cham (2018). https://doi.org/10.1007/978-3-030-01713-2_13
13. Fornara, N., Chiappa, A., Colombetti, M.: Using semantic web technologies and production rules for reasoning on obligations and permissions. In: Lujak, M. (ed.) AT 2018. LNCS (LNAI), vol. 11327, pp. 49–63. Springer, Cham (2019). https://doi.org/10.1007/978-3-030-17294-7_4
14. Gebser, M., Kaminski, R., König, A., Schaub, T.: Advances in *gringo* series 3. In: Delgrande, J.P., Faber, W. (eds.) LPNMR 2011. LNCS (LNAI), vol. 6645, pp. 345–351. Springer, Heidelberg (2011). https://doi.org/10.1007/978-3-642-20895-9_39
15. Gebser, M., Kaminski, R., Kaufmann, B., Schaub, T.: Clingo = ASP + control: preliminary report. CoRR, abs/1405.3694 (2014)
16. Gelfond, M., Lifschitz, V.: The stable model semantics for logic programming. In: Kowalski, R.A., Bowen, K.A. (eds.) Logic Programming, Proceedings of the Fifth International Conference and Symposium, Seattle, Washington, USA, 15–19 August 1988 (2 Volumes), pp. 1070–1080. MIT Press (1988). ISBN 0-262-61056-6
17. Gelfond, M., Lifschitz, V.: Classical negation in logic programs and disjunctive databases. New Gener. Comput. 9(3–4), 365–386 (1991)
18. Governatori, G., Hashmi, M., Lam, H.-P., Villata, S., Palmirani, M.: Semantic business process regulatory compliance checking using LegalRuleML. In: Blomqvist, E., Ciancarini, P., Poggi, F., Vitali, F. (eds.) EKAW 2016. LNCS (LNAI), vol. 10024, pp. 746–761. Springer, Cham (2016). https://doi.org/10.1007/978-3-319-49004-5_48
19. Information Commissioner's Office (ICO) UK: Getting ready for the GDPR (2017). https://ico.org.uk/for-organisations/resources-and-support/data-protection-self-assessment/getting-ready-for-the-gdpr. Accessed 1 May 2019
20. Jones, A., Sergot, M.: A formal characterisation of institutionalised power. Logic J. IGPL 4(3), 427–443 (1996)
21. Kagal, L., Finin, T.: A policy language for a pervasive computing environment. In: Proceedings POLICY 2003, IEEE 4th International Workshop on Policies for Distributed Systems and Networks (2003)

22. Lam, H.-P., Hashmi, M.: Enabling reasoning with LegalRuleML. Theor. Pract. Logic Program. **19**(1), 1–26 (2019)
23. Li, T., Balke, T., Vos, M.D., Padget, J.A., Satoh, K.: A model-based approach to the automatic revision of secondary legislation. In: Francesconi, E., Verheij, B. (eds.) International Conference on Artificial Intelligence and Law, ICAIL 2013, Rome, Italy, 10–14 June 2013, pp. 202–206. ACM (2013). ISBN 978-1-4503-2080-1, https://doi.org/10.1145/2514601.2514627
24. Microsoft Trust Center: Detailed GDPR Assessment (2017). http://aka.ms/gdprdetailed assessment. Accessed 1 May 2019
25. Motik, B., Horrocks, I., Rosati, R., Sattler, U.: Can OWL and logic programming live together happily ever after? In: Cruz, I., et al. (eds.) ISWC 2006. LNCS, vol. 4273, pp. 501–514. Springer, Heidelberg (2006). https://doi.org/10.1007/11926078_36
26. Nymity: GDPR Compliance Toolkit. https://www.nymity.com/gdpr-toolkit.aspx. Accessed 1 May 2019
27. Padget, J., ElDeen Elakehal, E., Li, T., De Vos, M.: InstAL: an institutional action language. In: Aldewereld, H., Boissier, O., Dignum, V., Noriega, P., Padget, J. (eds.) Social Coordination Frameworks for Social Technical Systems. LGTS, vol. 30, pp. 101–124. Springer, Cham (2016). https://doi.org/10.1007/978-3-319-33570-4_6
28. Padget, J., Vos, M.D., Page, C.A.: Deontic sensors. In: Proceedings of the Twenty-Seventh International Joint Conference on Artificial Intelligence, IJCAI-2018, pp. 475–481. International Joint Conferences on Artificial Intelligence Organization (2018). https://doi.org/10.24963/ijcai.2018/66
29. Palmirani, M., Governatori, G., Rotolo, A., Tabet, S., Boley, H., Paschke, A.: LegalRuleML: XML-based rules and norms. In: Olken, F., Palmirani, M., Sottara, D. (eds.) RuleML 2011. LNCS, vol. 7018, pp. 298–312. Springer, Heidelberg (2011). https://doi.org/10.1007/978-3-642-24908-2_30
30. Panasiuk, O., Steyskal, S., Havur, G., Fensel, A., Kirrane, S.: Modeling and reasoning over data licenses. In: Gangemi, A., et al. (eds.) ESWC 2018. LNCS, vol. 11155, pp. 218–222. Springer, Cham (2018). https://doi.org/10.1007/978-3-319-98192-5_41
31. Pandit, H.J., Fatema, K., O'Sullivan, D., Lewis, D.: GDPRtEXT - GDPR as a linked data resource. In: Gangemi, A., et al. (eds.) ESWC 2018. LNCS, vol. 10843, pp. 481–495. Springer, Cham (2018). https://doi.org/10.1007/978-3-319-93417-4_31
32. Steyskal, S., Kirrane, S.: If you can't enforce it, contract it: enforceability in policy-driven (linked) data markets. In: SEMANTiCS (Posters & Demos) (2015)
33. Steyskal, S., Polleres, A.: Defining expressive access policies for linked data using the ODRL ontology 2.0. In: Proceedings of the 10th International Conference on Semantic Systems (2014)
34. von Wright, G.: Deontic logic. Mind **60**(237), 1–15 (1951). ISSN 00264423, 14602113

Aligning, Interoperating, and Co-executing Air Traffic Control Rules Across PSOA RuleML and IDP

Marjolein Deryck[1]([⊠]), Theodoros Mitsikas[2]([⊠]), Sofia Almpani[2],
Petros Stefaneas[2], Panayiotis Frangos[2], Iakovos Ouranos[3], Harold Boley[4],
and Joost Vennekens[1]

[1] KU Leuven, Sint-Katelijne Waver, Belgium
{marjolein.deryck,joost.vennekens}@kuleuven.be
[2] National Technical University of Athens, Athens, Greece
{mitsikas,salmpani,pfrangos}@central.ntua.gr, petros@math.ntua.gr
[3] Hellenic Civil Aviation Authority, Heraklion Airport, Heraklion, Greece
iouranos@central.ntua.gr
[4] University of New Brunswick, Fredericton, Canada
harold.boley@unb.ca

Abstract. This paper studies Knowledge Bases (KBs) in PSOA RuleML and IDP, aligning, interoperating, and co-executing them for a use case of Air Traffic Control (ATC) regulations. We focus on the common core of facts and rules in both languages, explaining basic language features. The used knowledge sources are regulations specified in (legal) English, and an aircraft data schema. In the modeling process, inconsistencies in both sources were discovered. We present the discovery process utilizing both specification languages, and highlight their unique features. We introduce three extensions to this ATC KB core: (1) While the current PSOA RuleML does not distinguish the ontology separately from the instance level, IDP does. Hence, we specify a vocabulary-enriched version of ATC KB in IDP for knowledge validation. (2) While the current IDP uses relational modeling, PSOA additionally supports graph modeling. Hence, we specify a relationally interoperable graph version of ATC KB in PSOA. (3) The KB is extended to include optimization criteria to allow the determination of an optimal sequence of more than two aircraft.

Keywords: PSOA RuleML · IDP · Interoperation · Knowledge Base · Alignment · Co-execution · Regulations · Air Traffic Control

1 Introduction

Contributing to cross-fertilizations between, e.g., the Semantic Technologies and Decision Management Communities,[1] in this paper we use the Positional-Slotted

---

[1] For specific references see http://blog.ruleml.org/post/132677817-decisioncamp-and-ruleml-rr-will-meet-again-in-luxembourg.

© Springer Nature Switzerland AG 2019
P. Fodor et al. (Eds.): RuleML+RR 2019, LNCS 11784, pp. 52–66, 2019.
https://doi.org/10.1007/978-3-030-31095-0_4

Object-Applicative (PSOA) RuleML version of the ATC KB [14]² as a starting point for an IDP (Imperative/Declarative Programming) version, and explore the consequences for both. We compare Knowledge Bases (KBs) in IDP and PSOA RuleML to find how modeling the same knowledge in two languages can help us to improve our specification and to achieve an architecture that combines the best of both systems. Based on Air Traffic Control (ATC) regulations and data obtained from [6], the PSOA specification was created. From it, we derive the IDP KB and we study the similarities and differences between both specifications. In the first step, we try to align both systems by choosing similar ways of modeling from different possibilities. In doing this, we discovered that there were not only inconsistencies in the source aircraft characteristics data as discussed in [14], but also in one of the regulations. In the second step, we investigate how both systems can be interoperated by translating pieces of knowledge from one source to the other. In the third step, the co-execution of both systems is examined allowing us to validate results from both systems. The resulting KBs were then expanded: an optimization logic was formalized in IDP, while a perspectival graph version of the KB was created in PSOA. As the systems can be co-executed, an architecture in which the strength of each system is exploited can be envisaged.

Examples of ATC regulations formalization are [13] and [15]. The former presented an overview of a method for formal requirements capture and validation, in the domain of oceanic ATC. The obtained model focused on conflict prediction, while being compliant to the regulations governing aircraft separation in oceanic airspace. The presented examples are expressed in many-sorted first-order logic or in the Prolog notation, and include rules about conflict prediction and aircraft separation. Supplementary, the model was validated by automated processes, formal reasoning, and domain experts. [15] focuses on capturing ATC regulations valid in the airport area. The authors formalized the separation minima mandated by International Civil Aviation Organization (ICAO), Federal Aviation Administration (FAA), and FAA's "RECAT" regulations in POSL RuleML. It formed the foundations for further expansion that focuses on cases of conditional reduced separation minima. It was the basis for the development of [14], a PSOA RuleML version of ATC KB that in turn, served as the basis for this paper.

The paper is structured as follows. In the next section we introduce IDP and PSOA. Then the use case of ATC regulations is introduced in Sect. 3. The aligned KBs are presented in Sect. 4, while Sect. 5 discusses the interoperation and co-execution of the two systems and compares their results. Section 6 discusses inconsistencies found within the regulations, which is followed by the presentation of KBs' extensions in Sect. 7. Section 8 provides some final conclusions and directions for future work.

2   Knowledge Formalization and Reasoning

In this section we introduce the two specification languages, IDP and PSOA RuleML.

---

² See the PSOA ATC KB sources at http://users.ntua.gr/mitsikas/ATC_KB/.

2.1   IDP and the Knowledge Base Paradigm

The IDP system [3] adheres to the Knowledge Base Paradigm (KBP): it stresses
the distinction between domain knowledge *an sich*, and the different ways in
which this knowledge can be put into use [18]. The domain knowledge is formalized and centralized in a KB, which collects not only simple knowledge (e.g.,
data in a database), but also complex knowledge, such as definitions, implications, propositions, etc. One of the advantages of this separation of concerns of
knowledge versus problem solving, is the high maintainability of the KB because
it only contains descriptive information on the domain. Another advantage is the
flexibility to use this KB in different – often unforeseen – use cases.

The IDP system allows KBs to be written in the IDP language, which is
based on typed first-order logic, enriched with features such as aggregates and
inductive definitions. A KB typically consists of three kinds of components. A
vocabulary describes the logical symbols (types, constants, functions and predicates) that are used to formulate the domain knowledge. It represents the ontology of the domain. A *structure* for a vocabulary provides an interpretation for
each of the symbols of this vocabulary. Finally, a *theory* contains the actual
domain knowledge, represented as a set of formulas. A formula can be either a
classical first-order logic (FO) formula, or a *rule-based (inductive) definition*. For
instance, a theory can contain as a formula the following inductive definition of
the transitive closure T of a graph G:

$$\left\{ \begin{array}{c} \forall x \; y : T(x,y) \leftarrow G(x,y). \\ \forall x \; y : T(x,y) \leftarrow \exists z : T(x,z) \land T(z,y). \end{array} \right\}$$

The rules in such a definition are built using the *definitional implication* symbol
$\leftarrow$, which is to be distinguished from the *material implication* of classical logic,
the latter denoted as $\Leftarrow$ in IDP. The formal semantics of such a definition is given
by its *well-founded model*, because this coincides with the expected semantics
of an inductive definition [4]. The semantics of the FO formulas in a theory is
simply given by the standard satisfaction relation $\models$ of classical logic.

The IDP system allows a number of different inference tasks to be performed on
a KB. The most common is that of *Model Expansion (MX)*: for a given vocabulary
V, and theory T and structure S for some subvocabulary $V' \subseteq V$, MX constructs
a structure S' for the entire vocabulary V that extends S (i.e., $\sigma^S = \sigma^{S'}$ for all
symbols $\sigma \in V'$) and that is a model of T (i.e., $S' \models T$). Another useful inference
task is that of *Optimization*, which selects the most optimal structure (according
to some provided criterion) among all the possible model expansions of a given S
w.r.t. T. The optimality criterion is provided in the form of a term t which must be
minimized, i.e., the solution is the model expansion S' for which the value of $t^{S'}$ of
the term t in the structure S' is minimal.

2.2   PSOA RuleML for Graph-Relational Knowledge

PSOA RuleML generalizes RIF-BLD and POSL RuleML by a homogeneous
integration of table-like relationships and graph-like frames into **p**ositional-
slotted **o**bject-**a**pplicative (***psoa***) *terms*. The initially used *single-dependent-tuple*

independent-slot special case of psoa terms, oidless or oidful, has these forms [2, 20] (where $n \geq 0$ and $k \geq 0$, of which we focus on either $n = 0$ for – oidless – *frameships* and – oidful – *framepoints* or $k = 0$ for – oidless – *relationships*):

$$\text{Oidless:} \quad \text{f}(\text{t}_1 \ldots \text{t}_n \; \text{p}_1\text{->}\text{v}_1 \ldots \text{p}_k\text{->}\text{v}_k) \tag{1}$$

$$\text{Oidful:} \text{o\#f}(\text{t}_1 \ldots \text{t}_n \; \text{p}_1\text{->}\text{v}_1 \ldots \text{p}_k\text{->}\text{v}_k) \tag{2}$$

While (2) starts with an Object IDentifier (OID) o via a membership, o # f, of o in f (acting as a class), both (1) and (2) apply a function or predicate f (acting as a relator) to a tuple of arguments $\text{t}_1 \ldots \text{t}_n$ and to a bag of slots $\text{p}_j\text{->}\text{v}_j$, $j = 1, \ldots, k$, each pairing a slot name (attribute) p_j with a slot filler (value) v_j.

For example, in ATC KB, the term `:be91#:Aircraft(:mtow->9300.0)` is a single-slot framepoint atom, while `:AircraftIcaoCategory(:a388 icao:Super)` is a binary relationship. Both are *ground* atoms, i.e. variableless.

Variables in PSOA are '?'-prefixed names, e.g. ?x. The most common atomic formulas are psoa atoms in the form of (1) or (2). Compound formulas can be constructed using the Horn-like subset of first-order logic. A PSOA KB then consists of clauses that are ground facts and non-ground rules: while facts are – ground – psoa atoms, rules are defined – within **Forall** wrappers – using a Prolog-like *conclusion* `:-` *condition* syntax, where *conclusion* can be a psoa atom and *condition* can be a psoa atom or an **And**-prefixed conjunction of psoa atoms.

The reference implementation for deduction in PSOA RuleML is the open-source framework system PSOATransRun, currently in Version 1.4.2[3].

3 Air Traffic Control Regulations

Collision prevention in ATC is realized by ensuring a minimum distance between aircraft, a concept also called *separation minimum*. Separation of aircraft serves an additional role, which is the avoidance of wake turbulence. The separation minimum is defined for aircraft pairs depending on their wake turbulence category. The current FAA and ICAO regulations categorize aircraft according to their maximum takeoff weight/mass (MTOW/MTOM). MTOW/MTOM represent the wake turbulence of the leading aircraft, as well as how much a following aircraft is affected by the wake turbulence of the leader. Both agencies are in the process of a wake turbulence recategorization (RECAT), which recategorizes aircraft in six categories, taking into account the wingspan as an additional parameter.

For example, ICAO discerns four categories: Light (MTOM of 7000 kg or less), Medium (MTOM of greater than 7000 kg, but less than 136000 kg), Heavy (MTOM of 136000 kg or greater), and Super (a separate designation that currently only refers to the Airbus A380 with MTOM 575000 kg) [11,12]. The associated separation minima for flights under Instrument Flight Rules (IFR)[4] are defined in Table 1[5]. The Minimum Radar Separation (MRS), is 3 NM or 2.5 NM depending on operational conditions unrelated to wake turbulence (e.g. visibility) [12].

---

[3] http://psoa.ruleml.org/transrun/1.4.2/local/.

[4] Separation minima for flights on Visual Flight Rules (VFR) are time-based [8,11].

[5] The minima set out at Table 1 shall be applied when e.g. both aircraft are using the same runway, or parallel runways separated by less than 760 m (2 500 ft) [11].

Table 1. Current ICAO weight categories and associated separation minima [12]

ICAO separation standards (nautical miles (NM))				
		Follower			
		Super	Heavy	Medium	Light
Leader	Super	MRS	6	7	8
	Heavy	MRS	4	5	6
	Medium	MRS	MRS	MRS	5
	Light	MRS	MRS	MRS	MRS

4   Alignment

KB languages should be able to represent objects, facts, and relations in the knowledge domain. If two KBs are developed for the same knowledge domain, it is expected that they express the same information. Hence, it should be possible to align both. In this section we will discuss the way certain parts of knowledge are represented in both languages.

4.1   Common Core of the KBs

We performed an alignment for all PSOA and IDP constructs used in the ATC KBs. Typical parts of this PSOA-IDP alignment are shown below (aircraft-characterizing facts were obtained from [6]):

```
% PSOA KB fragment (from
% users.ntua.gr/mitsikas/
% ATC_KB/atc_kb-v201906.psoa)
```
```
// IDP KB fragment (from
// gitlab.com/mderyck/atc-kb-idp/)

vocabulary V {
    type Mtom isa int
    type Aircraft isa string
    MTOM(Aircraft,Mtom)
    ... }
theory T:V{
```

```
Forall ?a ?w (
   :AircraftIcaoCategory(?a icao:Light) :-
    And(?a#:Aircraft(:mtom->?w)
      math:lessEq(?w 7000))   )
```
```
!a[Aircraft] w[Mtom]:
AircraftIcaoCategory(a, Light) <=
    MTOM(a,w)
      & w =< 7000.
```

```
Forall ?a ?w (
   :AircraftIcaoCategory(?a icao:Medium) :-
    And(?a#:Aircraft(:mtom->?w)
      math:greaterThan(?w 7000)
      math:lessThan(?w 136000))   )
```
```
!a[Aircraft] w[Mtom]:
AircraftIcaoCategory(a, Medium) <=
    MTOM(a,w)
      & 7000 < w
      & w < 136000.
```

```
Forall ?a ?w (
   :AircraftIcaoCategory(?a icao:Heavy) :-
    And(?a#:Aircraft(:mtom->?w)
      math:greaterEq(?w 136000)
      not:Naf(:AircraftIcaoCategory(?a icao:Super))
    )
)
```
```
!a[Aircraft] w[Mtom]:
AircraftIcaoCategory(a, Heavy) <=
    MTOM(a,w)
      & 136000 =< w
      & a ~= a388
      & a ~= a38f.
```

```
:AircraftIcaoCategory(:a388 icao:Super)
:AircraftIcaoCategory(:a38f icao:Super)
```
```
AircraftIcaoCategory("a388", Super).
AircraftIcaoCategory("a38f", Super).
```

```
%% ICAO Separation example
```
```
// ICAO Separation Example
```

```
Forall ?l ?f (
   :icaoSeparation(:leader->?l
```
```
!l[Leader],f[Follower]:
```

```
              :follower->?f                    IcaoSeparation(l, f) = 8 <=
              :miles->8) :-
  And(:AircraftIcaoCategory(?l icao:Super)      AircraftIcaoCategory(l, Super)
     :AircraftIcaoCategory(?f icao:Light))      & AircraftIcaoCategory(f, Light).
  )                                             }

%% Sample Aircraft Facts %%               structure S1 : V {
                                          //specific value assignments:
:be91#:Aircraft(:mtom->4218.41            Leader = {a388}
              :mtow->9300.0               Follower = {be91}
              :wingspan->45.92            ...
              :appSpeed->100.0)
                                          //aircraft data
:a388#:Aircraft(:mtom->575000.0           MTOM = {be91, 4218; a388, 575000}
              :mtow->1267658.0            MTOW = {be91, 9300; a388, 1267658}
              :wingspan->261.65           WingSpan = {be91, 45; a388, 261}
              :appSpeed->145.0)           AppSpeed = {be91, 100; a388, 145}
                                          }
```

Vocabulary. In IDP the types/sorts that will be used in the knowledge base need to be explicitly declared in the *vocabulary*. It binds the use of types in relations that are appropriate for it. When instances of a type are (correctly) used in IDP, the types can be derived by the system, based on the place in which they occur. In PSOA there is no separate signature declaration.

Using the KBs. For the alignment we created specifications that are classically equivalent, in the sense that they both have the same class of possible worlds: an interpretation (called a structure in IDP) W satisfies the PSOA fragment P if and only if it satisfies the IDP fragment I: $W \models P \Leftrightarrow W \models I$.

As written above, both PSOA and IDP represent the categorization of aircraft as a set of implications: the :- symbol of PSOA and the <= symbol of IDP both denote the material implication of classical first-order logic. In other words, an interpretation W is a model of an implication F:- G in PSOA, or of F <= G in IDP, if and only if G holds in W or F is false in W. Accordingly, the class of models of the above IDP/PSOA specification is quite large, since every superset of a model is again a model; e.g., there are models in which the same aircraft belongs to all four categories at the same time.

The existence of these "extra" models is not a problem for PSOA, since this system uses the KB by means of the inference task of *query answering*, which looks for properties that hold in *all* models of the specification.

The IDP system offers the same inference task, allowing the above IDP specification to be used in precisely the same way (and, because of the equivalence of the two specifications, producing identical results). However, this is not an idiomatic use of IDP: IDP adheres to the KBP, which emphasizes that the same KB should be usable by different inference tasks. To guarantee that different inference tasks produce correct results, it is crucial that the KB is constructed in such a way that its models correspond one-to-one to possible worlds in reality. The above specification obviously does not have this property and therefore produces correct results *only* when the inference task of query answering is applied to it. When applying, e.g., the inference task of *model expansion*, we will obtain erroneous results, in which an aircraft is assigned some category even though it does not satisfy the condition for belonging to this category (such as :be91 belonging to the category Heavy).

The more idiomatic way of representing the above knowledge in IDP is by means of a *definition*, replacing each of the material implications <= by IDP's *definitional implication* symbol <-. Such a definitional implication entails the material implication (i.e., if F <- G then also F <= G), but in addition, it also implies that F can *only* be true if there is at least one rule of the form F <- G such that G is true. As mentioned above, such a set of definitional implications is interpreted under the well-founded semantics. Since the above specification is not recursive, this means that it is equivalent to its Clark's completion. In other words, it states that an aircraft belongs to a certain category if *and only if* it satisfied the corresponding condition. Therefore, the definitional IDP specification only has a single model, in which each aircraft is assigned a single category, namely that whose condition it satisfies. This specification is therefore considerably stronger than the PSOA specification. However, when we apply either *model expansion* (compute a single model) or *query answering* (compute facts that are true in all models—but there is only one model in this case) to the definitional IDP specification, we still obtain precisely the same answers as when we query the PSOA specification.

Expressing Relations. In the aligned KBs above, the purpose is to establish the relation between an aircraft and the ICAO regulation. In Sect. 6 we discuss the modeling choice that was made earlier to use aircraft type as an identifier, and the challenges put forward by this. In this part we assume that an aircraft can only be assigned to one category, as this is the case for every specific aircraft.

In both modeling languages, it is possible to employ relations in different ways, of which we chose a compatible subset for our KBs:

PSOA allows very general atoms [2], but here uses the *single-dependent-tuple independent-slot* special case of psoa terms (cf. Section 2.2). Specializing further, we need atoms that are oidful-slotted (*framepoints*) for the KB facts, oidless-tupled (*relationships*) for the aircraft categorization, and oidless-slotted (*frameships*) for the separation. A dependent-slotted version is discussed in Sect. 7.2.

IDP allows relations (which can be true or false) and functions (that have exactly one image). A 0-ary relation is a Boolean, a 0-ary function is a constant. Both also have unary and n-ary variants. A function is a special relation, in the sense that a function f(x) = y could also be written as a relation r(x,y), with the additional constraint that each argument x needs to have exactly one image y. As an aircraft can only belong to one category, the use of function represents the actual domain knowledge in the most appropriate way: AircraftIcaoCategory(Aircraft):Category. Alternatively, a relation can also be used, which is closer to the modeling in PSOA. The relation is expressed as: AircraftIcaoCategory(Aircraft, Category). A separate constraint can be formulated to express that an aircraft may belong to only one category: $\forall a[Aircraft] : \#c : AircraftIcaoCategory(a, c) = 1$.

Exceptions to the Regulations. ICAO regulations identify two specific types of aircraft—a388 and a38f—as belonging to the category *super*, even though their weight would normally put them in the category *heavy*. These two aircraft can be seen as exceptions to the general rule that aircraft with a weight over 136000 are "heavy". In the above IDP specification, we have represented these exceptions by excluding these two aircraft from the rule for "heavy" *by name*. Obviously, this is a poor representation, because it requires us to update both the rule for "heavy" and the rule for "super" if more aircraft are added to the "super" category[6]. In PSOA, we have an appealing alternative in the use of *negation as failure (naf)*: we can write not:Naf(:AircraftIcaoCategory(?a icao:Super)) in the body of the rule for "heavy". This atom will hold for any aircraft for which it *cannot be proven* that it belongs to *super* (which will be precisely all those aircraft that are not enumerated as being "super").

IDP does not have negation as failure and we therefore cannot adopt the same representation as long as we are using material implication. However, as discussed before, the idiomatic IDP representation would be to use definitional implications instead. Under this representation, we can simply write ~AircraftIcaoCategory(a, Super) in the body of the rule for "heavy". The ~ symbol represent simply classical negation, meaning that in any model in which a is not "super", it will be "heavy". Because we make use of definitional implications, there is only one model, in which a388 and a38f are the only two aircraft that are "super", and therefore this representation is correct. We therefore see that the combination of material implication with negation as failure in PSOA is functionally identical (though not formally equivalent, since the former has many more models than the latter) to the combination of definitional implementation with classical negation in IDP.

Comparing to [14], the newest PSOA version presented here does not need the workaround of the extra slot SpecialCase, as PSOATransRun now supports negation as failure.

5   Interoperation and Co-execution

Many of the commonalities and differences between the PSOA and IDP have been discussed in the previous section. In this section, we examine how both systems interoperate and co-execute.

---

[6] In the ATC domain the regulations are stable and new types of aircraft e.g. in the **Super** category are not currently in active development. Therefore, we do not consider this a major problem.

5.1 Syntactic Translation for Interoperation

PSOA is a rule-based system, whereas IDP is a generic constraint-based system (with rules as a special form of constraints). Their interoperation is only applicable to facts and rules. Also because of this, IDP features different reasoning tasks. This means that it might be more advantageous to go beyond the literal translation of the PSOA KB, to find the appropriate notation useful for all inferences.

Proceeding from the aligned KBs from Sect. 4.1, a partial translation can be realized:

Atoms: For *relationships*, there is a direct PSOA-IDP tuple correspondence. For *framepoints*, a slot name (e.g., :mtom) in PSOA is reflected by a *binary relation* (e.g., MTOM) in IDP, with the OID as the first argument and the filler as the second argument (the predicate name is already part of IDP's vocabulary declaration). For *frameships*, n-1 slots in PSOA can map to the argument tuple of a function in IDP, and 1 slot to its returned value.

Symbols: Some symbols can be directly translated from one language to the other, e.g. quantifiers (Forall $\leftrightarrow \forall$, Exists $\leftrightarrow \exists$) and implication (: $-\leftrightarrow \Leftarrow$).

Operators: PSOA uses prefix operators, while IDP uses infix operators (And vs. &, Or vs. |). *Externals* and *libraries* in PSOA are also prefixed, while in IDP are infixed (e.g., the comparison math:lessEq vs. =<).

Rules: These are wrapped into Forall/$\forall$ quantifiers, and built on atoms, possibly within an And/&, for both PSOA (: $-$) and IDP ($\Leftarrow$).

The interoperation between PSOA RuleML and IDP provides a link between the Semantic Technologies Community (e.g., N3 [1]) and the Decision Management Community (e.g., OMG DMN [16]). For example, it enables the interoperation path N3→PSOA↔IDP←DMN (for the link N3Basic→PSOA see [19] and for the link IDP←DMN see [5][7]).

5.2 Semantics-Preserving Co-execution

After having performed several experiments with the PSOA version developed from [14], we have also experimented with the new IDP version of the same KB.

For the common core of the KBs presented in Sect. 4.1 (assuming only two aircraft and omitting the namespaces to conserve space), the answers of PSOA queries are in accordance with the IDP least model, as shown below:

---

[7] The IDP language typically offers more expressivity than DMN decision tables. Current work focuses on an extension to DMN to strengthen the link IDP→DMN.

```
% PSOA queries                      // IDP least model

                                    structure  : V {
:AircraftIcaoCategory(?a ?c)          Leader = { "a388" }
                                      Follower = { "be91" }
Answer(s):                            ...
?a=<...#be91 ?c=<...#Light>           AircraftIcaoCategory = { "be91", Light;
?a=<...#a388 ?c=<...#Super>                                   "a388", Super}
                                      ...
:icaoSeparation(:leader->:a388
               :follower->:be91
               :miles->?d)
Answer(s):
?d=8                                  IcaoSeparation = { "a388","be91"->8 }
                                    }
```

In this example, as in all consistent cases (see Sect. 6), PSOA and IDP provide semantically compatible results. In general, PSOA/IDP co-execution benefits both systems for the following reasons:

1. We have compared and cross-validated the results from both systems. The inconsistencies that were discovered in the original regulations, using both systems, have been described in Sect. 6.
2. The top-down processing (backward-reasoning) of PSOATransRun is complementary to the bottom-up processing (forward-reasoning) of the IDP system. Since the ATC KB's required logical expressiveness is on the level of Datalog (function-free Horn logic), both methods are applicable, although there is the expected speed/memory trade-off.
3. Each system can be used for a task it is best suited to. For example, decimal-preserving numeric calculations are currently not supported by IDP, but are available in PSOA. Therefore, calculations involving decimal numbers are handled by PSOA. On the other hand, as discussed in Sect. 7.1, IDP uses constraint solving for efficiently optimizing a landing queue.

6   Inconsistencies Within Regulations

The process of aligning and co-executing several KBs does not only serve a theoretical purpose. These steps are especially useful in the construction of the KB. The detection of inconsistencies is an example of the added value of our approach. The KB validation for both PSOA and IDP aims to ensure the completeness (i.e. all aircraft will be categorized) and consistency (i.e. all individual aircraft are categorized in exactly one category for each applicable regulation). It serves a two-fold purpose. First, to ensure that the KB is in accordance with the regulations. Second, to ensure that the regulations and the source dataset are complete and consistent.

The PSOA KB in [14] considers that an aircraft is represented by its ICAO type designator and assigns the latter as an oid. This design choice, while efficient when the KB is used as a computational tool where individual aircraft would be handled by a front-end framework, can lead to problems in stand-alone execution: as a specific aircraft type can be assigned in more that one category due to variations (see e.g., [10]), an aircraft oid can be categorized in two different categories, as demonstrated by the following PSOA RuleML query:

```
And(:AircraftIcaoCategory(?a ?X) :AircraftIcaoCategory(?a ?Y)
    External(isopl:generic_not_eq(?X ?Y)))
Answer(s):
?a=<...#b350> ?X=<...#Light> ?Y=<...#Medium>
?a=<...#c207> ?X=<...#Light> ?Y=<...#Medium>
```

As explained in [14], the first result is a case where variants of the same type can be categorized in different categories, while the second result is an inconsistency of the source dataset.

Additional validation of the regulations can be realized by using PSOA RuleML or IDP. In [7], the categorization for categories **D** and **F** according to RECAT regulations is defined:

Category D. Aircraft capable of MTOW of less than 300,000 lb and wingspan greater than 125 ft and less than or equal to 175 ft; or aircraft with wingspan greater than 90 ft and less than or equal to 125 ft.

Category F. Aircraft capable of MTOW of less than 41,000 lb and wingspan less than or equal to 125 ft, or aircraft capable of MTOW less than 15,500 lb regardless of wingspan, or a powered sailplane.

According to the above, any aircraft capable of MTOW of less than 41,000 lb with wingspan greater than 90 ft and less than or equal to 125 ft would be categorized in both **D** and **F** categories. This inconsistency was discovered by both PSOA and IDP. In PSOA, appropriate non-ground queries can identify the problem, as shown below:

```
And(:AircraftRecatCategory(?a ?X) :AircraftRecatCategory(?a ?Y)
    External(isopl:generic_not_eq(?X ?Y)))
Answer(s):
?a=<...#dc3> ?X=<...#D> ?Y=<...#F>
?a=<...#dhc4> ...
```

In [14] the problem was not identified, as such non-ground queries were not possible to construct due to lack of the `generic_not_eq` built-in before PSOA-TransRun Version 1.4.1.

A later revision of the regulations attempted to correct the above problem and can be seen in [9, 10]:

Category D. Aircraft capable of MTOW of less than 300,000 lb and wingspan greater than 125 ft and less than or equal to 175 ft; or aircraft capable of a MTOW greater than 41,000 lb with a wingspan greater than 90 ft and less than or equal to 125 ft.

This definition leads to an incompleteness for aircraft capable of MTOW of exactly 41,000 lb with wingspan greater than 90 ft and less than or equal to 125 ft. While an aircraft with the above characteristics did not exist in the KB, the discovery of the incompleteness was made by PSOA by adding witness aircraft representing corner cases.

The behavior of the IDP system for these different inconsistencies depends on the modeling choice that have been made. In general, the more accurate the domain knowledge is represented, the more likely the system will behave as expected with inconsistencies. In this example, the appropriate way to model the categorization, would be to use a function `IcaoAircraftCategory(Aircraft):` `Category` and the definitional implication that was already discussed in Sect. 4. The definition assumes a closed world: a definition contains a set of sufficient and necessary conditions. In practice this means that no model can be found and an `unsatisfiable` message will be shown if not every case is defined, independent of the presence of an aircraft of 41,000 lb. It will act likewise if overlapping categories have been defined. It can be tedious to find the exact inconsistency in a theory. Specific inferences can be used to help identify the exact problem (e.g. `explainunsat` and `printunsatcore`). If we move away from the ideal model, e.g. by defining categories as separate definitions, using the *if and only if* operator, the missing definition will only be discovered if an aircraft of 41,000 lb is present in the database. Finally, the last way to model the categorization, is the use of material implication. If no category is explicitly assigned to aircraft of 41,000 lb, any category may be assigned. This is the most dangerous situation, as a random category will be assigned.

While both PSOA and IDP helped to identify the above inconsistency and incompleteness in the regulations, this was through different mechanisms: PSOA RuleML needs the construction of an appropriate query and an example in the KB (a "witness"). If the domain knowledge is appropriately modeled in IDP, inconsistencies or incompleteness will be found without such a witness. Thus, using both approaches to model safety-critical use cases, can benefit the final KB and can also help to identify such problems in safety-critical regulations.

7   Extensions of the KB

Both KBs are easily expandable, for example if new categories are added in the existing regulations, or if other regulations (e.g. RECAT) should be included in the same KB. Because of the strong typing of IDP, this does however ask attention to avoid overloading. If other regulations use the same category names, an additional letter or a prefix should be added in IDP to discern between the *Heavy* category of one regulation-set versus the *Heavy* category of the other regulations. The use of prefixes to handle different ICAO-FAA categories with the same name is also recommended in PSOA.

7.1   Optimization of Landing Order

As described in Sect. 2.1 IDP supports a variety of inferences that can be applied on the KB. An example of an application of this is the calculation of an optimal landing order of a number of given aircraft. There are multiple ways to define the optimal landing order. Typically this is either a time-based optimum (to minimize the time between consecutive aircraft landings) or a distance-based optimum. In the latter case, the purpose is to minimize the total separation distance for a series of aircraft, based on the pairwise separation minima. As we already formalized the pairwise calculation of the separation minimum, we will optimize the distance based metric.

An approach queue is constructed from a subset of aircraft, e.g.; `Waiting1` = `a319`; `Waiting2` = `a388`; `Waiting3` = `b788`; `Waiting4` = `be91`. The pairwise separation minima that are calculated with the main program are considered to be given for the example, e.g.; the pair leader a319 with follower a388 has 3 NM as separation minimum. We now want to come up with an order of aircraft: Leader, Follower1, Follower2, and Follower3; that minimizes the sum of the consecutive separation minima. A term `totalseparation` is created: $sum\{ac : Leader = ac|Follower1 = ac|Follower2 = ac|Follower3 = ac| : Separation(ac, Next(ac))\}$.

With the inference `Minimise` a term, in this case `totalseparation`, is minimized. A random order of aircraft could be: Leader be91; Follower1 b788; Follower2 a319; Follower3 a388. The total separation minimum, which is calculated as the sum of consecutive separation minima is 11NM. After minimization, the combination is Leader be91; Follower1 a319; Follower2 b788; Follower3 a388 with a separation minimum of 9.

7.2 Dependent-Slot ATC KB Version

PSOA RuleML explicitly specifies for each descriptor (tuple, slot) whether it is to be interpreted *dependent on* (under the perspective of) the predicate in whose scope it occurs. This dependency dimension refines the design space between oidless atoms with a single dependent tuple (relationships) and oidful atoms with only independent slots (framepoints): it also permits atoms with independent tuples and atoms with dependent slots, the latter denoted by "+>" (instead of "->" for independent slots, e.g. used in Sect. 4.1). This supports advanced data and knowledge representation where, for the same OID, a slot name can have different fillers depending on a predicate [2].

For the disambiguation of multi-valued slots, the ATC KB was enriched into a dependent-slot graph version. Examples of dependent-slot KB facts are shown below (the slot denoting the wake turbulence category, `wtc`, has two fillers, `icao:Super` vs. `faa:Super`, disambiguated, for `:a388` and `:a38f`, by the two perspective-providing predicates, `IcaoRegulated` vs. `FaaRegulated`[8]:

```
%% ICAO Wake Turbulence Categories, Super %%
:a388#:IcaoRegulated(wtc+>icao:Super)
:a38f#:IcaoRegulated(wtc+>icao:Super)

%% FAA Wake Turbulence Categories, Super %%
:a388#:FaaRegulated(wtc+>faa:Super)
:a38f#:FaaRegulated(wtc+>faa:Super)
:a225#:FaaRegulated(wtc+>faa:Super)
```

Interoperation between such dependence-enriched PSOA KB and IDP KB would require a dependence-to-independence reduction [2].

8   Conclusions and Future Work

In this paper, we presented the specification in IDP and PSOA of an Air Traffic Control use case. We discussed the alignment of both specifications and the implications of modeling choices that are involved in this. We demonstrated that a partial interoperation is possible for facts and rules. During the process of constructing and aligning the KBs, some inconsistencies in the original regulations were discovered. The discovery process was different for both systems, which points to their respective unique features and to their internal functioning. It also demonstrates the added value of combining two separate systems to formalize the same knowledge. As the systems can be co-executed, the advantages of each system can be exploited from within a combined application. Examples of this are the introduction of optimization, which is only efficient in the constraint-based system IDP, and the disambiguation of slots via their dependence, which is only possible in the graph-based system PSOA RuleML.

Future work includes the round-trippable translation between increasing subsets of the two languages. An application can be created in which both systems are connected through an API. Based on the PSOA/IDP cross-fertilization, both systems can be further developed, e.g. by support for a separated vocabulary in PSOA RuleML and for graph modeling in IDP. PSOA and IDP could be aligned for constructs used in additional KBs, ultimately defining the complete intersection of PSOA and IDP constructs. Conversely, additional languages could be aligned for formalizing the ATC KB, ultimately making ATC KB a standard use case. Future extensions to regulations could be easily incorporated into the existing KBs. ATC KB could become a shared resource of a multi-agent environment founded on [17].

Acknowledgments. The author of the Hellenic Civil Aviation Authority wants to thank his colleagues for discussions about the domain-expert knowledge. Any deficiencies are ours and we make this disclaimer: the described work constitutes an informative computational model of ATC regulations and is not intended for use in real aviation environments. The authors of KU Leuven are partially supported by the Flemish Agency for Innovation and Entrepreneurship (HBC.2017.0039).

References

1. Berners-Lee, T., Connolly, D., Kagal, L., Scharf, Y., Hendler, J.: N3Logic: a logical framework for the world wide web. Theor. Pract. Logic Program. (TPLP) **8**(3), 249–269 (2008)
2. Boley, H., Zou, G.: Perspectival knowledge in PSOA RuleML: representation, model theory, and translation. CoRR abs/1712.02869, v3 (2019)
3. de Cat, B., Bogaerts, B., Bruynooghe, M., Denecker, M.: Predicate logic as a modelling language: the IDP system. CoRR abs/1401.6312 (2014)
4. Denecker, M., Vennekens, J.: The well-founded semantics is the principle of inductive definition, revisited. In: Fourteenth International Conference on the Principles of Knowledge Representation and Reasoning, pp. 1–10 (2014)

5. Deryck, M., Hasić, F., Vanthienen, J., Vennekens, J.: A case-based inquiry into the decision model and notation (DMN) and the knowledge base (KB) paradigm. In: Benzmüller, C., Ricca, F., Parent, X., Roman, D. (eds.) RuleML+RR 2018. LNCS, vol. 11092, pp. 248–263. Springer, Cham (2018). https://doi.org/10.1007/978-3-319-99906-7_17

6. FAA: Aircraft characteristics database. https://www.faa.gov/airports/engineering/aircraft_char_database/. Accessed 31 May 2019

7. FAA: Advisory Circular 90–23G - Aircraft Wake Turbulence (2014)

8. FAA: ORDER JO 7110.65V, Air Traffic Control (2014)

9. FAA: Order JO 7110.659C, Wake Turbulence Recategorization (2016)

10. FAA: Order JO 7360.1C - Aircraft Type Designators (2017)

11. ICAO: Doc 4444-RAC/501, Procedures for Air Navigation Services - Rules of the Air and Air Traffic Services

12. Lang, S., et al.: Progress on an ICAO wake turbulence re-categorization effort. In: AIAA Atmospheric and Space Environments Conference (2010). https://doi.org/10.2514/6.2010-7682

13. McCluskey, T., Porteous, J., Naik, Y., Taylor, C., Jones, S.: A requirements capture method and its use in an air traffic control application. Softw.: Pract. Experience **25**(1), 47–71 (1995)

14. Mitsikas, T., Almpani, S., Stefaneas, P., Frangos, P., Ouranos, I.: Formalizing air traffic control regulations in PSOA RuleML. In: Proceedings of the Doctoral Consortium and Challenge@ RuleML+ RR 2018 Hosted by 2nd International Joint Conference on Rules and Reasoning, CEUR Workshop Proceedings, vol. 2204 (2018)

15. Mitsikas, T., Stefaneas, P., Ouranos, I.: A rule-based approach for air traffic control in the vicinity of the airport. In: Lambropoulou, S., Theodorou, D., Stefaneas, P., Kauffman, L. (eds.) AlModTopCom 2015, vol. 219, pp. 423–438. Springer, Cham (2017). https://doi.org/10.1007/978-3-319-68103-0_20

16. Object Management Group (OMG): Decision Model and Notation 1.2. (2019). https://www.omg.org/spec/DMN/1.2/

17. Valkanas, G., Natsiavas, P., Bassiliades, N.: A collision detection and resolution multi agent approach using utility functions. In: 2009 Fourth Balkan Conference in Informatics, BCI 2009, Thessaloniki, Greece, 17–19 September 2009, pp. 3–7. IEEE Computer Society (2009)

18. Van Hertum, P., Dasseville, I., Janssens, G., Denecker, M.: The KB paradigm and its application to interactive configuration. Theor. Pract. Logic Program. (TPLP) **17**(1), 91–117 (2017)

19. Zou, G.: Translators for Interoperating and Porting Object-Relational Knowledge. Ph.D. thesis, Faculty of Computer Science, University of New Brunswick, April 2018

20. Zou, G., Boley, H., Wood, D., Lea, K.: Port clearance rules in PSOA RuleML: from controlled-english regulation to object-relational logic. In: Proceedings of the RuleML+RR 2017 Challenge, vol. 1875. CEUR, July 2017

An ASP-based Solution for Operating Room Scheduling with Beds Management

Carmine Dodaro[1], Giuseppe Galatà[2], Muhammad Kamran Khan[3], Marco Maratea[3(✉)], and Ivan Porro[2]

[1] DEMACS, University of Calabria, Rende, Italy
dodaro@mat.unical.it
[2] SurgiQ srl, Genova, Italy
{giuseppe.galata,ivan.porro}@surgiq.com
[3] DIBRIS, University of Genova, Genova, Italy
muhammad.kamrankhan@edu.unige.it, marco.maratea@unige.it

Abstract. The Operating Room Scheduling (ORS) problem is the task of assigning patients to operating rooms, taking into account different specialties, lengths and priority scores of each planned surgery, operating room session durations, and the availability of beds for the entire length of stay both in the Intensive Care Unit and in the wards. A proper solution to the ORS problem is of utmost importance for the quality of the health-care and the satisfaction of patients in hospital environments. In this paper we present an improved solution to the problem based on Answer Set Programming (ASP) that, differently from a recent one, takes explictly into account beds management. Results of an experimental analysis, conducted on benchmarks with realistic sizes and parameters, show that ASP is a suitable solving methodology for solving also such improved problem version.

1 Introduction

The Operating Room Scheduling (ORS) [1,8,25,26] problem is the task of assigning patients to operating rooms, taking into account different specialties, surgery durations, and the availability of beds for the entire length of stay (LOS) both in the Intensive Care Unit (ICU) and in the wards. Given that patients may have priorities, the solution has to find an accommodation for the patients with highest priorities, and then to the other with lower priorities, if space is still available, at the same time taking into proper account beds availability. Recently, a solution based on Answer Set Programming (ASP) [10,11,21,22,27] was proposed and proved to be effective for solving ORS problems [15]. Nonetheless, such solution does not take into account beds management. In most modern hospitals, very long surgical waiting list are present and often worsened, if not altogether caused, by inefficiencies in operating room planning, and the availability of beds in the wards and, if necessary, in the Intensive Care Unit (ICU) for each patient for the entire duration of their stay, is a very important factor for such inefficiencies.

© Springer Nature Switzerland AG 2019
P. Fodor et al. (Eds.): RuleML+RR 2019, LNCS 11784, pp. 67–81, 2019.
https://doi.org/10.1007/978-3-030-31095-0_5

In this paper we thus propose an improved solution based on ASP that takes explictly into account beds management. In such solution, problem specifications related to beds management are modularly added as ASP rules to the previous encoding of the basic version of the problem where beds management was not considered, and then efficient ASP solvers are used to solve the resulting ASP program. We have then generated ORS benchmarks with realistic sizes and parameters inspired by those of small-medium Italian hospitals, and run an experimental analysis on such benchmarks using the ASP solver CLINGO [19]. Benchmarks have been organized in two scenarios: a first scenario is characterized by an abundance of available beds, so that the constraining resource becomes the OR time, while for the second scenario the number of beds is the constrained resource. Overall, results show that ASP is a suitable solving methodology for ORS also when beds management is taken into account, on both scenario, given that our solution is able to utilize efficiently whichever resource is more constrained; moreover, this is obtained in short timings in line with the needs of the application.

To summarize, the main contributions of this paper are the following:

- We provide an ASP encoding for solving the complete ORS problem (Sects. 4 and 5).
- We run an experimental analysis assessing the good performance of our ASP solution (Sect. 6).
- We analyze related literature (Sect. 7), with focus on beds management.

The paper is completed by Sect. 2, which contains needed preliminaries about ASP, by an informal description of the ORS problem in Sect. 3, and by conclusions and possible topics for future research in Sect. 8.

2   Background on ASP

Answer Set Programming (ASP) [11] is a programming paradigm developed in the field of nonmonotonic reasoning and logic programming. In this section we overview the language of ASP. More detailed descriptions and a more formal account of ASP, including the features of the language employed in this paper, can be found in [11,13]. Hereafter, we assume the reader is familiar with logic programming conventions.

Syntax. The syntax of ASP is similar to the one of Prolog. Variables are strings starting with uppercase letter and constants are non-negative integers or strings starting with lowercase letters. A *term* is either a variable or a constant. A *standard atom* is an expression $p(t_1, \ldots, t_n)$, where p is a *predicate* of arity n and $t_1, \ldots, t_n$ are terms. An atom $p(t_1, \ldots, t_n)$ is ground if $t_1, \ldots, t_n$ are constants. A *ground set* is a set of pairs of the form $\langle consts : conj \rangle$, where $consts$ is a list of constants and $conj$ is a conjunction of ground standard atoms. A *symbolic set* is a set specified syntactically as $\{Terms_1 : Conj_1; \cdots ; Terms_t : Conj_t\}$, where $t > 0$, and for all $i \in [1, t]$, each $Terms_i$ is a list of terms such that $|Terms_i| =$

$k > 0$, and each $Conj_i$ is a conjunction of standard atoms. A *set term* is either a symbolic set or a ground set. Intuitively, a set term $\{X : a(X, c), p(X); Y : b(Y, m)\}$ stands for the union of two sets: the first one contains the X-values making the conjunction $a(X, c), p(X)$ true, and the second one contains the Y-values making the conjunction $b(Y, m)$ true. An *aggregate function* is of the form $f(S)$, where S is a set term, and f is an *aggregate function symbol*. Basically, aggregate functions map multisets of constants to a constant. The most common functions implemented in ASP systems are the following:

- *#count*, number of terms;
- *#sum*, sum of integers.

An *aggregate atom* is of the form $f(S) \prec T$, where $f(S)$ is an aggregate function, $\prec \in \{<, \leq, >, \geq, \neq, =\}$ is a comparison operator, and T is a term called guard. An aggregate atom $f(S) \prec T$ is ground if T is a constant and S is a ground set. An *atom* is either a standard atom or an aggregate atom. A *rule r* has the following form:

$$a_1 \vee \ldots \vee a_n :- b_1, \ldots, b_k, not\ b_{k+1}, \ldots, not\ b_m.$$

where $a_1, \ldots, a_n$ are standard atoms, $b_1, \ldots, b_k$ are atoms, $b_{k+1}, \ldots, b_m$ are standard atoms, and $n, k, m \geq 0$. A literal is either a standard atom a or its negation *not a*. The disjunction $a_1 \vee \ldots \vee a_n$ is the *head* of r, while the conjunction $b_1, \ldots, b_k, not\ b_{k+1}, \ldots, not\ b_m$ is its *body*. Rules with empty body are called *facts*. Rules with empty head are called *constraints*. A variable that appears uniquely in set terms of a rule r is said to be *local* in r, otherwise it is a *global* variable of r. An ASP program is a set of *safe* rules, where a rule r is *safe* if the following conditions hold: *(i)* for each global variable X of r there is a positive standard atom ℓ in the body of r such that X appears in ℓ; and *(ii)* each local variable of r appearing in a symbolic set $\{Terms : Conj\}$ also appears in a positive atom in $Conj$.

A *weak constraint* [12] ω is of the form:

$$:\sim b_1, \ldots, b_k, not\ b_{k+1}, \ldots, not\ b_m. \ [w@l]$$

where w and l are the weight and level of ω, respectively. (Intuitively, $[w@l]$ is read "as weight w at level l", where weight is the "cost" of violating the condition in the body of w, whereas levels can be specified for defining a priority among preference criteria). An ASP program with weak constraints is $\Pi = \langle P, W \rangle$, where P is a program and W is a set of weak constraints.

A standard atom, a literal, a rule, a program or a weak constraint is *ground* if no variables appear in it.

Semantics. Let P be an ASP program. The *Herbrand universe* U_P and the *Herbrand base* B_P of P are defined as usual. The ground instantiation G_P of P is the set of all the ground instances of rules of P that can be obtained by substituting variables with constants from U_P.

An *interpretation* I for P is a subset I of B_P. A ground literal ℓ (resp., *not* ℓ) is true w.r.t. I if $\ell \in I$ (resp., $\ell \notin I$), and false (resp., true) otherwise. An aggregate atom is true w.r.t. I if the evaluation of its aggregate function (i.e., the result of the application of f on the multiset S) with respect to I satisfies the guard; otherwise, it is false.

A ground rule r is *satisfied* by I if at least one atom in the head is true w.r.t. I whenever all conjuncts of the body of r are true w.r.t. I.

A model is an interpretation that satisfies all rules of a program. Given a ground program G_P and an interpretation I, the *reduct* [17] of G_P w.r.t. I is the subset G_P^I of G_P obtained by deleting from G_P the rules in which a body literal is false w.r.t. I. An interpretation I for P is an *answer set* (or stable model) for P if I is a minimal model (under subset inclusion) of G_P^I (i.e., I is a minimal model for G_P^I) [17].

Given a program with weak constraints $\Pi = \langle P, W \rangle$, the semantics of Π extends from the basic case defined above. Thus, let $G_\Pi = \langle G_P, G_W \rangle$ be the instantiation of Π; a constraint $\omega \in G_W$ is violated by an interpretation I if all the literals in ω are true w.r.t. I. An *optimum answer set* for Π is an answer set of G_P that minimizes the sum of the weights of the violated weak constraints in G_W in a prioritized way.

Syntactic Shortcuts. In the following, we also use *choice rules* of the form $\{p\}$, where p is an atom. Choice rules can be viewed as a syntactic shortcut for the rule $p \vee p'$, where p' is a fresh new atom not appearing elsewhere in the program, meaning that the atom p can be chosen as true.

3   Problem Description

In this section we provide an informal description of the ORS problem and its requirements.

As we already said in the introduction, most modern hospitals are characterized by a very long surgical waiting list, often worsened, if not altogether caused, by inefficiencies in operating room planning. A very important factor is represented by the availability of beds in the wards and, if necessary, in the Intensive Care Unit for each patient for the entire duration of their stay.

This means that hospital planners have to balance the need to use the OR time with the maximum efficiency with an often reduced beds availability.

In this paper, the elements of the waiting list are called *registrations*. Each registration links a particular surgical procedure, with a predicted surgery duration and length of stay in the ward and in the ICU, to a patient.

The overall goal of the ORS problem is to assign the maximum number of registrations to the operating rooms (ORs), taking into account the availability of beds in the associated wards and in the ICU. This approach entails that the resource optimized is the one, between the OR time and the beds, that represents the bottleneck in the particular scenario analyzed.

As first requirement of the ORS problem, the assignments must guarantee that the sum of the predicted duration of surgeries assigned to a particular OR session does not exceed the length of the session itself: this is referred in the following as *surgery requirement*. Moreover, registrations are not all equal: they can be related to different medical conditions and can be in the waiting list for different periods of time. These two factors are unified in one concept: *priority*. Registrations are classified according to three different priority categories, namely P_1, P_2 and P_3. The first one gathers either very urgent registrations or the ones that have been in the waiting list for a long period of time; it is required that these registrations are all assigned to an OR. Then, the registrations of the other two categories are assigned to the top of the ORs capacity, prioritizing the P_2 over the P_3 ones (*minimization*).

Regarding the bed management part of the problem, we have to ensure that a registration can be assigned to an OR only if there is a bed available for the patient for the entire LOS. In particular, we have considered the situation where each specialty is related to a ward with a variable number of available beds exclusively dedicated to the patients associated to the specialty. This is referred in the following as *ward bed requirement*. The ICU is a particular type of ward that is accessible to patients from any specialty. However, only a small percentage of patients is expected to need to stay in the ICU. This requirement will be referred as the *ICU bed requirement*. Obviously, during their stay in the ICU, the patient does not occupy a bed in the specialty's ward.

In our model, a patient's LOS has been subdivided in the following phases:

- a LOS in the ward before surgery, in case the admission is programmed a day (or more) before the surgery takes place;
- the LOS after surgery, which can be further subdivided into the ICU LOS and the following ward LOS.

The encoding described in Sects. 4 and 5 supports the generation of an optimized schedule of the surgeries either in the case where the bottleneck is represented by the OR time or by the beds availability.

4   ASP Encoding for the Basic ORS Problem

Starting from the specifications in the previous section, in this section the scheduling problem, limited to the assignments of the registrations to the ORs, is described in the ASP language, in particular following the input language of CLINGO.

4.1   OR scheduling

Data Model. The input data is specified by means of the following atoms:

- Instances of *registration(R,P,SU,LOS,SP,ICU,A)* represent the registrations, characterized by an id (*R*), a priority score (*P*), a surgery duration (*SU*) in minutes, the overall length of stay both in the ward and the ICU after the surgery (*LOS*) in days, the id of the specialty (*SP*) it belongs to, a length

$$\{x(R, P, O, S, D)\} \;:\!\!-\; registration(R, P, \_, \_, SP, \_, \_), mss(O, S, SP, D). \qquad (r_1)$$

$$:\!\!-\; x(R, P, O, S1, \_), x(R, P, O, S2, \_), S1! = S2. \qquad (r_2)$$

$$:\!\!-\; x(R, P, O1, \_, \_), x(R, P, O2, \_, \_), O1! = O2. \qquad (r_3)$$

$$surgery(R, SU, O, S) \;:\!\!-\; x(R, \_, O, S, \_), registration(R, \_, SU, \_, \_, \_, \_). \qquad (r_4)$$

$$:\!\!-\; x(\_, \_, O, S, \_), duration(N, O, S),$$
$$\#sum\{SU, R : surgery(R, SU, O, S)\} > N \qquad (r_5)$$

$$:\!\!-\; N = totRegsP1 - \#count\{R : x(R, 1, \_, \_, \_)\}, \; N > 0. \quad (r_6)$$

$$:\!\!\sim\; N = totRegsP2 - \#count\{R : x(R, 2, \_, \_, \_)\}. \; [N@3] \quad (r_7)$$

$$:\!\!\sim\; N = totRegsP3 - \#count\{R : x(R, 3, \_, \_, \_)\}. \; [N@2] \quad (r_8)$$

Fig. 1. ASP encoding of the ORS problem, excluding the bed management

of stay in the ICU (ICU) in days, and finally a parameter representing the number of days in advance (A) the patient is admitted to the ward before the surgery. It must be noted that the variables LOS, ICU and A become relevant for the beds management (see Sect. 5).

- Instances of $mss(O,S,SP,D)$ link each operating room (O) to a session (S) for each specialty and planning day (D) as established by the hospital Master Surgical Schedule (MSS).
- The OR sessions are represented by the instances of the predicate $duration(N,O,S)$, where N is the session duration.

The output is an assignment represented by atoms of the form $x(R,P,O,S,D)$, where the intuitive meaning is that the registration R with priority P is assigned to the OR O during the session S and the day D.

Encoding. The related encoding is shown in Fig. 1, and is described in the following. Rule (r_1) guesses an assignment for the registrations to an OR in a given day and session among the ones permitted by the MSS for the particular specialty the registration belongs to.

The same registration should not be assigned more than once, in different OR or sessions. This is assured by the constraints (r_2) and (r_3). Note that in our setting there is no requirement that every registration must actually be assigned.

Surgery requirement. With rules (r_4) and (r_5), we impose that the total length of surgery durations assigned to a session is less than or equal to the session duration.

Minimization. We remind that we want to be sure that every registration having priority 1 is assigned, then we assign as much as possible of the others, giving precedence to registrations having priority 2 over those having priority 3. This is accomplished through constraint (r_6) for priority 1 and the weak constraints (r_7) and (r_8) for priority 2 and 3, respectively, where *totRegsP1*, *totRegsP2* and *totRegsP3* are constants representing the total number of registrations having priority 1, 2 and 3, respectively.

Minimizing the number of unassigned registrations could cause an implicit preference towards the assignments of the registrations with shorter surgery durations. To avoid this effect, one can consider to minimize the idle time, however this is in general slower from a computational point of view and often unnecessary, since the preference towards shorter surgeries is already mitigated by our three-tiered priority schema.

5 ASP Encoding for ORS with Beds Management

This section is devoted to the beds management task of the ORS problem; the ASP rules and data model described here are added to those presented in the previous section.

5.1 OR scheduling with beds

Data Model. In order to deal with the beds management for the wards and the ICU, the data model outlined in Sect. 4.1 must be supplemented to include data about the availability of beds in each day of the planning and for each ward associated to the specialties and the ICU.

Instances of *beds(SP,AV,D)* represent the number of available beds (AV) for the beds associated to the specialty SP in the day D. The ICU is represented by giving the value 0 to SP.

Encoding. The related encoding is shown in Fig. 2, and is described in the following. Rule (r_9) assigns a bed in the ward to each registration assigned to an OR, for the days before the surgery. Rule (r_{10}) assigns a ward bed for the period after the patient was dismissed from the ICU and transferred to the ward. Rule (r_{11}) assigns a bed in the ICU.

Ward Bed Requirement. Rule (r_{12}) ensures that the number of patients occupying a bed in each ward for each day is never larger than the number of available beds.

ICU bed requirement. Finally, rule (r_{13}) performs a similar check as the one in rule (r_{12}), but for the ICU.

Remark. We note that, given that the MSS is fixed, our problem and encoding can be decomposed by considering each specialty separately in case the beds are not a constrained resource, as will be the case for one of our scenario. We decided not to use this property because (i) this is the description of a practical application that is expected to be extended over time and to correctly work even if the problem becomes non-decomposable, e.g. a (simple but significant)

$$stay(R, D - A..D - 1, SP) :- registration(R, \_, \_, LOS, SP, \_, A),$$
$$x(R, \_, \_, \_, D), A > 0. \qquad (r_9)$$
$$stay(R, D + ICU..D + LOS - 1, SP) :- registration(R, \_, \_, LOS, SP, ICU, \_),$$
$$x(R, \_, \_, \_, D), LOS > ICU. \qquad (r_{10})$$
$$stayICU(R, D..ICU + D - 1) :- registration(R, \_, \_, \_, \_, ICU, \_),$$
$$x(R, \_, \_, \_, D), ICU > 0. \qquad (r_{11})$$
$$:- \#count\{R : stay(R, D, SP)\} > AV,$$
$$SP > 0, beds(SP, AV, D). \qquad (r_{12})$$
$$:- \#count\{R : stayICU(R, D)\} > AV,$$
$$beds(0, AV, D). \qquad (r_{13})$$

Fig. 2. ASP encoding of the bed management portion of the ORS problem

Table 1. Beds availability for each specialty and in each day in scenario A.

Specialty	Monday	Tuesday	Wednesday	Thursday	Friday
0 (ICU)	40	40	40	40	40
1	80	80	80	80	80
2	58	58	58	58	58
3	65	65	65	65	65
4	57	57	57	57	57
5	40	40	40	40	40

extension in which a room is shared among specialties brings to a problem which is not anymore decomposable in all cases, and (*ii*) it is not applicable to all of our scenarios. Additionaly, even not considering this property at the level of encoding, the experimental analysis that we will present is already satisfying for our use case.

6   Experimental Results

In this section we report about the results of an empirical analysis of the ORS problem. Data have been randomly generated but having parameters and sizes inspired by real data. Both experiments were run on a Intel Core i7-7500U CPU @ 2.70GHz with 7.6 GB of physical RAM. The ASP system used was CLINGO [18], version 5.5.2.

6.1   ORS Benchmarks

The final encoding employed in our analysis is composed by the ASP rules $(r_1), \ldots, (r_{13})$. The test cases we have assembled are based on the requirements of a typical small-medium size Italian hospital, with five surgical specialties to be

Table 2. Beds availability for each specialty and in each day in scenario B.

Specialty	Monday	Tuesday	Wednesday	Thursday	Friday
0 (ICU)	4	4	5	5	6
1	20	30	40	45	50
2	10	15	23	30	35
3	10	14	21	30	35
4	8	10	14	16	18
5	10	14	20	23	25

Table 3. Parameters for the random generation of the scheduler input.

Specialty	Reg.	ORs	Surgery duration (min) mean (std)	LOS (d) mean (std)	ICU %	ICU LOS (d) mean (std)	LOS (d) before surgery
1	80	3	124 (59.52)	7.91 (2)	10	1 (1)	1
2	70	2	99 (17.82)	9.81 (2)	10	1 (1)	1
3	70	2	134 (25.46)	11.06 (3)	10	1 (1)	1
4	60	1	95 (19.95)	6.36 (1)	10	1 (1)	0
5	70	2	105 (30.45)	2.48 (1)	10	1 (1)	0
Total	350	10					

managed over the widely used 5-days planning period. Two different scenarios were assembled. The first one (scenario A) is characterized by an abundance of available beds, so that the constraining resource becomes the OR time. For the second one (scenario B), we severely reduced the number of beds, in order to test the encoding in a situation with plenty of OR time but few available beds. Each scenario was tested 10 times with different randomly generated inputs. The characteristics of the tests are the following:

- 2 different benchmarks, comprising a planning period of 5 working days, and different numbers of available beds, as reported in Tables 1 and 2 for scenario A and B, respectively;
- 10 ORs, unevenly distributed among the specialties;
- 5 h long morning and afternoon sessions for each OR, summing up to a total of 500 h of ORs available time for the 2 benchmarks;
- 350 generated registrations, from which the scheduler will draw the assignments. In this way, we simulate the common situation where a hospital manager takes an ordered, w.r.t. priorities, waiting list and tries to assign as many elements as possible to each OR.

Table 4. Scheduling results for the scenario A benchmark.

Assigned registrations				OR time efficiency	Bed occupation efficiency
Priority 1	Priority 2	Priority 3	Total		
62/62	132/150	72/138	266/350	96.6%	52.0%
72/72	128/145	64/133	264/350	95.6%	51.0%
71/71	132/132	69/147	272/350	96.7%	53.0%
66/66	138/142	57/142	261/350	96.2%	50.7%
79/79	119/130	67/141	265/350	96.0%	51.9%
67/67	131/131	66/152	264/350	96.6%	53.8%
66/66	121/132	69/152	256/350	96.0%	49.8%
69/69	130/135	68/146	267/350	96.8%	51.6%
60/60	139/153	59/137	258/350	96.0%	50.8%
68/68	138/142	57/139	263/350	95.2%	51.3%

Table 5. Scheduling results for the scenario B benchmark.

Assigned registrations				OR time efficiency	Bed occupation efficiency
Priority 1	Priority 2	Priority 3	Total		
62/62	106/150	13/138	181/350	66.3%	92.7%
72/72	77/145	43/133	192/350	67.5%	94.2%
71/71	80/132	38/147	189/350	68.2%	96.1%
66/66	81/142	41/142	188/350	71.4%	93.4%
79/79	90/130	20/141	189/350	69.0%	94.1%
67/67	95/131	25/152	187/350	66.5%	93.9%
66/66	92/132	30/152	188/350	71.8%	94.1%
69/69	84/135	36/146	189/350	68.7%	92.7%
60/60	91/153	34/137	185/350	69.7%	94.1%
68/68	82/142	35/139	185/350	69.3%	95.1%

The surgery durations have been generated assuming a normal distribution, while the priorities have been generated from a uneven distribution of three possible values (with weights respectively of 0.20, 0.40 and 0.40 for registrations having priority 1, 2 and 3, respectively). The lengths of stay (total LOS after surgery and ICU LOS) have been generated using a truncated normal distribution, in order to avoid values less than 1. In particular for the ICU, only a small percentage of patients have been generated with a predicted LOS while the large majority do not need to pass through the ICU and their value for the ICU LOS is fixed to 0. Finally, since the LOS after surgery includes both the LOS in the wards and in the ICU, the value generated for the ICU LOS must be less than or equal to the total LOS after surgery. The parameters of the test have been summed up in Table 3. In particular, for each specialty (1 to 5), we reported the number of registrations generated, the number of ORs assigned to the specialty, the mean duration of surgeries with its standard deviation, the mean LOS after the surgery with its standard deviation, the percentage of patients that need to stay in the ICU, the mean LOS in the ICU with its standard deviation and, finally, the LOS before the surgery (i.e. the number of days, constant for each specialty, the patient is admitted before the planned surgery is executed).

6.2   Results

Results of the experiments are reported for scenario A in Table 4 and for scenario B in Table 5, respectively. A time limit of 60 seconds was given in view of a practical use of the program, each scenario was run 10 times with different input registrations. For each of the 10 runs executed, the tables report in the first three columns the number of the assigned registrations out of the generated ones for each priority, and in the remaining two columns a measure of the total time occupied by the assigned registrations as a percentage of the total OR time available (indicated as OR time Efficiency in the Table) and the ratio between the beds occupied after the planning to the available ones before the planning (labeled as Bed Occupation Efficiency in the tables). As a general observation, these results show that our solution is able to utilize efficiently whichever resource is more constrained: scenario A runs manage to reach a very high efficiency, over 95%, in the use of OR time, while scenario B achieves an efficiency of bed occupation between 92% and 95%. Note that to better be able to confront the results, for each run the bed configurations of the two scenarios were applied to the same generated registrations. Taking into consideration a practical use of this solution, the user would be able to individuate and quantify the resources that are more constraining and take the appropriate actions. This means that the solution can also be used to test and evaluate "what if" scenarios.

Finally, in Fig. 3 we (partially) present the results achieved on one instance (i.e., the first instance of Tables 4 and 5) with 350 registrations for 5 days. Each bar represents the total number of available beds for specialty 1, as reported in Table 1 for the plot at the top and Table 2 for the bottom one, for each day of the week, from Monday through Friday. The colored part of the bars indicates the amount of occupied beds while the gray part the beds left unoccupied by our planning.

7   Related Work

In this section we review related literature, organized into two paragraphs. The first paragraph is devoted to outlining different techniques for solving the ORS problem, with focus on the inclusion of beds management, while in the second paragraph we report about other scheduling problems where ASP has been employed.

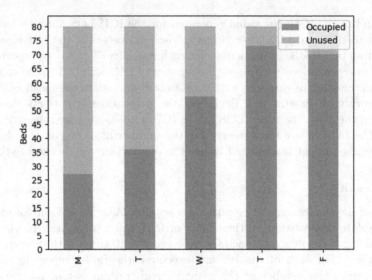

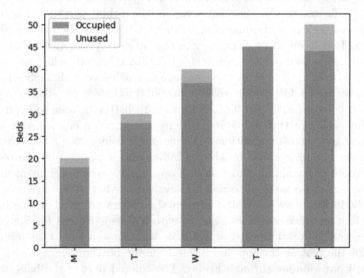

Fig. 3. Example of bed occupation of the ward corresponding to specialty 1 for 5 days scheduling. The plot on the top corresponds to the first instance of scenario A, while the one on the bottom to the first instance of scenario B.

Solving ORS Problems. Aringhieri et al. [8] addressed the joint OR planning (MSS) and scheduling problem, described as the allocation of OR time blocks to specialties together with the subsets of patients to be scheduled within each time block over a one week planning horizon. They developed a 0–1 linear programming formulation of the problem and used a two-level meta-heuristic to solve it.

Its effectiveness was demonstrated through numerical experiments carried out on a set of instances based on real data and resulted, for benchmarks of 80–100 assigned registrations, in a 95–98% average OR utilization rate, for a number of ORs ranging from 4 to 8. The execution times were around 30–40 seconds. In [25], the same authors introduced a hybrid two-phase optimization algorithm which exploits neighborhood search techniques combined with Monte Carlo simulation, in order to solve the joint advance and allocation scheduling problem, taking into account the inherent uncertainty of surgery durations. In both the previous works, the authors solve the beds management problem limited to weekend beds, while assuming that each specialty has its own post-surgery beds from Monday to Friday with no availability restriction. In [9], some of the previous authors face the beds management problem for all the days of the week, with the aim to level the post-surgery ward bed occupancies during the days, using a Variable Neighbourhood Search approach.

Other relevant approaches are: Abedini et al. [1], that developed a bin packing model with a multi-step approach and a priority-type-duration rule; Molina-Pariente et al. [26], that tackled the problem of assigning an intervention date and an operating room to a set of surgeries on the waiting list, minimizing access time for patients with diverse clinical priority values; and Zhang et al. [29], that addressed the problem of OR planning with different demands from both elective patients and non-elective ones, with priorities in accordance with urgency levels and waiting times. However, beds management is not considered in this last three mentioned approaches.

ASP in Scheduling Problems. We already mentioned in the introduction that ASP has been already successfully used for solving hard combinatorial and application problems in several research areas. Concerning scheduling problems other than ORS, ASP encodings were proposed for the following problems: *Incremental Scheduling Problem* [14,20], where the goal is to assign jobs to devices such that their executions do not overlap one another; *Team Building Problem* [28], where the goal is to allocate the available personnel of a seaport for serving the incoming ships; and *Nurse Scheduling Problem* [2,3,16], where the goal is to create a scheduling for nurses working in hospital units. Other relevant problems are *Interdependent Scheduling Games* [5], which requires interdependent services among players, that control only a limited number of services and schedule independently, the *Conference Paper Assignment Problem* [7], which deals with the problem of assigning reviewers in the PC to submitted conference papers, and the *Stable Roommates Problem* [6], which is a modified version of the well-known Stable Marriage Problem.

8 Conclusions

In this paper we have employed ASP for solving to the ORS problem with beds management, given ASP has already proved to be a viable tool for solving scheduling problems due to the readability of the encoding, and availability of efficient solvers. Specifications of the problem are modularly expressed as rules in the ASP encoding, and ASP solver CLINGO has been used. We finally presented the results of an experimental analysis on ORS benchmarks with realistic sizes and parameters on two scenario, that reveal that our solution is able to utilize efficiently whichever resource is more constrained, being either the OR time or the beds. Moreover, for the planning length of 5 days usually used in small-medium Italian hospitals, this is obtained in short timings in line with the needs of the application. Future work includes the design and analysis of a re-scheduling solution, in case the off-line solution proposed in this paper can not be fully implemented for circumstances such as canceled registrations, and the evaluation of heuristics and optimization techniques (see, e.g., [4,23,24]) for further improving the effectiveness of our solution.

All materials presented in this work, including benchmarks, encodings and results, can be found at: http://www.star.dist.unige.it/~marco/RuleMLRR19/ material.zip.

References

1. Abedini, A., Ye, H., Li, W.: Operating room planning under surgery type and priority constraints. Procedia Manufact. **5**, 15–25 (2016)
2. Alviano, M., Dodaro, C., Maratea, M.: An advanced answer set programming encoding for nurse scheduling. In: Esposito, F., Basili, R., Ferilli, S., Lisi, F. (eds.) AI*IA 2017 Advances in Artificial Intelligence. LNCS, pp. 468–482. Springer, Cham (2017). https://doi.org/10.1007/978-3-319-70169-1_35
3. Alviano, M., Dodaro, C., Maratea, M.: Nurse (re)scheduling via answer set programming. Intelligenza Artificiale **12**(2), 109–124 (2018)
4. Alviano, M., Dodaro, C., Marques-Silva, J., Ricca, F.: Optimum stable model search: Algorithms and implementation. J. Log. Comput. https://doi.org/10.1093/logcom/exv061 (in press)
5. Amendola, G.: Preliminary results on modeling interdependent scheduling games via answer set programming. In: RiCeRcA@AI*IA CEUR Workshop Proceedings, vol. 2272. CEUR-WS.org (2018)
6. Amendola, G.: Solving the stable roommates problem using incoherent answer set programs. In: RiCeRcA@AI*IA CEUR Workshop Proceedings, vol. 2272. CEUR-WS.org (2018)
7. Amendola, G., Dodaro, C., Leone, N., Ricca, F.: On the application of answer set programming to the conference paper assignment problem. In: Adorni, G., Cagnoni, S., Gori, M., Maratea, M. (eds.) AI*IA 2016. LNCS (LNAI), vol. 10037, pp. 164–178. Springer, Cham (2016). https://doi.org/10.1007/978-3-319-49130-1_13
8. Aringhieri, R., Landa, P., Soriano, P., Tànfani, E., Testi, A.: A two level metaheuristic for the operating room scheduling and assignment problem. Comput. Oper. Res. **54**, 21–34 (2015)
9. Aringhieri, R., Landa, P., Tànfani, E.: Assigning surgery cases to operating rooms: A VNS approach for leveling ward beds occupancies. Electron. Notes Discrete Math. **47**, 173–180 (2015). https://doi.org/10.1016/j.endm.2014.11.023
10. Baral, C.: Knowledge Representation, Reasoning and Declarative Problem Solving. Cambridge University Press, Cambridge (2003). https://doi.org/10.1017/CBO9780511543357

11. Brewka, G., Eiter, T., Truszczynski, M.: Answer set programming at a glance. Commun. ACM **54**(12), 92–103 (2011)
12. Buccafurri, F., Leone, N., Rullo, P.: Enhancing disjunctive datalog by constraints. IEEE Trans. Knowl. Data Eng. **12**(5), 845–860 (2000)
13. Calimeri, F., et al.: ASP-Core-2 Input Language Format (2013). https://www.mat. unical.it/aspcomp2013/files/ASP-CORE-2.01c.pdf
14. Calimeri, F., Gebser, M., Maratea, M., Ricca, F.: Design and results of the fifth answer set programming competition. Artif. Intell. **231**, 151–181 (2016)
15. Dodaro, C., Galatà, G., Maratea, M., Porro, I.: Operating room scheduling via answer set programming. In: Ghidini, C., Magnini, B., Passerini, A., Traverso, P. (eds.) AI*IA 2018. LNCS (LNAI), vol. 11298, pp. 445–459. Springer, Cham (2018). https://doi.org/10.1007/978-3-030-03840-3_33
16. Dodaro, C., Maratea, M.: Nurse scheduling via answer set programming. In: Balduccini, M., Janhunen, T. (eds.) LPNMR 2017. LNCS (LNAI), vol. 10377, pp. 301–307. Springer, Cham (2017). https://doi.org/10.1007/978-3-319-61660-5_27
17. Faber, W., Pfeifer, G., Leone, N.: Semantics and complexity of recursive aggregates in answer set programming. Artif. Intell. **175**(1), 278–298 (2011)
18. Gebser, M., Kaminski, R., Kaufmann, B., Ostrowski, M., Schaub, T., Wanko, P.: Theory solving made easy with clingo 5. In: ICLP (Technical Communications). OASICS, vol. 52, pp. 2:1–2:15. Schloss Dagstuhl - Leibniz-Zentrum fuer Informatik (2016)
19. Gebser, M., Kaufmann, B., Schaub, T.: Conflict-driven answer set solving: from theory to practice. Artif. Intell. **187**, 52–89 (2012)
20. Gebser, M., Maratea, M., Ricca, F.: The sixth answer set programming competition. J. Artif. Intell. Res. **60**, 41–95 (2017)
21. Gelfond, M., Lifschitz, V.: The stable model semantics for logic programming. In: Proceedings of the Fifth International Conference and Symposium, Seattle, Washington, 15–19 August 1988, vol. 2, pp. 1070–1080. MIT Press (1988)
22. Gelfond, M., Lifschitz, V.: Classical negation in logic programs and disjunctive databases. New Gener. Comput. **9**(3/4), 365–386 (1991)
23. Giunchiglia, E., Maratea, M., Tacchella, A.: Dependent and independent variables in propositional satisfiability. In: Flesca, S., Greco, S., Ianni, G., Leone, N. (eds.) JELIA 2002. LNCS (LNAI), vol. 2424, pp. 296–307. Springer, Heidelberg (2002). https://doi.org/10.1007/3-540-45757-7_25
24. Giunchiglia, E., Maratea, M., Tacchella, A.: (In)Effectiveness of look-ahead techniques in a modern SAT solver. In: Rossi, F. (ed.) CP 2003. LNCS, vol. 2833, pp. 842–846. Springer, Heidelberg (2003). https://doi.org/10.1007/978-3-540-45193-8_64
25. Landa, P., Aringhieri, R., Soriano, P., Tànfani, E., Testi, A.: A hybrid optimization algorithm for surgeries scheduling. Oper. Res. Health Care **8**, 103–114 (2016)
26. Molina-Pariente, J.M., Hans, E.W., Framinan, J.M., Gomez-Cia, T.: New heuristics for planning operating rooms. Comput. Ind. Eng. **90**, 429–443 (2015)
27. Niemelä, I.: Logic programs with stable model semantics as a constraint programming paradigm. Ann. Math. Artif. Intell. **25**(3–4), 241–273 (1999)
28. Ricca, F., et al.: Team-building with answer set programming in the Gioia-Tauro seaport. Theory Pract. Logic Program. **12**(3), 361–381 (2012)
29. Zhang, J., Dridi, M., El Moudni, A.: A stochastic shortest-path MDP model with dead ends for operating rooms planning. In: ICAC, pp. 1–6. IEEE (2017)

EASE: Enabling Hardware Assertion Synthesis from English

Rahul Krishnamurthy[✉] and Michael S. Hsiao

Department of Electrical and Computer Engineering, Virginia Tech Blacksburg,
Blacksburg, VA 24061, USA
{rahulk4,mhsiao}@vt.edu

Abstract. In this paper, we present EASE (Enabling hardware Assertion Synthesis from English) which translates hardware design specifications written in English to a formal assertion language. Existing natural language processing (NLP) tools for hardware verification utilize the vocabulary and grammar of a few specification documents only. Hence, they lack the ability to provide linguistic variations in parsing and writing natural language assertions. The grammar used in EASE does not follow a strict English syntax for writing design specifications. Our grammar incorporates dependency rules for syntactic categories which are coupled with semantic category dependencies that allow users to specify the same design specification using different word sequences in a sentence. Our NLP engine consists of interleaving operations of semantic and syntactic analyses to understand the input sentences and map differently worded sentences with the same meaning to the same logical form. Moreover, our approach also provides semantically driven suggestions for sentences that are not understood by the system. EASE has been tested on natural language requirements extracted from memory controller, UART and AMBA AXI protocol specification documents. The system has been tested for imperative, declarative and conditional types of specifications. The results show that the proposed approach can handle a more diverse set of linguistic variations than existing methods.

Keywords: Natural Language Processing · Hardware verification · Natural language programming

1 Introduction

Automation of design verification from natural language specifications has the advantage of reducing design life cycle, design errors and identification of incoherent specifications in an early design stage [1]. Motivated by these advantages, various approaches have been proposed [2–6] to automate design verification by generating assertions from its available natural language specifications.

A template-based approach to automatically generate assertions has been proposed in [2], in which natural language assertions are categorized based on

© Springer Nature Switzerland AG 2019
P. Fodor et al. (Eds.): RuleML+RR 2019, LNCS 11784, pp. 82–96, 2019.
https://doi.org/10.1007/978-3-030-31095-0_6

their syntactic and semantic structures and then mapped to a System Verilog Assertion (SVA) template. However, such an approach is not fully automated and require the assistance of a verification engineer in the creation of these templates. In [3], translation rules are manually extracted from properties defined in specification documents to automatically generate a formal model. In addition, it also synthesized LTL formulas from natural language requirements when sufficient information about design variables are available.

In [4,5], an attribute grammar is constructed which represents vocabulary and linguistic variations of the corpus containing assertions written in English. A parser based on this grammar enables the generation of formal assertions from natural language properties specified in the corpus. However, a new grammar has to be constructed manually for another design specification document with a different writing style. The work in [6], aims to alleviate the manual creation of the grammar by automatically learning the grammar using a training set of natural language assertions and their corresponding formal representations.

All of the above grammar based approaches require low-level design assertions as input from the specification document. System requirements in these documents describe high-level design behavior and include charts, diagrams and tables [1]. Moreover, for a complex System on chip (SoC) design, each module may have a separate specification document. Consequently, it becomes difficult to specify the overall design properties in a specification document for a design that span multiple design blocks. Because of this, we may not even find design properties in the specification document for a complex SoC design [7], which may result in a lack of data for the creation of grammar and translation rules.

In the absence of design properties in specification documents, an engineer has to understand the design intent from the corresponding high level specifications and then write it in a formal assertion language. Manually writing such formal assertions for complex designs can be error-prone and time-consuming.

Natural language based automation can relieve the user from manually writing executable assertions from specifications. However, as mentioned earlier, specifications are either in the form of high level description in tables, charts or written in unrestricted natural language which are imprecise, ambiguous and incomplete [8,9]. Hence, it is not possible to automatically parse these specifications and accurately translate them to a formal language. Our research objective is to design a tool that can assist the user in writing and parsing design specifications unambiguously at the beginning stage of the design. These design specifications can then be parsed to create executable specifications which can be used to automatically verify the design.

Recently, Controlled Natural Language (CNL) tools have been developed that translate English sentences to executable code [10]. In [11], a CNL framework is proposed to translate specifications written in natural language to a formal verification language. However, their architecture perform syntactic and semantic analysis separately and hence unable to use the partial semantic understanding for disambiguation of natural language specification. Moreover, their CNL gives suggestion only when it detects ambiguity in the specification after completely parsing the sentence.

We propose an architecture called EASE which stands for Enabling hardware Assertion Synthesis from English. EASE guides users in writing design specifications in English that can be automatically processed by the system and translated to a formal language. Our objective is to minimize restrictions on sequence of categories in a CNL grammar so that the user can express his ideas more freely and get the maximum benefits from natural language based automation. In order to achieve this objective, we have created a grammar based on syntactic and semantic dependencies instead of a grammar with sequential structure of English categories. Another purpose for creating such a grammar was to create parse trees with dependency links that are more suited for Natural Language Understanding (NLU) tasks [12]. We propose a joint syntactic-semantic parsing of dependency trees at run-time to understand any partially written sentence and provide suggestion to user on completing this sentence. Moreover, our interleaving syntactic and semantic analysis is incremental in nature and also considers context to understand the written sentence. This approach creates the same semantic expression for a design intent written with a different sequence of words.

Our contributions can be summarized as follows:

- EASE is independent of grammar and linguistic variations of a specification document which is contrary to earlier approaches that are mostly document specific. We introduce semantic categories of hardware verification domain to reduce ambiguity in parsing a specification written in unrestricted English. Semantic categories in our grammar assists in disambiguation of semantic roles for various words in the input sentence.
- A dependency grammar allows flexibility in writing a sentence with different word orders. However, a dependency grammar with only dependency links does not contain explicit semantic interpretation of syntactic rules. To improve semantic interpretation of sentences we propose a joint syntactic and semantic analysis using dynamic programming in our natural language parsing stage. Such an analysis ensures that any partially written sentence is correctly understood and also assists in providing suggestions to user on completing a partially written sentence.
- Our natural language understanding analysis is an incremental algorithm and considers the context with which a word is written/parsed while assigning semantic expressions to a complete sentence. This contextual based understanding enables the mapping of the same semantic expression to sentences with the same meaning but written with different word order.
- We provide semantically driven suggestions to the user in completing the sentence. This suggestion framework has the capability of giving feedback to the user on sentences that are semantically wrong even if they are syntactically correct.

The rest of the paper is organized as follows: In Sect. 2, we will describe EASE and its components. In Sect. 3, we will discuss our experimental evaluation and results. Finally, in Sect. 4, we present our Conclusion.

2   EASE Architecture

The architecture of EASE is illustrated in Fig. 1. An overview of our framework is as follows: Our approach dynamically updates the semantic expressions of the sentence based on words written (or parsed) thus far. In the beginning, a word is pre-processed to ensure that it complies with the input requirements. After pre-processing, each word is tagged by syntactic and semantic categories that are defined in our grammar. We create an initial clause relationship tree (CRT), and each input word is appropriately placed in the CRT. The CRT is created to understand complex English sentences containing connecting words like coordinating conjunction and conditional words. This CRT is updated based on every incoming words. Updating the CRT involves several sub-tasks, as shown in Fig. 1. First, we dynamically detect a clause and place it in a leaf node of the CRT. Then, a dependency parse tree for this clause is created and stored in the CRT. Semantic contribution of each incoming word in a dependency tree is computed, and the overall semantic expression of a clause is updated. The process of updating the CRT, dependency parse trees, and semantic expressions continues until the last word is processed in a design specification. The overall meaning of the sentence is understood by the collective analysis of the semantic expressions of all the clauses. Finally, the formal representation that is generated from the CRT can be translated to any of the hardware verification languages like System Verilog Assertion (SVA) or Property Specification Language (PSL). SVA or PSL based assertion can be used to verify the hardware design. In the subsequent subsections, detail of these stages is presented.

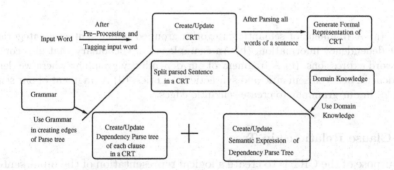

Fig. 1. EASE Architecture dynamically analyse each input word and generates a formal representation of design specification.

2.1   Dependency Grammar and Lexicon

The categories in our grammar are commonly used in writing hardware verification functions. Figure 2(a), shows a fragment of semantic classes available in our grammar. These categories are grouped under syntactic categories. Based on syntactic information of these categories, we have defined their dependency relations with other syntactic classes like verbs, prepositions, and coordinate conjunctions. A small set of words belonging to these categories are shown in Fig. 2(b).

A fragment of our dependency grammar is shown in Fig. 2(c). In Fig. 2(c), the verbs are shown to be dependent on a dummy node 'root'. The main verbs are the action words, for example 'set', 'de-assert' and 'change'. The predicate-argument structure in dependency parse trees may not be explicit for some syntactic constructions like coordinate conjunctions and prepositional phrases. We attempt to capture even these semantic relationships by adding semantic information to our dependency grammar. Figure 2(d) shows a fragment of our rule-based semantic information, which creates a direct relationship between semantic categories.

Adj	Noun	Num
Rel	Signal	Value
Delay	Data_word	Data
Consec	Value_word	
	Clock_cycle	

(a)

Category	Lexicon
Rel	greater, lower, equal ,..
Data_word	data, number of bits ,...
Clock_cycle	cd cycle, same cycle ,...

(b)

Category 1	Connector	Category 2	Direc	Sem_edge
Consec	[for,*]	Clock_cycle	Same	consec_clock
Data	[of,*]	Data_word	Reverse	data_is
Data_word	[of,in,*]	Register	Same	data_of
Rel	[than,to]	Value	Same	rel_obj
Value_word	[of,on,*]	Signal	Same	value_of

(d)

Head	Dependents
root	main verb, linking verb, modal verb
main verb	signal, register, clock_cycle, value_word, data_word
value_word	signal, register, value, prep
clock_cycle	consec, delay, prep

(c)

Fig. 2. (a) A fragment of semantic categories grouped under syntactic categories to extract dependency information. (b) An example set of vocabulary that also contains multi-word expressions. (c) A fragment of our dependency grammar where we defined dependencies between semantic and syntactic categories. (d) A fragment of our semantic tuples in our grammar to create semantic edges.

2.2   Clause Relationship Tree

The purpose of the CRT is to create a logical representation of the input sentence in the form of a tree where the semantic expression of clauses is connected by conditional words or by coordinating conjunctions. As illustrated in Fig. 3, the CRT is a binary tree. All the leaf nodes in this tree are clauses, and non-leaf nodes are words which connect these clauses in a sentence. The structure of a clause and list of head-words are shown in Fig. 3. The CRT is updated dynamically based on the input word received for processing.

A partial sentence is split into clauses and stored in leaf nodes of CRT only if a head-word separates these clauses. The dependency parse tree and corresponding semantic expression of a clause are also stored in the leaf node of the CRT. The semantic expressions of all the clauses are combined to generate an executable

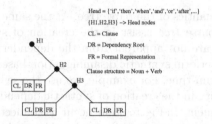

Fig. 3. A clause relationship tree with 3 types of data stored in a leaf node. The head node words are sentence connectors or conditional words as mentioned in Head list.

assertion. Figure 4 illustrates the creation and reconfiguration of the CRT based on the words received from the input sentence. The figure also shows that we create the dependency parse trees of the clauses taken from the leaf nodes of this CRT.

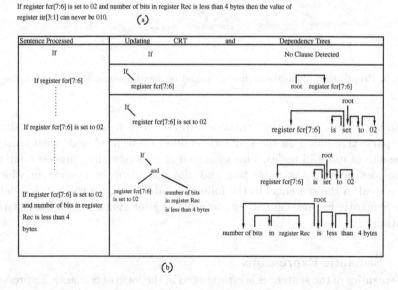

Fig. 4. (a) An example design specification for which CRT creation is shown. (b)Figure shows the creation and update performed in a CRT. Also each leaf node of CRT is parsed to create dependency parse trees.

2.3   Syntactic-Semantic Parsing

In our EASE architecture, syntactic and semantic parsing are coupled in a complementary fashion to disambiguate any clarity issues in the English sentence. The parsing process begins from the current leaf node of CRT where a new word is received from the user. The syntactic edges in our parse tree assist in

understanding the semantics of the parse tree. At the same time, partial semantic expressions of a parse tree assist in the creation of syntactic edges when
dependency relations are not apparent from the dependency grammar. In this
architecture, we can perform syntactic disambiguation based on semantic understanding of words at run-time. For example, we use the semantic understanding
to resolve the ambiguity in the creation of syntactic dependency links for prepositional phrase attachment. In Fig. 5, node 'occur' is connected to the preposition
'within' instead of connecting node 'command active' to this preposition. Such a
connection was made because of the disambiguation based on the partial semantic expression created on the node 'occur' where 'command active' is considered
as an 'occur_when_after' parameter. According to the rules, we would have connected 'command active' to 'within' if 'command active' had a semantic role of
'occur_what'.

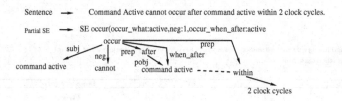

Fig. 5. Preposition phrase attachment based on partial semantic understanding.

Our algorithm learns the semantic expression of a head-node in a dependency parse tree based on the syntactic connection of head-node with semantic
expressions of its child nodes. This incremental understanding process continues
till the root node of the parse tree and also considers the context in which a
word is used in the sentence. In the following subsection, we will provide details
of our Semantic Expression (SE) structure and joint syntactic-semantic parsing
algorithm.

2.3.1   Semantic Expressions

The Semantics of the sentences is represented in the form of Semantic Expressions
(SE) in our framework. These SEs are analogous to frame semantics used in NLP.
In our approach, we define an SE as a set of parameters where each parameter is
a placeholder for either a function or an argument of a function. For example, the
phrase 'assigning a 1 to a register' describe a Value function where 'register' and
'1' are arguments. These SE's are created based on the syntactic and semantic
relations that exist between words in a dependency parse tree.

Figure 6a, illustrates a Value SE with three parameters, namely, value_of,
value_is and value_when. Also, in this figure, we have shown the type of values
that can be assigned to value_of and value_when parameters. As shown in this
figure, value_of can be assigned nodes of category type 'Signal' or 'Register'.

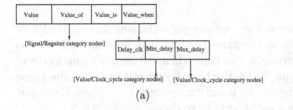

(a)

exp(value) : par(value_of) & par(value_is) & par(value_when)

par(value_is) : cat(value) & edge(same)

par(value_of) : (cat(register) I cat(signal)) & edge(value_of)

exp(delay_clock) : par(min_delay) & par(max_delay)

par(min_delay) : (cat(value) I cat(clock_cycle)) & edge(delay)

cat(clock_cycle) : cd cycle I same cycle I next cycle

(b)

Fig. 6. (a) For a 'Value' semantic expression, we have shown how 'Value_of' and 'Value_when' parameter can be filled. (b) This figure shows rules that are defined to fill Value SE and delay_clock SE. In this figure, keyword 'exp' means semantic expression, 'par' is a slot in expression, and 'cat' is the category of a word that can fill a parameter.

Whereas, the Value_when parameter can take another SE named Delay_clk as its value.

A fragment of our rules used to fill Value SE is shown in Fig. 6b. In this figure, we have shown a set of rules that creates SE based on the edges between words in a dependency parse tree. For example, the parameter(value_of) is assigned either a signal or a register category node, and the edge between the SE and this node should be value_of.

Algorithm 1. Syntactic-Semantic Parser

```
 1: procedure PARSE(CRT_leaf_node data)
 2:     Push data to buffer, Push root to stack          ▷ Initializing Buffer and Stack
 3:     while Buffer is not empty do
 4:         move_dict ← find_move(buffer,stack)
 5:         if move_dict = 'shift' then
 6:             stack.append(buffer_top) , buffer.remove(buffer_top)
 7:         end if
 8:         if move_dict = 'reduce' then
 9:             stack.remove(stack_top)
10:         end if
11:         if move_dict = 'left_arc' then
12:             apply_arc(head = buffer_top , child = stack_top)
13:             traverse_with_semantic_update(buffer_top,'down')
14:         end if
15:         if move_dict = 'right_arc' then
16:             apply_arc(head = stack_top , child = buffer_top)
17:             traverse_with_semantic_update(buffer_top,'up')
18:         end if
19:     end while
20: end procedure
```

2.3.2   Algorithm

For each edge created in our parse tree, we compute the semantic influence of this edge on the overall semantics of the parse tree. Such an approach allow us to detect an error in the input sentence at run-time when the semantics understood from the new word is not compatible with the existing semantics of the parse tree. We adopted such an approach to avoid the necessity to parse the input sentence according to the strict English syntax to detect ambiguity.

We maintain a list of semantic expressions at each node, which represents the meaning contributed by this node and its child nodes. In a parse tree, the leaf node consists of all possible semantic expressions which can be inferred based on the word in this leaf node. While parsing from a leaf node to the root node in a parse tree, we select semantic expressions at each node that describes the context in which this node and its child-node word is being used. In this process, the ambiguity in understanding the input sentence is detected if we are left with multiple SE's at the root node. Then according to the available SE's at the root node, we provide an example sentence to the user for each these SE's. This helps in explaining to the user on how to express these SE's correctly in an English sentence.

Head SE	Child SE	Connecting Edge	Semantic Update
rel	value	rel_obj	rel(rel_obj: value SE)
occur	command	subj	occur(occur_what: command SE)
assign	delay_clk	when_after	assign(when_after: delay_clk)

Fig. 7. Rules for creating semantic expressions for a head node from its child nodes semantic expressions based on the semantic edge connecting these nodes.

Algorithm 1 presents our Syntactic-Semantic parser. A list of words from a leaf node of CRT is passed as input to the parser. This algorithm is based on the shift-reduce dependency parsing algorithm of [13]. We added steps to update semantics in this algorithm whenever we are creating an edge in our parse tree, as shown in line 13 and 17 of Algorithm 1. A new head of parse tree is formed on the creation of a left arc. SE of the new head is created based on its child node SE's by traversing in a downward direction from the head node to its child nodes, as shown in line 13 of Algorithm 1. Similarly, for the creation of the right arc, we will traverse the parse tree from the new word added as a child node. We continue to propagate the effects of this child node semantics to the root node of the parse tree, as shown in Algorithm 2 from line 9–13. In this algorithm, we update the SE of nodes based on its child node SE's using the rules. Figure 7 shows only a fragment of these rules. In this figure, the column 'Connecting edge' is a semantic edge that is inferred based on connecting syntactic edge or a series of syntactic edges for a long range semantic dependency. Figure 8 illustrates the working of our syntactic-semantic parser.

Algorithm 2. Traversal with Semantic update

```
1: procedure TRAVERSE_WITH_SEMANTIC_UPDATE(node,direc)
2:     if node == root then                              ▷ root is a dummy node
3:         return
4:     end if
5:     if direc == 'down' then
6:         For all child_nodes of node
7:             Update_SE_node(node,child_node)
8:     end if
9:     if direc == 'up' then
10:        head_node ← node.dep_head          ▷ head of the node in dependency tree
11:        Update_SE_node(head_node,node)
12:        traverse_with_semantic_update(head_node,'up')
13:    end if
14: end procedure
```

2.4 Suggestion Generation

Suggestion mechanism relies on the partial semantic understanding to generate additional text to complete the sentence. Our syntactic-semantic parser ensures that at every stage of parsing, we will have a list of partial semantic expressions that could be inferred from the group words written by the user. This list of partial semantic expressions is filtered down to a specific semantic expression when the user completes the sentence. Earlier work done in the area of grammar based CNL authoring tools like [14], lacks the ability to consider semantics while giving suggestions based on the partially written sentence. Hence, the previous work cannot give feedback to the user for a syntactically correct partial sentence that is semantically wrong.

We propose a semantically driven suggestion framework by utilizing the list of SE's for partial sentence created in the syntactic-semantic parser stage. A fragment of our rules to generate suggestions is shown in Fig. 9. Suggestions are generated after a 'triggered word' is detected by the system. Triggered words comprise of prepositions, linking verbs(lv), modal verbs, main verb and conditional words like if-then, when. In this figure, the column 'Partial sentence Meaning' represents the semantic expression that is understood from the partially written sentence that appears before the trigger word. As shown in the Fig. 9, a combination of the meaning based on the SE and the trigger word is used to generate suggestions for completing the sentence. The column 'Suggestions' in this figure consist of categories like 'AW' which refers to 'action words' and lexicon like 'set' that belongs to this category. The SE's in the 'Suggestion' column indicates the intent of the suggestions. The sentence structure of the suggestion is generated by mapping these SE's to the sentence template tuned according to the parameters in these SE's.

Clause Processed	Updating Dependency parse tree and semantic understanding	Explanation
register fcr[7:6]	root ↓ register fcr[7:6] SE value(value_of:fcr[7:6]) SE data(data_of:fcr[7:6])	Multiple semantic expression are used to generate suggestion to help user complete the sentence
register fcr[7:6] is	root ↓ is SE value(value_of:fcr[7:6]) subj⟋ SE data(data_of:fcr[7:6]) ↑Semantic Propagation register fcr[7:6] SE value(value_of:fcr[7:6]) SE data(data_of:fcr[7:6])	Dependency parse tree is updated when system detects the word 'is'. The semantic expressions of leaf node 'register' is propagated to head node 'is'. This is shown as 'Semantic Propagation' in the figure
register fcr[7:6] is set	root ↓ set SE value(value_of:fcr[7:6]) aux⟋ SE data(data_of:fcr[7:6]) ↙ subj ↑Semantic Propagation is register fcr[7:6] SE value(value_of:fcr[7:6]) SE data(data_of:fcr[7:6])	'set' being the main verb becomes the new root of the tree. In the next figure we have shown a tree after processing two words 'to' and 02.
register fcr[7:6] is set to 02	SE value(value_of:fcr[7:6], value_is:02) set subj⟋ ↘prep ↑Semantic propagation aux↙ ↘ SE value(value_is:02) is to register fcr[7:6] ↓ pobj SE value(value_of:fcr[7:6]) 02 ↑Semantic Propagation SE data(data_of:fcr[7:6]) SE value(value_is:02)	Based on newly created syntactic edges, the semantics of root word is updated

Fig. 8. Words in a clause are processed one after the other. A syntactic parsing is followed by semantic parsing on partial parse trees to generate semantic expression of the clause.

3   Evaluation

EASE was implemented in Python. The grammar had three syntactic categories (verb, prepositions, conjunction) and 12 semantic categories for the hardware verification domain. We extracted a total of 80 assertions for memory controller design [16] and UART architecture [15]. We followed the industry standard practices as given in [17] for writing assertions. We wrote these assertions according to our vocabulary and grammar to test our algorithm. The assertions were translated to CRTs which represent the logical forms of these assertions. These CRTs were then translated to SVA based on the underlying rules. A sample of the sentences that we tested are as follows:

1. if signal sel_n[3] is de-asserted and signal oe_n falls from 1 to 0 then after 16 clock cycles and before 900 cycles signal oe_n rises from 0 to 1.
2. Signal AWREADY is asserted within 5 clock cycles after signal AWVALID being asserted.
3. if signal sel_n[3] is de-asserted and signal oe_n falls from 1 to 0 then only after 15 clock cycles signal oe_n rises from 0 to 1.
4. if signal sel_n[2] and signal oe_n are low and signal we_n is high then after 1 clock cycle register addr data should be stable and in the same clock cycle data in register Acc should be available.

5. If command active is initiated then command read or command write cannot occur within 3 cycles. If command active is initiated then command read or command write must be issued within 10 clocks.
6. signal ads should not be asserted for consecutive 2 clocks.

Partial Sentence Meaning	Triggered word	Suggestions
SE value(empty)	lv | modal	<AW(set)> for < register | signal> | <value> for <register> | ...
SE value(value_of)	lv | modal	<AW(set | assigned)> 'a' <value> | <rel(greater | lower) than <value>
SE count(count_of)	lv | modal	<rel(greater | lower)> than <data> | <data> | <data> on <system func.>
SE Value(value_of, value_is)	conditional words	SE value(value_of,value_is) | SE rel(rel_of,rel_is,rel_obj) | ....
SE Value(value_of, value_is)	prep (after)	SE value(when: <specific clock>) | <specific_clock> | .....
SE Value(value_of, value_is, value_when)	conj (and)	SE value(when : <specific_clk>) | SE value() | ....

Fig. 9. Rules for generating suggestions based on Semantic Expressions (SE). We have used | to denote that any one of the element will suffice.

The above sentences demonstrate the ability of EASE in handling various features of hardware verification assertions. For example, the first specification shows that various operations ranging from 'de-assertions', 'transition' and clock events can be handled in a single sentence. The second sentence illustrates a conditional assertion without using explicit conditional words like 'if-then' or 'when'. Sentence 3 is a safety property where an event should not happen before 15 clock cycles. Sentence 6 is an example of an imperative specification.

In terms of varying the way the spec is written, the following variations of the earlier sentences were also successfully translated to the corresponding SVA logic. These sentences have the same semantic but were written in different word and/or phrase order:

(a) If command read is issued then command write can occur only after 3 clocks.
(b) Command write cannot occur after command read before 3 clocks.
(c) Command write should be executed only after 3 clocks when command read is issued.

We then evaluated EASE on assertions that are available in specification documents [18]. A list of design variables were first added in the vocabulary and were treated as nouns subsequently. We have compared EASE with previous approaches in Table 1. The work in [4] and [6] cannot handle multi-line and sequential specifications. Sequential specifications are assertions that span multiple clock cycles. The column 'Architectures' indicates the underlying hardware architectures for which assertions were automatically generated from specifications. The last column shows some example sentences that cannot be handled by the previous work. For example, the work in [6] did not show the translation of specifications of AMBA AXI protocol that span multiple clock cycles. Such specification is shown by sentence 2 in above given sample sentences. Our approach is different from [11] since we are using syntactic-semantic parser which assists in providing suggestions to complete the sentence.

Table 1. Comparison with previous works on Hardware Assertions Synthesis.

Paper	Feedback	Sequential specs	Architectures	Multi-line tested	Which sentences could not handle
[4]	No	No	AMBA AXI3	No	1, 2, 3, 4
[6]	No	No	AMBA AXI3, AMBA AXI4	No	2
This work	Auto-completion	Yes	Memory controller, UART, AMBA AXI3	Yes	-

We also evaluated the proposed suggestion mechanism and ambiguity detection on specifications. Figure 10 illustrates how the suggestion mechanism responded to a partial sentence based on SEs. Consider the first example of suggestion when the word is 'value'. The system assigns a partial semantics of 'SE value' to the word 'value'. When the system detects the word 'of', it checks the partial semantics of the written sentence. Two suggestions templates are displayed based on the combination of the anticipated SE and the trigger word. The suggestion shows the category of words and the lexicon that user can use in completing the sentence. Although the suggestions are not the complete list of possibilities, they are useful to help the users avoid potential syntax and semantic errors.

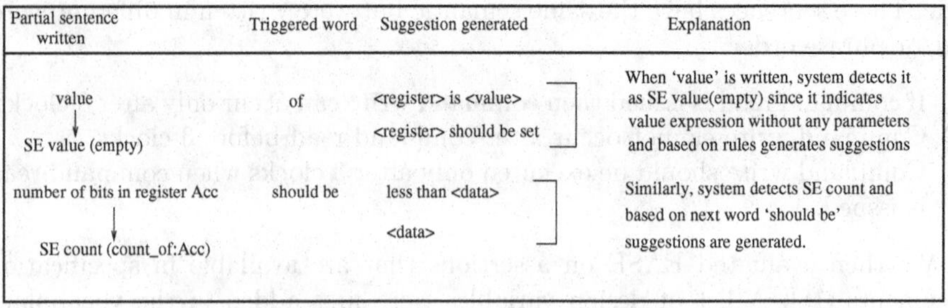

Partial sentence written	Triggered word	Suggestion generated	Explanation
value ↓ SE value (empty)	of	<register> is <value> <register> should be set	When 'value' is written, system detects it as SE value(empty) since it indicates value expression without any parameters and based on rules generates suggestions
number of bits in register Acc ↓ SE count (count_of:Acc)	should be	less than <data> <data>	Similarly, system detects SE count and based on next word 'should be' suggestions are generated.

Fig. 10. Suggestion mechanism responds with some small set of possible categories and lexicon that can be used to complete the sentence.

In Fig. 11, the mechanism for the detection of ambiguity is tested for the sentence: 'Command active to command active cannot come within 3 clocks'. When the user writes the phrase 'Command active to command active', the understanding algorithm halts as it is unable to propagate the SE of command active to the root of the partial sentence which is the Noun 'command active' in this case. An error message is displayed and explains the incompatibility of the SEs connected by the preposition 'to' and suggests the use of a connector like 'after' which can connect the two SEs. We have broadly classified three types of errors that we have encountered in our framework. The first type of error appears when we do not find at least one pair of design variable and verb in a clause. Identifying a set of design variables and verb is crucial to splitting the sentence into clauses and hence in the overall analysis of the sentence. The second type of errors occurs when some words of the input sentence are left floating in the parse tree without any head or child nodes. This error occurs due to the unavailability of syntactic or semantic dependency rules in our grammar. The third type of error appears when we detect semantic expressions connected by ambiguous words which we did not anticipate while creating the SE update rules. Overcoming these errors is a continuous process and can be achieved when we take sentences from different users which give us the needed data to create rules for efficient working of EASE.

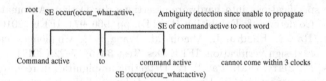

Fig. 11. Ambiguity detection is based on the compatibility check between SE's. SE's in this figure could not be combined due to the ambiguous nature of preposition 'to' used to connect two commands.

In terms of limitations, we are currently unable to create an SVA for high level design functions which required additional information about design variables. For example in AMBA AXI document [18], a sentence like 'A sequence of locked transaction must use a single ID' cannot be mapped to an SVA without knowing the name of signals or storage elements involved in this assertion.

4 Conclusion

We have presented the architecture of EASE to generate formal assertions automatically from design specifications. The framework was first evaluated on specifications that we wrote according to the vocabulary and grammar of EASE. EASE was also tested on existing specifications taken from the spec documents. A syntactic-semantic parser enabled understanding of sentences based on dependency grammar. Moreover, partial semantic analysis also assisted in providing suggestions to the user in writing specifications. In the future, we would like to add a reasoning framework which can detect inconsistencies in a list of specifications.

References

1. Ray, S., Harris, I.G., Fey, G., Soeken, M.: Multilevel design understanding: from specification to logic. In: 2016 IEEE/ACM International Conference on Computer-Aided Design (ICCAD), pp. 1–6. IEEE (2016)
2. Soeken, M., Harris, C.B., Abdessaied, N., Harris, I.G., Drechsler, R.: Automating the translation of assertions using natural language processing techniques. In: Proceedings of the 2014 Forum on Specification and Design Languages (FDL), Munich, pp. 1–8 (2014)
3. Ghosh, S., Elenius, D., Li, W., Lincoln, P., Shankar, N., Steiner, W.: ARSE-NAL: automatic requirements specification extraction from natural language. In: Rayadurgam, S., Tkachuk, O. (eds.) NFM 2016. LNCS, vol. 9690, pp. 41–46. Springer, Cham (2016). https://doi.org/10.1007/978-3-319-40648-0_4
4. Harris, I.G.: Capturing assertions from natural language descriptions. In: 2013 1st International Workshop on Natural Language Analysis in Software Engineering (NaturaLiSE), pp. 17–24. IEEE (2013)
5. Harris, C.B., Harris, I.G.: Generating formal hardware verification properties from natural language documentation. In: 2015 IEEE International Conference on Semantic Computing (ICSC), pp. 49–56. IEEE (2015)
6. Harris, C.B., Harris, I.G.: Glast: learning formal grammars to translate natural language specifications into hardware assertions. In: Design, Automation & Test in Europe Conference & Exhibition (DATE), pp. 966–971. IEEE (2016)
7. Chen, W., Ray, S., Bhadra, J., Abadir, M., Wang, L.C.: Challenges and trends in modern SoC design verification. IEEE Des. Test **34**(5), 7–22 (2017)
8. Kamsties, E., Berry, D.M., Paech, B.: Detecting ambiguities in requirements documents using inspections. In: Proceedings of the First Workshop on Inspection in Software Engineering (WISE01), pp. 68–80 (2001)
9. Berry, D.M.: Ambiguity in natural language requirements documents. In: Paech, B., Martell, C. (eds.) Monterey Workshop 2007. LNCS, vol. 5320, pp. 1–7. Springer, Heidelberg (2008). https://doi.org/10.1007/978-3-540-89778-1_1
10. Hsiao, M.S.: Automated program synthesis from object-oriented natural language for computer games. In: The International Workshop on Controlled Natural Language, August 2018
11. Krishnamurthy, R., Hsiao, M.S.: Controlled natural language framework for generating assertions from hardware specifications. In: 2019 IEEE 13th International Conference on Semantic Computing (ICSC), pp. 367–370. IEEE (2019)
12. Nivre, J.: Dependency grammar and dependency parsing. MSI report 5133, no. 1959, pp. 1–32 (2005)
13. Nivre, J.: An efficient algorithm for projective dependency parsing. In: Proceedings of the 8th International Workshop on Parsing Technologies (IWPT) (2003)
14. Kuhn, T., Schwitter, R.: Writing support for controlled natural languages. In: Proceedings of the Australasian Language Technology Association Workshop 2008, pp. 46–54 (2008)
15. Gorban, J.: UART IP core specification. Architecture **15**, 1 (2002)
16. Micron SDR SRAM. https://www.micron.com/~/media/documents/products/data-sheet/dram/512mb_sdr.pdf
17. Vijayaraghavan, S., Meyyappan, R.: A Practical Guide for SystemVerilog Assertions. Springer, New York (2005). https://doi.org/10.1007/b137011
18. AMBA 3 AXI Protocol Checker User Guide, r0p1 ed., ARM, June 2009

Formalizing Object-Ontological Mapping Using F-logic

Martin Ledvinka[(⊠)] and Petr Křemen

Department of Computer Science, Faculty of Electrical Engineering,
Czech Technical University in Prague, Prague 6, Dejvice, Czech Republic
{martin.ledvinka,petr.kremen}@fel.cvut.cz

Abstract. Ontologies can represent a significant asset of domain-specific information systems, written predominantly using the object-oriented paradigm. However, to be able to work with ontological data in this paradigm, a mapping must ensure transformation between the ontology and the object world. While software libraries provide such a mapping, they lack standardization or formal guarantees of its semantics. In this paper, we provide a formalism for mapping ontologies between description logics and F-logic, a formal language for representing structural aspects of object-oriented programming languages. This formalism allows to precisely specify the semantics of the object-ontological mapping and thus ensure a predictable shape and behavior of the object model.

Keywords: Object-ontological mapping · $\mathcal{SROIQ}$ · F-logic

1 Introduction

The object-oriented paradigm (OOP) has been a dominant software development technique in the past two decades, mainly due to its ability to represent the underlying domains in a natural and understandable way [4]. Ontologies, on the other hand, can significantly increase the capabilities of information systems, especially due to their formal semantics (in this paper, we consider description logic (DL) [2] as the language backing the formal semantics of ontologies), elements with shared meaning and global identification, and inference enabled by expressive languages. Yet, to be able to fully embrace the benefits of ontologies in OOP, a *mapping* is needed to transform data between the two worlds. Many software libraries provide such functionality, however, without sufficient guarantees as to the semantics of the mapping.

One differentiating aspect between an object model and a DL ontology[1] is the *open world assumption* of the latter – DL ontologies assume incomplete

---

[1] In the sequel, we consider mapping of DL ontology specifications since the most widespread ontology-related standards (e.g., OWL) and relevant tools are based on the description logic formalism.

© Springer Nature Switzerland AG 2019
P. Fodor et al. (Eds.): RuleML+RR 2019, LNCS 11784, pp. 97–112, 2019.
https://doi.org/10.1007/978-3-030-31095-0_7

knowledge of the domain. However, most domain-specific information systems assume data completeness and thus, if a fact cannot be derived from the existing data, it is considered false. To overcome this mismatch, *integrity constraints* can be used to place restrictions on the knowledge base. Consider a simple vocabulary management system which needs to keep track of authors of the vocabularies. In description logics, this is represented by placing existential quantification on the vocabulary records. Unfortunately, this merely ensures that *some* author exists for each vocabulary. But they may remain unknown which is hardly sufficient for ensuring integrity of data created by the system. Integrity constraints can be used to enforce that a known author is explicitly assigned to each record.

Thus, the goal of this paper is to provide a formalism for *object-ontological mapping* (OOM) moderated by integrity constraints. We chose F-logic as a vessel for this formalism. The main reasons are that it is a logic-based language, it has been used to describe ontologies and it is specifically designed to represent the structural aspects of object-oriented languages. Its syntax allows to concisely represent the most common constructs needed by object-oriented domain models – class hierarchies, local restrictions on property value types, possible cardinality restrictions and individual assertions.

1.1   Running Example

We shall use the following example DL ontology throughout this paper to illustrate the mapping. $\mathcal{T}$ represents the ontology schema, $\mathcal{A}$ are the actual data and $\mathcal{IC}$ are *integrity constraints* placed on the ontology. All the corresponding notions will be explained in Sect. 2.

In the example, we declare an asset and specify that it has to have an author and it may have a last editor. This generic ontology is restricted by integrity constraints for a system working with vocabularies, which are kinds of assets. The constraints specify the same cardinalities of both author and last editor as $\mathcal{T}$, but require their values to be users of the system.

$$\mathcal{T} = \{Asset \sqsubseteq =1author.\top, Asset \sqsubseteq \leqslant 1lastEditor.\top,$$
$$Vocabulary \sqsubseteq Asset, author \sqsubseteq editor, lastEditor \sqsubseteq editor\}$$
$$\mathcal{A} = \{User(Tom), User(Sarah), Vocabulary(MetropolitanPlan)\}$$
$$\mathcal{IC} = \{Vocabulary \sqsubseteq \forall author.User, Vocabulary \sqsubseteq =1author.User,$$
$$Vocabulary \sqsubseteq \forall lastEditor.User, Vocabulary \sqsubseteq \leqslant 1lastEditor.User\}$$

Figure 1 then illustrates how the resulting integrity constraints-based application model may look like in terms of a UML class diagram.

The paper is structured as follows: Sect. 2 provides the necessary theoretical background, Sect. 3 presents the object-ontological mapping, while Sect. 4 introduces the mapping of integrity constraints. Section 5 discusses related work and Sect. 6 concludes the paper.

Fig. 1. UML class diagram of a model based on the running example ICs.

2 Background

This section presents the most important notions of the description logic $\mathcal{SROIQ}$, application of *integrity constraints* to ontologies, and F-logic.

2.1 $\mathcal{SROIQ}$

$\mathcal{SROIQ}$ [15] is an expressive description logic (DL), i.e., a decidable sub-language of the first order logic (FOL), used to describe ontologies. Each $\mathcal{SROIQ}$ ontology $\mathcal{O}$ is comprised of a *terminology* (TBox and RBox), which describes the schema of the ontology, and a set of individual assertions representing actual data (ABox)[2]. TBox consists of a *concept hierarchy* where concepts can be either *atomic* or *concept descriptions* of the following forms: $\neg C$, $C \sqcap D$, $C \sqcup D$, $\geqslant nR.C$, $\leqslant nR.C$, $\exists R.Self$, $\{a\}$, $\forall R.C$, $\exists R.C$, where C, D are concepts, R is a role, n is a non-negative integer and a is an individual. RBox consists of a hierarchy of roles and axioms stating their properties, for instance, $Sym(R)$ denoting a symmetric role, or $Dis(R, Q)$ denoting disjoint roles. The schema also contains built-in concepts $\top$, $\bot$ and a built-in universal role R_U.

Individual assertions are of the form $C(a)$, $R(a, b)$, $a = b$ and $a \neq b$, where a and b are individuals, C is a concept and R is a role. The set N_C represents concept names, N_R role names, and N_I denotes the set of individual names.

The semantics of a $\mathcal{SROIQ}$ ontology $\mathcal{O}$ is given by an *interpretation* $\mathcal{I} = (\Delta^{\mathcal{I}}, \cdot^{\mathcal{I}})$, where $\Delta^{\mathcal{I}}$ is the *domain* of the interpretation and $\cdot^{\mathcal{I}}$ is the *interpretation function*. This function assigns to every atomic concept A a set $A^{\mathcal{I}} \subseteq \Delta^{\mathcal{I}}$, to every atomic role R a binary relation $R^{\mathcal{I}} \subseteq \Delta^{\mathcal{I}} \times \Delta^{\mathcal{I}}$ and to every individual an element of $\Delta^{\mathcal{I}}$. $\top^{\mathcal{I}}$ is $\Delta^{\mathcal{I}}$, $\bot^{\mathcal{I}}$ is the empty set $\emptyset$ and $R_U^{\mathcal{I}}$ is $\Delta^{\mathcal{I}} \times \Delta^{\mathcal{I}}$. $\mathcal{I}$ is a model of an ontology $\mathcal{O}$ consisting of a TBox $\mathcal{T}$, an RBox $\mathcal{R}$, and an ABox $\mathcal{A}$ ($\mathcal{I} \models \mathcal{O} = \mathcal{T} \cup \mathcal{R} \cup \mathcal{A}$) if it satisfies all the axioms in $\mathcal{O}$. A set of axioms Θ *logically entails* an axiom θ ($\Theta \models \theta$) if and only if all models of Θ are also models of θ. Concrete rules for interpretation of concept descriptions and axioms are described in [15] and we omit them here for the lack of space.

2.2 Integrity Constraints

The intention of integrity constraints (ICs) in the area of application access to DL ontologies is mostly to restrict the open-world nature of (a portion of) an

---

[2] $\mathcal{SROIQ}$ allows expressing individual assertions using TBox axioms with nominals. However, ABox assertions provide a natural, easy to read syntax which we will use throughout this paper.

ontology. While OWL does allow to express certain constraints, its expressiveness in this regard is limited. For example, a DL ontology might require that every asset must have a unique author (*Asset* $\sqsubseteq$ 1*author*.$\top$). If an asset v does not have one, the reasoner will infer an anonymous individual for the sake of ontology consistency. However, a vocabulary management system requires a stronger condition to be satisfied – every vocabulary (a special kind of asset) must have a *known* author of type user. Such a constraint cannot be enforced in OWL with standard semantics. However, there are approaches which allow this type of restrictions. [10] introduces *Minimal Knowledge and Negation as Failure* logics, while [25] uses minimal Herbrand models. [29] points out that these approaches can lead to counterintuitive results and presents an alternative solution. A more recent effort in [27] discusses the flaws of all of the aforementioned solutions and proposes the use of DBox-based *completely specified* concepts and roles. However, not even this approach is immune to debatable results. Consider the following example:

$$\mathcal{T} = \{Employee \sqsubseteq Person, Flight \sqsubseteq \exists hasPassenger.Person\}$$
$$\mathcal{A} = \{Flight(c), Flight(d)\}$$
$$\mathcal{DB} = \{Person(a), Employee(b), hasPassenger(c, a), hasPassenger(d, b)\}$$

We put *hasPassenger*, *Person* and *Employee* into the DBox, so that no unexpected instances are generated. However, this will cause an IC violation, because *Person*(b) will be inferred for a completely specified concept *Person*. The approach of Tao et al. [29] does not suffer from such issues, because, while it does work only with named individuals, it does not prevent inference of types/roles. We will be using it for this work as we consider it the most suitable for OOM.

2.3 F-logic

F-logic [1,17,18] is a formalism rooted in FOL which can be used to describe structural aspects of object-oriented or frame-based languages. It has model-theoretic semantics and a sound and complete proof theory. In the discussion of F-logic syntax, we use the revised version of [1] and w.l.o.g. omit the distinction between inheritable and non-inheritable methods. We use a restricted variant of F-logic, which is suitable for mapping of DL, but does not contain, for instance, methods with arbitrary arity (we use only *attributes* – parameterless methods). We use *sorted* F-logic, so that (atomic) classes are disjoint from individuals and methods (much like classes, individuals and properties are disjoint in DL).

F-logic Syntax. The alphabet of an F-logic language $\mathcal{L}$ consists of

- A set of *object constructors* $\mathcal{F} = \mathcal{C} \cup \mathcal{R} \cup \mathcal{E} \cup \mathcal{A}$, where $\mathcal{C}$ is a set of class names (0-ary function symbols), $\mathcal{R}$ is a set of methods (0-ary function symbols), $\mathcal{E}$ is a set of instances (0-ary function symbols), and $\mathcal{A}$ is a set of function symbols (it essentially allows us to *parameterize* concept constructors, as will be seen in Sect. 3). $\mathcal{C}$, $\mathcal{R}$, $\mathcal{E}$, and $\mathcal{A}$ are mutually disjoint,

- A set of predicate symbols $\mathcal{P}$,
- An infinite set of variables $\mathcal{V}$,
- Auxiliary symbols like (,), [,], $\rightarrow$ etc.,
- Logical connectives and quantifiers $\wedge$, $\vee$, $\neg$, $\forall$, $\exists$.

An *id-term* is a first-order term composed of an object constructor and variables. A variable-free object constructor is called a *ground id-term* and the set of all ground id-terms is denoted $U(\mathcal{F})$. Formulas in F-logic can be either *molecular formulas*, or complex formulas consisting of other formulas connected using logical connectives and quantifiers. Molecular formulas can be:

1. *Is-a* assertions of the form $A::B$ or $o:A$, where o, A, B are id-terms,
2. *Object molecules* of the form $O[$ a ';' separated list of method expressions]. Where method expressions can be:

- *data expressions* of the form $m \rightarrow v$, where m and v are id-terms (v is the attribute value),
- *Signature expressions* of the form $m \Rightarrow (T_1, ..., T_n)$, where $n \geq 1$ and m and T_i are id-terms (T_i are the return types).

In short, data expressions represent attribute values, whereas signature expressions represent their return types.

F-logic Semantics. Semantics of F-logic is specified using *F-structures*. Before we define an F-structure, we need several additional notions.

For a pair of sets U, V, $Total(U,V)$ denotes the set of all *total* functions from U to V. Similarly, $Partial(U,V)$ denotes the set of all *partial* functions from U to V. We use $\mathcal{P}(U)$ to express the power set of U. $\mathcal{P}_\uparrow(U)$ is the set of all *upward-closed* subsets of U. A set $V \subseteq U$ is upward closed if for $v \in V$, $u \in U$, $v \prec_U u$ implies $u \in V$, where $\prec_U$ is an irreflexive partial order on U (see below). $PartialAM_{\prec_U}(U, \mathcal{P}_\uparrow(U))$ denotes the set of all partial *anti-monotonic* functions from U to $\mathcal{P}_\uparrow(U)$. A function f is *partial anti-monotonic* if for vectors $\boldsymbol{u}, \boldsymbol{v} \in U^k$, $\boldsymbol{v} \prec_U \boldsymbol{u}$, if $f(\boldsymbol{u})$ is defined, then $f(\boldsymbol{v})$ is also defined and $f(\boldsymbol{u}) \subseteq f(\boldsymbol{v})$.

An F-structure is then a tuple $\mathbf{I} = \langle U, \prec_U, \in_U, I_F, I_\mathcal{P}, I_\rightarrow, I_\Rightarrow \rangle$, where:

- U is the domain of $\mathbf{I}$ consisting of disjoint subdomains $U_\mathcal{E}$, $U_\mathcal{C}$, $U_\mathcal{R}$, $U_\mathcal{A}$,
- $\prec_U$ is an irreflexive partial order on $U_{\mathcal{C}\cup\mathcal{A}}$ – the subclass relationship,[3]
- $\in_U$ is a binary relationship on $U_\mathcal{E} \times U_{\mathcal{C}\cup\mathcal{A}}$ – instance membership in classes,
- $I_F : \mathcal{F} \rightarrow \bigcup_{k=0}^{\infty} Total(U^k, U)$ is a mapping which represents function symbols from $\mathcal{F}$ by functions from U^k to U. For $k = 0$, $I_F(f)$ can be identified with an element of U. I_F maps names to their respective subdomains, e.g., class names from $\mathcal{C}$ to $U_\mathcal{C}$,
- $I_\mathcal{P}(p) \subseteq U^n$ for any n-ary predicate symbol $p \in \mathcal{P}$,
- $I_\rightarrow : U_\mathcal{R} \rightarrow Partial(U_\mathcal{E}, \mathcal{P}(U_\mathcal{E}))$,
- $I_\Rightarrow : U_\mathcal{R} \rightarrow PartialAM_{\prec_U}(U_{\mathcal{C}\cup\mathcal{A}\cup\mathcal{E}}, \mathcal{P}_\uparrow(U_{\mathcal{C}\cup\mathcal{A}}))$.

---

[3] $U_{\mathcal{C}\cup\mathcal{A}}$ is an abbreviation for $U_\mathcal{C} \cup U_\mathcal{A}$.

Remarks. The use of *upward-closed* sets is important for class hierarchies – it means that along with each class, the set also contains all its superclasses. The relationship between $I_\rightarrow$ and $I_\Rightarrow$ is such that $I_\Rightarrow$ defines the target (range) type of an attribute, whereas $I_\rightarrow$ defines particular values.

A variable assignment ν is a mapping from the set of variables, $\mathcal{V}$, to the domain U, which extends to id-terms as follows: $\nu(d) = I_\mathcal{F}(d)$ if $d \in \mathcal{F}$ has arity 0 and $\nu(f(...,t,...)) = I_\mathcal{F}(f)(...,\nu(t),...)$. Intuitively, given an F-structure $\mathbf{I}$ and a variable assignment ν, a molecule $t[...]$ is *true* under $\mathbf{I}$ w.r.t. to ν, written $\mathbf{I} \vDash_\nu t[...]$, iff the object $\nu(t)$ has the properties defined by the F-molecule. For example, $\mathbf{I} \vDash_\nu (O{:}P)$ iff $\nu(O) \preceq_U \nu(P)$. For attributes, this means that there exist functions interpreting them and they have the right return values (types), e.g., $\mathbf{I} \vDash_\nu q[m \rightarrow v]$ iff $I_\rightarrow(\nu(m))(\nu(q))$ is defined and contains $\nu(v)$. An object molecule is a conjunction of method expressions. Precise definitions of logical implication in F-logic can be found in [18], Sec. 5.2. Satisfaction of complex formulas is defined in the usual first-order sense. An F-logic theory $\mathcal{S}$ logically implies an axiom α ($\mathcal{S} \vDash_\nu \alpha$) iff all models of $\mathcal{S}$ are also models of α. Since we will be working with closed formulas only, we can omit the variable assignment identifier. Instead, we shall denote F-logic semantic implication by $\vDash^F$ to distinguish it from DL entailment. We omit discussion of properties of F-structures here due to the lack of space. Nevertheless, since these properties do affect the formalization, the reader should refer to [18], Sec. 7, if necessary.

Queries. An F-logic *query* Q is a molecule. The set of answers to Q w.r.t. a set of formulas P is the smallest set of molecules that:

- contains all instances of Q (variable assignments for all variables in Q) that are found in the model of P,
- is closed under $\vDash^F$ (see [18], Sec. 12.1.2).

3   Mapping

We begin by introducing the mapping of concept descriptions and ontological axioms. We then show that it preserves entailment in both directions.

The mapping is inspired by [3], but supports a more expressive DL. The use of sorted F-logic and proofs of entailment equivalence are based on [6]. While the version provided here is for $\mathcal{SROIQ}$, the latest version of F-logic supports also datatypes [1], so it could be easily extended to $\mathcal{SROIQ(D)}$. Table 1 shows mapping of concept descriptions. Similar to [3], several new function symbols are introduced – *Not, AtLeast, AtMost, HasSelf, Nom, All, Some* $\in \mathcal{A}$ – that allow us to represent $\mathcal{SROIQ}$ concept constructs which cannot be directly mapped to F-logic. For instance, $\geq nR.C$ does not correspond to $[R_\mathcal{R} \Rightarrow_{\{n:*\}} C_\mathcal{C}]$ because the $\mathcal{SROIQ}$ version admits also R-fillers of other types than C, whereas the F-logic signature expression would require all $R_\mathcal{R}$-fillers to belong to $C_\mathcal{C}$. Also, signature expressions cannot be used to infer the type of values of the corresponding data expressions. The relationship between signature and data expressions becomes relevant under the well-typing conditions. For each of the new function symbols, we specify a condition on the underlying F-structures to ensure correct semantics w.r.t. their $\mathcal{SROIQ}$ counterparts.

Table 1. Mapping of concept descriptions. By default, all variables are universally quantified over $\mathcal{E}$. X_C (X_R) represents a concept (method) name, i.e., a function symbol from C (R). $AtMost$ is defined analogously to $AtLeast$ and corresponds to $\leqslant nR.C$.

$\mathcal{SROIQ}$	F-logic	F-logic Semantics
A	A_C	
$\neg C$	$Not(C_C)$	$\mathbf{I} \models^F x\!:\!Not(C_C)$ iff $I_{\mathcal{F}}(x) \notin_U I_F(C_C)$
$C \sqcap D$	C_C and D_C	
$C \sqcup D$	C_C or D_C	
$\geqslant nR.C$	$AtLeast(n, R_R, C_C)$	$\mathbf{I} \models^F x\!:\!AtLeast(n, R_R, C_C)$ iff
		$\exists y_1...y_n \in U_{\mathcal{E}}$ s.t. $y_i \in I_{\to}(I_F(R_R))(I_F(x))$
		$\wedge y_i \in_U I_F(C_C)$, for $\neg(y_i = y_j)$
$\exists R.Self$	$HasSelf(R_R)$	$\mathbf{I} \models^F x\!:\!HasSelf(R_R)$ iff $I_F(x) \in I_{\to}(I_F(R_R))(I_F(x))$
$\{a\}$	$Nom(a_{\mathcal{E}})$	$\mathbf{I} \models^F x\!:\!Nom(a_{\mathcal{E}})$ iff $I_{\mathcal{F}}(x) = I_{\mathcal{F}}(a_{\mathcal{E}})$
$\forall R.C$	$All(R_R, C_C)$	$\mathbf{I} \models^F x\!:\!All(R_R, C_C)$ iff
		$\forall y \in U_{\mathcal{E}}$ s.t. $y \in I_{\to}(I_F(R_R))(I_F(x)) \Rightarrow y \in_U I_F(C_C)$
$\exists R.C$	$Some(R_R, C_C)$	$\mathbf{I} \models^F x\!:\!Some(R_R, C_C)$ iff
		$\exists y \in U_{\mathcal{E}}$ s.t. $y \in I_{\to}(I_F(R_R))(I_F(x)) \wedge y \in_U I_F(C_C)$

$\mathcal{SROIQ}$ top (bottom) concept $\top$ ($\bot$) is mapped to F-logic concept $\top_C$ ($\bot_C$) for which it must hold $\forall x \in U_{\mathcal{E}}$, $x \in_U I_F(\top_C)$ ($\forall x \in U_{\mathcal{E}}$, $x \notin_U I_F(\bot_C)$). Similarly, $\mathcal{SROIQ}$ universal role R_U is mapped to an F-logic method M_R such that $\forall x, y \in U_{\mathcal{E}}$, $y \in I_{\to}(I_F(M_R))(x)$.

TBox and RBox axiom mapping is shown in Table 2. We make use of F-logic predicates and define conditions under which they are true.

ABox individual assertions are mapped straightforwardly, $C(a)$ as an is-a assertion $a_{\mathcal{E}} : C_C$, $R(a,b)$ as a data expression $a_{\mathcal{E}}[R_R \to b_{\mathcal{E}}]$ and (in)equality $a = b$ ($a \neq b$) as $a_{\mathcal{E}} = b_{\mathcal{E}}$ ($\neg(a_{\mathcal{E}} = b_{\mathcal{E}})$).

Running Example. To illustrate the mapping, we revisit the running example. A corresponding F-logic ontology looks as follows:

$$\mathcal{T}^F = \{Asset_C\!:\!Some(author_R, \top_C), Asset_C\!:\!AtMost(1, author_R, \top_C),$$
$$Asset_C\!:\!AtMost(1, lastEditor_R, \top_C), Vocabulary\!:\!Asset,$$
$$subPropertyOf_P(author_R, editor_R),$$
$$subPropertyOf_P(lastEditor_R, editor_R)\}$$
$$\mathcal{A}^F = \{Tom_{\mathcal{E}}\!:\!User_C, Sarah_{\mathcal{E}}\!:\!User_C, MetropolitanPlan_{\mathcal{E}}\!:\!Vocabulary_C\}$$

Table 2. Mapping of $TBox$ and $RBox$ axioms. $RBox$ axioms are mapped to predicates, for which we provide satisfaction conditions on the F-structure $\mathbf{I}$. $\Rightarrow$ outside F-molecules represents regular logical implication. Variables are universally quantified over $U_{\mathcal{E}}$.

$\mathcal{SROIQ}$	F-logic	Condition on $\mathbf{I}$
$C \sqsubseteq D$	$C_C :: D_D$	$I_F(C_C) \preceq_U I_F(D_C)$
$R \sqsubseteq S$	$subPropertyOf_P(R_{\mathcal{R}}, S_{\mathcal{R}})$	$y \in I_{\rightarrow}(I_F(R_{\mathcal{R}}))(x) \Rightarrow y \in I_{\rightarrow}(I_F(S_{\mathcal{R}}))(x)$
$Sym(R)$	$Sym_P(R_{\mathcal{R}})$	$y \in I_{\rightarrow}(I_F(R_{\mathcal{R}}))(x) \Rightarrow x \in I_{\rightarrow}(I_F(S_{\mathcal{R}}))(y)$
$Asy(R)$	$Asy_P(R_{\mathcal{R}})$	$y \in I_{\rightarrow}(I_F(R_{\mathcal{R}}))(x) \Rightarrow x \notin I_{\rightarrow}(I_F(S_{\mathcal{R}}))(y)$
$Tra(R)$	$Tra_P(R_{\mathcal{R}})$	$y \in I_{\rightarrow}(I_F(R_{\mathcal{R}}))(x) \wedge z \in I_{\rightarrow}(I_F(R_{\mathcal{R}}))(y) \Rightarrow$ $z \in I_{\rightarrow}(I_F(R_{\mathcal{R}}))(x)$
$Ref(R)$	$Ref_P(R_{\mathcal{R}})$	$x \in I_{\rightarrow}(I_F(R_{\mathcal{R}}))(x)$
$Irr(R)$	$Irr_P(R_{\mathcal{R}})$	$x \notin I_{\rightarrow}(I_F(R_{\mathcal{R}}))(x)$
$Dis(R,S)$	$Dis_P(R_{\mathcal{R}}, S_{\mathcal{R}})$	$y \notin I_{\rightarrow}(I_F(R_{\mathcal{R}}))(x) \vee y \notin I_{\rightarrow}(I_F(S_{\mathcal{R}}))(x)$

Now we have to show that the mapping preservers entailment. First, we show that a formula θ is satisfiable in a $\mathcal{SROIQ}$ language $\mathcal{L}^{DL}$ if and only if a corresponding formula θ^F is satisfiable in a corresponding F-logic language $\mathcal{L}^F$.

Lemma 1. *Let θ be a formula in $\mathcal{L}^{DL}$ and θ^F a corresponding F-logic formula in an F-logic language $\mathcal{L}^F$. Then θ is satisfiable in some interpretation $\mathcal{I}$ of $\mathcal{L}^{DL}$ if and only if θ^F is satisfiable in some F-structure $\mathbf{I}$ of $\mathcal{L}^F$.*

Proof (Sketch). The lemma is proven by showing how an F-structure $\mathbf{I}$ can be constructed for a $\mathcal{SROIQ}$ interpretation $\mathcal{I}$ and vice versa. The interpretation correspondence is shown for RBox and TBox axioms, ABox axioms are *internalized* using TBox. The full proof can be found in the technical report [23]. □

The lemma allows us to show that entailment is preserved by the mapping.

Theorem 1. *Let Θ and Θ^F be corresponding theories in $\mathcal{L}^{DL}$ and $\mathcal{L}^F$. For any formula θ in $\mathcal{L}^{DL}$ holds:*

$$\Theta \models \theta \ iff \ \Theta^F \models^F \theta^F,$$

where $\models^F$ represents F-logic entailment.

Proof. The proof relies on Lemma 1 and the fact that entailment checking can be reduced to satisfiability checking. □

4 Mapping Integrity Constraints

Integrity constraint mapping between $\mathcal{SROIQ}$ and F-logic consists of two parts: (1) IC semantics with a closed-world view of the data; (2) means of their validation. Integrity constraint mapping is important because, while data are mapped using the ABox mapping shown above, the application object model is based on ICs.

4.1   Integrity Constraint Semantics

IC semantics allows to impose a closed-world view on a portion of the knowledge base $\mathcal{K}$ affected by the integrity constraints. We follow the approach of Tao et al. [29] and define an augmented F-structure $\mathbf{I}^{IC}$ with IC semantics. This approach has the advantage of not introducing additional syntactic constructs and giving the IC axioms a natural, easy-to-understand meaning. The semantics uses the notion of *minimal equality models* (Mod_{ME}) to support a *weak* form of unique name assumption. The original definition of Mod_{ME} from [29] can be carried over to F-logic as follows:

Consider a knowledge base $\mathcal{K}^F$ and let $E_{\mathbf{I}}$ be the set of equality relations satisfied by $\mathbf{I}$, i.e., $E_{\mathbf{I}} = \{\langle a,b \rangle \mid a, b \in \mathcal{E}$ s.t. $\mathbf{I} \models^F I_F(a) = I_F(b)\}$. A relation $\mathbf{I} \prec_=^F \mathbf{J}$, where $\mathbf{I}$ and $\mathbf{J}$ are F-structures, holds iff:

- $\forall C \in \mathcal{C} \cup \mathcal{A}$, if $\mathbf{I} \models^F a\!:\!C$, then $\mathbf{J} \models^F a\!:\!C$,
- $\forall R \in \mathcal{R}$, if $\mathbf{I} \models^F a[R \rightarrow b]$, then $\mathbf{J} \models^F a[R \rightarrow b]$,
- $E_{\mathbf{I}} \subset E_{\mathbf{J}}$.

$Mod_{ME}^F(\mathcal{K}^F)$ is then defined as $Mod_{ME}^F(\mathcal{K}^F) = \{\mathbf{I} \mid \mathbf{I}$ is a model of $\mathcal{K}^F$ s.t. $\nexists \mathbf{J}\ E_{\mathbf{J}} \prec_=^F E_{\mathbf{I}}\}$.

The augmented F-structure with IC semantics is a tuple $\mathbf{I}^{IC} = \langle U, \prec_U^{IC}, \in_U^{IC}, I_F, I_{\mathcal{P}}^{IC}, I_{\rightarrow}^{IC}, I_{\Rightarrow} \rangle$, where:

- $\prec_U^{IC} = \{\langle (I_F(x), I_F)(y) \rangle \mid x, y \in \mathcal{C}$ s.t. $\forall \mathcal{J} \in Mod_{ME}^F(\mathcal{K}^F), \mathcal{J} \models^F I_F(x) \prec_U I_F(y)\}$
- $\in_U^{IC} = \{\langle (I_F(x), I_F(y) \rangle \mid x \in \mathcal{E}, y \in \mathcal{C} \cup \mathcal{A}$ s.t. $\forall \mathcal{J} \in Mod_{ME}^F(\mathcal{K}^F), \mathcal{J} \models^F I_F(x) \in_U I_F(y)\}$
- $I_F(y) \in I_{\rightarrow}^{IC}(I_F(z))(I_F(x))$ iff $x, y \in \mathcal{E}, z \in \mathcal{R} \wedge \forall \mathcal{J} \in Mod_{ME}^F(\mathcal{K}^F), \mathcal{J} \models^F I_F(y) \in I_{\rightarrow}^{IC}(I_F(z))(I_F(x))$
- $I_{\mathcal{P}}^{IC}(p) = \{\langle I_F(y_1), ..., I_F(y_n) \rangle \mid y_i \in \mathcal{F}$ s.t. $\forall \mathcal{J} \in Mod_{ME}^F(\mathcal{K}^F), \mathcal{J} \models^F \langle I_F(y_1), ..., I_F(y_n) \rangle \in I_{\mathcal{P}}^{IC}(p)\}$, where n is the arity of p,
- And the other parts of $\mathbf{I}^{IC}$ are the same as in a regular F-structure.

Based on $\mathbf{I}^{IC}$, we can now define the IC semantics of concept descriptions. This is done in Table 3. One modification is the switch from $All(R_{\mathcal{R}}, C_{\mathcal{C}})$ to a signature expression $[R_{\mathcal{R}} \Rightarrow C_{\mathcal{C}}]$. This can be done thanks to the notion of *typing*, which requires data expressions to correspond to a signature expression declaring their types, e.g., for a signature $C_{\mathcal{C}}[R_{\mathcal{R}} \Rightarrow D_{\mathcal{C}}]$, typing requires d from $c\!:\!C_{\mathcal{C}}[R_{\mathcal{R}} \rightarrow d]$ to be of type $D_{\mathcal{C}}$. Typing is an optional, non-monotonic part of F-logic. We utilize it for IC declaration for its nice, succinct, frame-based syntax. IC semantics of axioms should follow trivially from definitions in Table 3.

Running Example. Reviewing our running example, the biggest change is the use of method signatures with cardinality constraints. This significantly reduces the verbosity of the ICs and makes them arguable easier to understand.

$$\mathcal{IC}^F = \{Vocabulary_C[author_{\mathcal{R}} \Rightarrow_{\{1:1\}} User_C; lastEditor_C \Rightarrow_{\{0:1\}} User_C]\}$$

Table 3. Integrity constraint semantics of F-logic concept descriptions. The right hand column specifies a condition under which an individual x is an instance of the concept specified in the left hand column under the IC semantics.

Concept	$\mathbf{I}^{IC} \models^F x : Concept$ iff
$Not(C_C)$	$x \in \mathcal{E} \wedge I_F(x) \notin_U^{IC} I_F(C_C)$
C_C and D_C	$x \in \mathcal{E} \wedge I_F(x) \in_U^{IC} I_F(C_C) \wedge I_F(x) \in_U^{IC} I_F(D_C)$
C_C or D_C	$x \in \mathcal{E} \wedge I_F(x) \in_U^{IC} I_F(C_C) \vee I_F(x) \in_U^{IC} I_F(D_C)$
$AtLeast(n, R_R, C_C)$	$x \in \mathcal{E} \wedge \exists y_1, ... y_n \in \mathcal{E}$ s.t. $I_F(y_i) \in I_{\rightarrow}^{IC}(I_F(R_R))(I_F(x))$
	$\wedge I_F(y_i) \in_U^{IC} I_F(C_C) \wedge \neg(I_F(y_i) = I_F(y_j))$
$HasSelf(R_R)$	$x \in \mathcal{E} \wedge I_F(x) \in I_{\rightarrow}^{IC}(I_F(R_R))(I_F(x))$
$Nom(a_\mathcal{E})$	$x \in \mathcal{E} \wedge I_F(x) = I_F(a_\mathcal{E})$
$[R_R \Rightarrow D_C]$	$\mathbf{I}^{IC}$ is a *typed* F-structure [18] (Sec. 13)
$Some(R_R, C_C)$	$x \in \mathcal{E} \wedge \exists y \in \mathcal{E}$ s.t. $I_F(y) \in I_{\rightarrow}^{IC}(I_F(R_R))(I_F(x))$
	$\wedge I_F(y) \in_U^{IC} I_F(C_C)$

4.2   Integrity Constraint Validation

Now that one is able to define integrity constraints for an ontology, it is necessary to be able to validate them as well. IC semantics is a convenient construct, but because no corresponding implementation exists, it is impractical. Thus, integrity constraint validation in F-logic utilizes the built-in possibility to execute queries over the underlying ontology. F-logic is a full-fledged logic programming language, so it allows to define rules and pose queries to the knowledge base.

Since ICs represent a close-world view of the ontology, *negation as failure* (NAF) is necessary to be able to represent it in the queries. Like most logic programming languages [24], we introduce the NAF operator **not**, whose semantics is $\mathcal{K}^F \models \mathbf{not}(\alpha)$ iff $\mathcal{K}^F \not\models \alpha$, where α is an F-formula.

We now show how the IC axioms can be translated into F-logic queries with **not**. The rationale is that if the knowledge base entails the query, there is an IC violation. We again follow the line of reasoning of [29], which introduces two operators for translating integrity constraints to validation queries: $\mathcal{T}_C$ for concepts and $\mathcal{T}$ for axioms. Their definitions are in Tables 4 and 5, respectively.

The universal role restriction concept comes with a little twist. Instead of representing a standalone concept, a corresponding signature expression is attached to the target concept, i.e., instead of mapping a GCI axiom $C \sqsubseteq \forall R.D$, we have directly $C_C[R_R \Rightarrow D_C]$. A validation query is created by verifying that data expressions of all instances of C_C comply with the signature expression, i.e., $\mathcal{T}_C(x[R_R \Rightarrow D_C])$. This can be also seen in the running example below.

Running Example. Since a signature expression with cardinality constraints is essentially a combination of multiple concept descriptions, it results in multiple

Table 4. Integrity constraint validation transformation rules for concepts. C_A is an atomic class name.

Concept	$\mathcal{T}_C$
$\mathcal{T}_C(x\!:\!C_A)$	$x\!:\!C_A$
$\mathcal{T}_C(x\!:\!Not(C))$	$\textbf{not}(x\!:\!\mathcal{T}_C(C))$
$\mathcal{T}_C(x\!:\!(C_1 \text{ and } C_2))$	$x\!:\!\mathcal{T}_C(C_1) \wedge x\!:\!\mathcal{T}_C(C_2)$
$\mathcal{T}_C(x\!:\!(C_1 \text{ or } C_2))$	$x\!:\!\mathcal{T}_C(C_1) \vee x\!:\!\mathcal{T}_C(C_2)$
$\mathcal{T}_C(x\!:\!AtLeast(n,R,C))$	$\bigwedge_{1\leq i\leq n} x[R \to y_i] \wedge y_i\!:\!\mathcal{T}_C(C) \bigwedge_{1\leq i\leq j\leq n} \textbf{not}(y_i = y_j)$
$\mathcal{T}_C(x\!:\!HasSelf(R))$	$x[R \to x]$
$\mathcal{T}_C(x\!:\!Nom(a))$	$x = a$
$\mathcal{T}_C(x[R \Rightarrow C])$	$x[R \to y] \Rightarrow y\!:\!\mathcal{T}_C(C)$
$\mathcal{T}_C(x\!:\!Some(R,C))$	$x[R \to y] \wedge y\!:\!\mathcal{T}_C(C)$

Table 5. Integrity constraint validation transformation rules for axioms. C_i is a concept, R_i is a role and x, y_i are variables.

Axiom	$\mathcal{T}$
$\mathcal{T}(C_1 :: C_2)$	$\mathcal{T}_C(x\!:\!C_1) \wedge \textbf{not}(\mathcal{T}_C(x\!:\!C_2))$
$\mathcal{T}(subPropertyOf_\mathcal{P}(R_1,R_2))$	$x[R_1 \to y] \wedge \textbf{not}(x[R_2 \to y])$
$\mathcal{T}(Sym_\mathcal{P}(R))$	$x[R \to y] \wedge \textbf{not}(y[R \to x])$
$\mathcal{T}(Asy_\mathcal{P}(R))$	$x[R \to y] \wedge y[R \to x]$
$\mathcal{T}(Tra_\mathcal{P}(R))$	$x[R \to y] \wedge y[R \to z] \wedge \textbf{not}(x[R \to z])$
$\mathcal{T}(Ref_\mathcal{P}(R))$	$\textbf{not}(x[R \to x])$
$\mathcal{T}(Irr_\mathcal{P}(R))$	$x[R \to x]$
$\mathcal{T}(Dis_\mathcal{P}(R_1,R_2))$	$x[R_1 \to y] \wedge x[R_2 \to y]$

validation queries. The queries below can be executed in F-logic implementations with **not**, e.g., FLORA-2.[4] The constraints are violated by the lack of an explicit author of $MetropolitanPlan$, manifested in the second query. Asserting an author, e.g., $MetropolitanPlan_\mathcal{E}[author_\mathcal{R} \to Tom_\mathcal{E}]$, would fix the IC violation.

$$\mathcal{T} = \{x\!:\!Vocabulary \wedge x[author_\mathcal{R} \to y] \wedge \textbf{not}(y\!:\!User_\mathcal{C}),$$
$$x\!:\!Vocabulary_\mathcal{C} \wedge \textbf{not}(x[author_\mathcal{R} \to y] \wedge y\!:\!User_\mathcal{C}),$$
$$x\!:\!Vocabulary_\mathcal{C} \wedge x[author_\mathcal{R} \to \{y_1,y_2\}] \bigwedge_{1\leq i\leq 2} y_i\!:\!User_\mathcal{C} \wedge \textbf{not}(y_1 = y_2)$$
$$x\!:\!Vocabulary \wedge x[lastEditor_\mathcal{R} \to y] \wedge \textbf{not}(y\!:\!User_\mathcal{C}),$$
$$x\!:\!Vocabulary_\mathcal{C} \wedge x[lastEditor_\mathcal{R} \to \{y_1,y_2\}] \bigwedge_{1\leq i\leq 2} y_i\!:\!User_\mathcal{C} \wedge \textbf{not}(y_1 = y_2)\}$$

---

[4] http://flora.sourceforge.net/, accessed 2019-04-10.

Finally, we have to show that the validation queries faithfully represent the IC semantics, i.e., that validation queries generated from IC axioms return results whenever any of the IC axioms are violated by the knowledge base.

Theorem 2. *Consider a knowledge base $\mathcal{K}$, a set of integrity constraint axioms $\mathcal{IC}$ and a set of IC validation queries $\mathcal{Q}$, constructed by applying the translation operator $\mathcal{T}$ on each IC axiom α in $\mathcal{IC}$. If $\mathcal{K}$ violates any of the IC axioms in $\mathcal{IC}$, then $\exists\ q \in \mathcal{Q}$ such that $\mathcal{K} \models^F q$.*

Proof (Sketch). The proof shows for RBox and TBox GCI integrity constraint axioms that if there is a model which violates an IC axiom, it satisfies the corresponding IC validation query. Thus, integrity constraint checking can be reduced to query answering in F-logic. A full proof can be again found in the technical report [23]. □

5 Related Work

This section reviews works concerning application access to DL ontologies, provides a comparison of approaches to mapping between description logics and F-logic and discusses methods of closed-world reasoning in DL ontologies.

5.1 Application Access to Ontologies

There exists a number of application libraries which provide programmatic access to ontologies. They can be roughly divided into two groups [21]:

Domain-independent APIs. Jena [8], OWL API [14], or RDF4J [5] are low-level libraries which represent ontological data on axiom level.
Domain-specific APIs. ActiveRDF [26], Empire [13], and KOMMA [30], allow the application to access ontological data in a frame-based manner.[5]

Libraries of type 1 are suitable for generic applications like ontology editors or vocabulary explorers, but their use in domain-specific applications is cumbersome, because they require a lot of boilerplate code to allow dealing with higher-level business objects. Libraries of type 2 employ some kind of object-ontological mapping (sometimes also called object-triple mapping), so that they map ontological concepts to programming language reference types, properties to attributes etc. The problem with these libraries is that they often do not take into account the open-world nature of ontologies. They do not deal with inferred knowledge (an inferred assertion cannot be directly removed), and the mapping is done without any formal basis. These libraries rarely support knowledge outside the mapped object model and tend to have issues with individual identity. For instance, given an OWL ontology $\mathcal{O} = \{Vocabulary \sqsubseteq Asset, Vocabulary(a)\}$, Empire, when asked to retrieve a twice - as an *Asset* and as a *Vocabulary*, will return two different objects. The consequences of such behavior can be anywhere between overwriting updates and deletion of an object that is being used.

---

[5] A detailed comparison of these libraries can be found in [22].

5.2 Mapping Between Description Logics and F-logic

The relationship between description logics and F-logic can be approached from different directions. One, which has been investigated in [12] or [16], enriches a DL knowledge base with (F-logic) rules to provide additional or more efficient inferences ([12] considers logic programming languages in general).

The other direction tries to develop a mapping between the two languages. [7] exploits the fact that DLs are a subset of the first order logic and maps them to the FOL flavor of F-logic, i.e., concepts to unary predicates and roles to binary predicates. On the other hand, Balaban [3] attempts to map DL constructs to F-logic frames. However, his article deals only with less expressive DLs ($\mathcal{ALC}$). Close to our approach is also the work of de Bruijn and Heymans [6] which maps $\mathcal{SHIQ}$ to F-logic by first translating it to predicate-based FOL and then mapping it to F-logic. Compared to Balaban, our mapping deals with more expressive languages and considers the mapping of integrity constraints. de Bruijn and Heymans' work presents, in our opinion, a less readable, although arguably more straightforward, approach to the mapping. The authors of F-logic themselves discuss its potential as ontology-modeling language in [17,31]. In [17], they provide an example of an ontology for describing Web services.

5.3 Closed-World Reasoning

Application of integrity constraints to DL ontologies, as discussed in Sect. 2.2, is closely related to *(local) closed-world reasoning*. Significant amount of work has been done in this area in connection with rule-based languages. They often split the knowledge base into a DL-based OWA part and a rule-based CWA part with stable [9,11] or well-founded [9,19] model semantics.

Another approach similar to [27] is based on *grounded circumscription* where selected concepts and roles are closed and minimized, i.e., they contain only the minimum necessary *known* individuals [28].

6 Conclusions

We have introduced a novel formalism for object-ontological mapping based on the description logic $\mathcal{SROIQ}$ and F-logic. The formalism maps both a DL ontology and integrity constraints which provide a closed-world view of (a portion of) the ontology. We have shown that the mapping preserves entailment and presented means of validating the integrity constraints. As has been shown in [20], integrity constraints represent the basis of the contract between an ontology and an object model and are used to define the object model.

However, the presented work is just the first step. The mapping represents a static structure of the model and the data. The next step should be defining operations over the data in terms of the formalism. With such definitions, ontological operations like data retrieval or modifications would have predictable and well defined results.

Another step is the actual translation of the F-logic intermediate model into an object model in a mainstream object-oriented programming language like Java. Finally, the operations need to be implemented according to the definitions.

Acknowledgment. This work was supported by grant No. SGS19/110/OHK3/2T/13 Efficient Vocabularies Management Using Ontologies of the Czech Technical University in Prague.

References

1. Angele, J., Kifer, M., Lausen, G.: Ontologies in F-logic. In: Staab, S., Studer, R. (eds.) Handbook on Ontologies, pp. 45–70. Springer, Heidelberg (2009). https://doi.org/10.1007/978-3-540-92673-3_2
2. Baader, F., Calvanese, D., McGuinness, D.L., Nardi, D., Patel-Schneider, P.F. (eds.): The Description Logic Handbook: Theory, Implementation, and Applications. Cambridge University Press, New York (2003)
3. Balaban, M.: The F-logic approach for description languages. Ann. Math. Artif. Intell. **15**(1), 19–60 (1995). https://doi.org/10.1007/BF01535840
4. Booch, G.: Object-oriented Analysis and Design with Applications, 2nd edn. Benjamin-Cummings Publishing Co. Inc., Redwood City (1994)
5. Broekstra, J., Kampman, A., van Harmelen, F.: Sesame: a generic architecture for storing and querying RDF and RDF schema. In: Horrocks, I., Hendler, J. (eds.) ISWC 2002. LNCS, vol. 2342, pp. 54–68. Springer, Heidelberg (2002). https://doi.org/10.1007/3-540-48005-6_7
6. de Bruijn, J., Heymans, S.: Translating ontologies from predicate-based to frame-based languages. In: Proceedings of the 2nd International Conference on Rules and Rule Markup Languages for the Semantic Web (RuleML2006) (2006)
7. de Bruijn, J., Lara, R., Polleres, A., Fensel, D.: OWL DL vs. OWL flight: conceptual modeling and reasoning for the semantic web. In: Proceedings of the 14th International Conference on World Wide Web, WWW 2005. ACM (2005). https://doi.org/10.1145/1060745.1060836
8. Carroll, J.J., Dickinson, I., Dollin, C., Reynolds, D., Seaborne, A., Wilkinson, K.: Jena: implementing the semantic web recommendations. In: Proceedings of the 13th International World Wide Web Conference (Alternate Track Papers & Posters), pp. 74–83 (2004)
9. Damásio, C.V., Analyti, A., Antoniou, G., Wagner, G.: Supporting open and closed world reasoning on the web. In: Alferes, J.J., Bailey, J., May, W., Schwertel, U. (eds.) PPSWR 2006. LNCS, vol. 4187, pp. 149–163. Springer, Heidelberg (2006). https://doi.org/10.1007/11853107_11
10. Donini, F.M., Nardi, D., Rosati, R.: Description Logics of Minimal Knowledge and Negation as Failure. ACM Trans. Comput. Logic **3**(2), 177–225 (2002). https://doi.org/10.1145/505372.505373
11. Eiter, T., Ianni, G., Lukasiewicz, T., Schindlauer, R., Tompits, H.: Combining answer set programming with description logics for the Semantic Web. IEEE Trans. Knowl. Data Eng. **22**(11), 1577–1592 (2010). https://doi.org/10.1109/TKDE.2010.111
12. Grosof, B.N., Horrocks, I., Volz, R., Decker, S.: Description logic programs: combining logic programs with description logic. In: Proceedings of the 12th International Conference on World Wide Web, WWW 2003, pp. 48–57. ACM, New York (2003). https://doi.org/10.1145/775152.775160

13. Grove, M.: Empire: RDF & SPARQL Meet JPA. semanticweb.com, April 2010. http://semanticweb.com/empire-rdf-sparql-meet-jpa_b15617
14. Horridge, M., Bechhofer, S.: The OWL API: A Java API for OWL ontologies. Semantic Web - Interoperability, Usability, Applicability (2011)
15. Horrocks, I., Kutz, O., Sattler, U.: The even more irresistible $\mathcal{SROIQ}$. In: Proceedings of the 10th International Conference on Principles of Knowledge Representation and Reasoning (KR 2006), pp. 57–67 (2006)
16. Kattenstroth, H., May, W., Schenk, F.: Combining OWL with F-logic rules and defaults. In: Proceedings of the ICLP'07 Workshop on Applications of Logic Programming to the Web, Semantic Web and Semantic Web Services, ALPSWS 2007 (2007)
17. Kifer, M.: Rules and ontologies in F-logic. In: Eisinger, N., Małuszyński, J. (eds.) Reasoning Web. LNCS, vol. 3564, pp. 22–34. Springer, Heidelberg (2005). https://doi.org/10.1007/11526988_2
18. Kifer, M., Lausen, G., Wu, J.: Logical foundations of object-oriented and frame-based languages. J. ACM **42**(4), 741–843 (1995). https://doi.org/10.1145/210332.210335
19. Knorr, M., Alferes, J.J., Hitzler, P.: Local closed world reasoning with description logics under the well-founded semantics. Artif. Intell. **175**(9–10), 1528–1554 (2011). https://doi.org/10.1016/j.artint.2011.01.007
20. Křemen, P.: Building Ontology-Based Information Systems. Ph.D. thesis, Czech Technical University, Prague (2012)
21. Křemen, P., Kouba, Z.: Ontology-driven information system design. IEEE Trans. Syst. Man Cybern. Part C **42**(3), 334–344 (2012)
22. Ledvinka, M., Křemen, P.: A comparison of object-triple mapping libraries. In: Semantic Web, p. 43, February 2019. https://doi.org/10.3233/SW-190345
23. Ledvinka, M., Křemen, P.: Formalizing object-ontological mapping using F-logic. Technical report SGS19/110/OHK3/2T/13-1, Department of Computer Science, CTU in Prague (2019). https://kbss.felk.cvut.cz/reports/2019/19ruleml-report.pdf. Accessed 30 May 2019
24. Lloyd, J.W.: Foundations of Logic Programming. Springer, Heidelberg (1984). https://doi.org/10.1007/978-3-642-83189-8
25. Motik, B., Horrocks, I., Sattler, U.: Bridging the gap between OWL and relational databases. Web Semant. Sci. Serv. Agents World Wide Web **7**(2), 74–89 (2009)
26. Oren, E., Heitmann, B., Decker, S.: ActiveRDF: Embedding Semantic Web data into object-oriented languages. Web Semant. Sci. Serv. Agents World Wide Web **6**(3), 191–202 (2008)
27. Patel-Schneider, P.F., Franconi, E.: Ontology constraints in incomplete and complete data. In: Cudré-Mauroux, P., et al. (eds.) ISWC 2012. LNCS, vol. 7649, pp. 444–459. Springer, Heidelberg (2012). https://doi.org/10.1007/978-3-642-35176-1_28
28. Sengupta, K., Krisnadhi, A.A., Hitzler, P.: Local closed world semantics: grounded circumscription for OWL. In: Aroyo, L., Welty, C., Alani, H., Taylor, J., Bernstein, A., Kagal, L., Noy, N., Blomqvist, E. (eds.) ISWC 2011. LNCS, vol. 7031, pp. 617–632. Springer, Heidelberg (2011). https://doi.org/10.1007/978-3-642-25073-6_39
29. Tao, J., Sirin, E., Bao, J., McGuinness, D.L.: Integrity constraints in OWL. In: Fox, M., Poole, D. (eds.) AAAI. AAAI Press (2010)

30. Wenzel, K.: KOMMA: an application framework for ontology-based software systems. In: Semantic Web - Interoperability, Usability, Applicability (2010)
31. Yang, G., Kifer, M.: Reasoning about anonymous resources and meta statements on the semantic web. In: Spaccapietra, S., March, S., Aberer, K. (eds.) Journal on Data Semantics I. LNCS, vol. 2800, pp. 69–97. Springer, Heidelberg (2003). https://doi.org/10.1007/978-3-540-39733-5_4

Alternating Fixpoint Operator for Hybrid MKNF Knowledge Bases as an Approximator of AFT

Fangfang Liu[1]($\boxtimes$) and Jia-Huai You[2]

[1] School of Computer Engineering and Science, Shanghai University,
99 Shangda Road, BaoShan District, Shanghai 20444, People's Republic of China
ffliu@shu.edu.cn

[2] Department of Computing Science, University of Alberta, Edmonton, Canada

Abstract. Approximation fixpoint theory (AFT) provides an algebraic framework for the study of fixpoints of operators on bilattices and has found its applications in characterizing semantics for various types of logic programs and nonmonotonic languages. In this paper, we show one more application of this kind: the alternating fixpoint operator by Knorr et al. [8] for the study of well-founded semantics for hybrid MKNF knowledge bases is in fact an approximator of AFT in disguise, which, thanks to the power of abstraction of AFT, characterizes not only the well-founded semantics but also two-valued as well as three-valued semantics for hybrid MKNF knowledge bases. Furthermore, we show an improved approximator for these knowledge bases, of which the least stable fixpoint is information richer than the one formulated from Knorr et al.'s construction. This leads to an improved computation for the well-founded semantics.

1 Introduction

AFT is a framework for the study of semantics of nonmonotonic logics based on operators and their fixpoints [6]. Under this theory, the semantics of a logic theory is defined in terms of respective stable fixpoints of an *approximator* on a bilattice. The approach is highly general as it only depends on mild conditions on approximators, and highly abstract as well since the semantics is given in terms of an algebraic structure. As different approximators may represent different structural intuitions, AFT provides an elegant way to treat semantics uniformly and allows to explore alternatives by different approximators. A major advantage is that we can understand some properties of a semantics even without a concrete approximator. For example, the well-founded fixpoint approximates all other fixpoints, and mathematically, this property holds for all approximators.

AFT has been applied in the study of the semantics of logic programs with aggregates [13] and disjunctive HEX programs [1]. Vennekens et al. [17] used AFT in a modularity study for a number of nonmonotonic logics, and by applying AFT, Strass [14] showed that many semantics from Dung's argumentation

© Springer Nature Switzerland AG 2019
P. Fodor et al. (Eds.): RuleML+RR 2019, LNCS 11784, pp. 113–127, 2019.
https://doi.org/10.1007/978-3-030-31095-0_8

frameworks and abstract dialectical frameworks can be obtained rather directly. More recently, AFT has been applied to study database revision by characterizing the semantics for active integrity constraints [3].

In this paper, we add one more application to the above collection for hybrid MKNF. Recall that hybrid MKNF (minimal knowledge and negation as failure) was proposed by Motik and Rosati [12] for integrating nonmonotonic rules with description logics (DLs). A hybrid MKNF knowledge base $\mathcal{K}$ consists of two components, $\mathcal{K} = (\mathcal{O}, \mathcal{P})$, where $\mathcal{O}$ is a DL knowledge base, which corresponds to a decidable first-order theory, and $\mathcal{P}$ is a collection of MKNF rules based on the stable model semantics. In [8], Knorr et al. formulated a three-valued extension of MKNF and defined three-valued MKNF models where the least one is called the *well-founded MKNF model.* An alternating fixpoint operator was then formulated for the computation of the well-founded MKNF model for (nondisjunctive) hybrid MKNF knowledge bases. In this paper, our primary goal is to show that this alternating fixpoint operator is in fact an approximator of AFT. Due to the abstract power of AFT, it follows that Knorr et al.'s alternating fixpoint operator provides a uniform characterization of all semantics based on various kinds of three-valued MKNF models, including two-valued MKNF models of [12].

As shown in [8,10], not all hybrid MKNF knowledge bases possess a well-founded MKNF model, and in general, deciding the existence of a well-founded MKNF model is intractable [10]. On the other hand, we also know that alternating fixpoint construction provides a tractable means to compute the well-founded MKNF model for a subset of hybrid MKNF knowledge bases. A question then is whether this subset can be enlarged. In this paper, we answer this question positively by formulating a *more precise* approximator.

The paper is organized as follows. The next section introduces approximation fixpoint theory; in particular, we give a relaxation of the original definition of approximators in order to accommodate inconsistency. Section 3 gives a review of three-valued MKNF and hybrid MKNF knowledge bases along with the underlying semantics. Then, in Sect. 4 we show how Knorr et al.'s alternating fixpoint operator may be recast as an approximator and provide semantic characterizations and in Sect. 5, we show an improved approximator. Section 6 is about related work and final remarks.

2   Approximation Fixpoint Theory

Briefly, recall that a *lattice* $\langle L, \leq \rangle$ is a *partially ordered set* (poset) in which every two elements have a *least upper bound* (lub) and a *greatest lower bound* (glb). A *complete lattice* is a lattice where every subset of L has a least upper bound and a greatest lower bound. An operator O on L is *monotone* if for all $x, y \in L$, $x \leq y$ implies $O(x) \leq O(y)$. The Knaster-Tarski fixpoint theory [15] tells us that a monotone operator on a complete lattice has fixpoints and a least fixpoint (denoted $lfp(O)$).

Given a complete lattice $\langle L, \leq \rangle$, AFT deals with the structure $\langle L^2, \leq_p \rangle$, which is the induced (product) bilattice, where $\leq_p$ is called the *precision order* and defined as: for all $x, y, x', y' \in L$, $(x, y) \leq_p (x', y')$ if $x \leq x'$ and $y' \leq y$. The $\leq_p$ ordering is a complete lattice ordering on L^2.

We define two *projection functions* for pairs: $(x, y)_1 = x$ and $(x, y)_2 = y$. A pair $(x, y) \in L^2$ is *consistent* if $x \leq y$, *inconsistent* otherwise, and *exact* if $x = y$. A consistent pair (x, y) in L defines an *interval*, denoted $[x, y]$, which is identified by the set $\{z \mid x \leq z \leq y\}$. We therefore also use an interval to denote the corresponding set. A consistent pair (x, y) in L can be seen as an approximation of every $z \in L$ such that $z \in [x, y]$. In this sense, the precision order $\leq_p$ corresponds to the precision of approximation, while an exact pair approximates the only element in it. We denote by L^c the set of consistent pairs in L^2.

AFT studies fixpoints of operators O on L through operators approximating O. An *approximator* is a $\leq_p$-monotone operator on L^2. We denote by $Appx(L^2)$ the set of all approximators on L^2. An *approximator for O* has the additional property that $A(x, x) = (O(x), O(x))$, for all $x \in L$. Since $(L^2, \leq_p)$ is a complete lattice, an approximator A has a least fixpoint, which is called *Kripke-Kleene fixpoint* of A. Our main interest in this paper is in *stable fixpoints* of A, which are the fixpoints of a *stable revision operator* $St_A : L^2 \to L^2$, defined as: $St_A(u, v) = (lfp(A(\cdot, v)_1), lfp(A(u, \cdot)_2))$, where $A(\cdot, v)_1$ and $A(u, \cdot)_2$ are the component operators on L.

AFT was first developed for consistent approximations, where an approximator is consistent if it maps consistent pairs to consistent pairs and has the property that $A(x, x)_1 = A(x, x)_2$, for all $x \in L$. Let us denote by $Appx(L^c)$ the set of all consistent approximators. Then, the notion of approximator is generalized to *symmetric approximators*, which are $\leq_p$-monotone operators A on L^2 such that $A(x, y)_1 = A(y, x)_2$, for all $x, y \in L$. Note that a symmetric approximator is defined for all pairs in L^2. It is easy to show that an operator A being symmetric implies that $A(x, x)$ yields an exact pair, for all $x \in L$. As remarked in [6], this generalization is motivated by operators occurring in knowledge representation that are symmetric. The authors also point out that it is possible to develop a generalization of the theory without the symmetry assumption. Apparently, such a generalization needs to relax the definition of *"approximator for O"*, and let us define: $A : L^2 \to L^2$ is an *approximator* for O if A is $\leq_p$-monotone and for all $x \in L$, if $A(x, x)$ is consistent then $A(x, x) = (O(x), O(x))$. That is, we make the notion of approximation for O *partial* - $A(x, x)$ captures O only when $A(x, x)$ is consistent.[1]

3 Hybrid MKNF Knowledge Bases

3.1 Minimal Knowledge and Negation as Failure

The logic of minimal knowledge and negation as failure (MKNF) [9] is based on a first-order language $\mathcal{L}$ (possibly with equality $\approx$) with two modal operators, **K**,

---

[1] We can in addition require that an approximator A be consistent for at least one exact pair. This will eliminate the undesired situation that if a $\leq_p$-monotone operator A is inconsistent on each exact pair, then it approximates every operator O, trivially.

for minimal knowledge, and **not**, for negation as failure. In MKNF, *first-order atom* is defined as usual and *MKNF formulas* are first-order formulas with **K** and **not**. An MKNF formula is *ground* if it contains no variables. Given a first-order formula φ, $\mathbf{K}\varphi$ is called a (modal) **K**-*atom* and **not** φ called a (modal) **not**-*atom*.

A *first-order interpretation* is understood as in first-order logic. The universe of a first-order interpretation I is denoted by $|I|$. A *first-order structure* is a nonempty set M of first-order interpretations with the universe $|I|$ for some fixed $I \in M$. An *MKNF structure* is a triple (I, M, N), where M and N are sets of first-order interpretations with the universe $|I|$. We define the *satisfaction relation* $\models$ between an MKNF structure (I, M, N) and an MKNF formula F. Then we extend the language $\mathcal{L}$ by object constants representing all elements of $|I|$ and call these constants *names*:

$$(I, M, N) \models A \ (A \text{ is a first-order atom}) \text{ if } A \text{ is true in } I,$$
$$(I, M, N) \models \neg F \text{ if } (I, M, N) \not\models F,$$
$$(I, M, N) \models F \wedge G \text{ if } (I, M, N) \models F \text{ and } (I, M, N) \models G,$$
$$(I, M, N) \models \exists x F \text{ if } (I, M, N) \models F[\alpha/x] \text{ for some name } \alpha,$$
$$(I, M, N) \models \mathbf{K} F \text{ if } (J, M, N) \models F \text{ for all } J \in M,$$
$$(I, M, N) \models \mathbf{not} F \text{ if } (J, M, N) \not\models F \text{ for some } J \in N.$$

The symbols $\top$, $\bot$, $\vee$, $\forall$, and $\supset$ are interpreted as usual.

An *MKNF interpretation M* is a nonempty set of first-order interpretations over the universe $|I|$ for some $I \in M$. In MKNF, a notion called *standard name assumption* is imposed on top of MKNF to avoid unintended behaviors [12]. This requires an interpretation to be a Herbrand interpretation with a countably infinite number of additional constants, and the predicate $\approx$ to be a congruence relation. Intuitively, given the assumption that each individual in the universe of an interpretation is denoted by a constant and the countability it implies, the standard name assumption becomes a convenient normalized representation of interpretations since each interpretation is isomorphic to the quotient (w.r.t. $\approx$) of a Herbrand interpretation and each quotient of a Herbrand interpretation is an interpretation.

An MKNF interpretation M *satisfies* an MKNF formula F, written $M \models_{MKNF} F$, if $(I, M, M) \models F$ for each $I \in M$.

Following [8], a *three-valued MKNF structure*, $(I, \mathcal{M}, \mathcal{N})$, consists of a first-order interpretation, I, and two pairs, $\mathcal{M} = \langle M, M_1 \rangle$ and $\mathcal{N} = \langle N, N_1 \rangle$, of sets of first-order interpretations, where $M_1 \subseteq M$ and $N_1 \subseteq N$. From $\langle M, M_1 \rangle$, we can identify three truth values for modal **K**-atoms in the following way: $\mathbf{K}\varphi$ is true w.r.t. $\langle M, M_1 \rangle$ if φ is true in all interpretations in M; it is false if it is false in at least one interpretation in M_1; and it is undefined otherwise. For **not**-atoms, a symmetric treatment w.r.t. $\langle N, N_1 \rangle$ is adopted. Let $\{\mathbf{t}, \mathbf{u}, \mathbf{f}\}$ be the set of truth values *true*, *undefined*, and *false* with the order $\mathbf{f} < \mathbf{u} < \mathbf{t}$, and let the operator *max* (resp. *min*) choose the greatest (resp. the least) element with respect to this ordering. Table 1 shows three-valued evaluation of MKNF formulas (where $F[t/x]$ denotes the formula obtained from F by replacing all free occurrences of the variable x with term t).

A *(three-valued) MKNF interpretation pair* (M, N) consists of two MKNF interpretations, M and N, with $\emptyset \subset N \subseteq M$. An MKNF interpretation pair satisfies an MKNF formula φ, denoted $(M, N) \models \varphi$, iff $(I, \langle M, N \rangle, \langle M, N \rangle)(\varphi) = \mathbf{t}$ for each $I \in M$. If $M = N$, the MKNF interpretation pair is called *total*. If there exists an MKNF interpretation pair satisfying a formula φ, then φ is said to be *consistent*; otherwise φ is *inconsistent*. If φ is a first-order formula, we also say that φ is *satisfiable* when φ is consistent.

Table 1. Evaluation in three-valued MKNF structure

$(I, \mathcal{M}, \mathcal{N})(P(t_1, \ldots, t_n)) =$	$\mathbf{t}$	iff $(t_1^I, \ldots, t_n^I) \in P^I$
	$\mathbf{f}$	iff $(t_1^I, \ldots, t_n^I) \notin P^I$
$(I, \mathcal{M}, \mathcal{N})(\neg\varphi) =$	$\mathbf{t}$	iff $(I, \mathcal{M}, \mathcal{N})(\varphi) = \mathbf{f}$
	$\mathbf{u}$	iff $(I, \mathcal{M}, \mathcal{N})(\varphi) = \mathbf{u}$
	$\mathbf{f}$	iff $(I, \mathcal{M}, \mathcal{N})(\varphi) = \mathbf{t}$
$(I, \mathcal{M}, \mathcal{N})(\varphi_1 \wedge \varphi_2) =$	$\min\{(I, \mathcal{M}, \mathcal{N})(\varphi_1), (I, \mathcal{M}, \mathcal{N})(\varphi_2)\}$	
$(I, \mathcal{M}, \mathcal{N})(\varphi_1 \supset \varphi_2) =$	$\mathbf{t}$	iff $(I, \mathcal{M}, \mathcal{N})(\varphi_2) \geq (I, \mathcal{M}, \mathcal{N})(\varphi_1)$
	$\mathbf{f}$	otherwise
$(I, \mathcal{M}, \mathcal{N})(\exists x : \varphi) =$	$\max\{(I, \mathcal{M}, \mathcal{N})(\varphi[n_\alpha/x]) \mid \alpha \in \Delta\}$	
$(I, \mathcal{M}, \mathcal{N})(\mathbf{K}\varphi) =$	$\mathbf{t}$	iff $(J, \langle M, M_1 \rangle, \mathcal{N})(\varphi) = \mathbf{t}$ for all $J \in M$
	$\mathbf{f}$	iff $(J, \langle M, M_1 \rangle, \mathcal{N})(\varphi) = \mathbf{f}$ for some $J \in M_1$
	$\mathbf{u}$	otherwise
$(I, \mathcal{M}, \mathcal{N})(\mathbf{not}\ \varphi) =$	$\mathbf{t}$	iff $(J, \mathcal{M}, \langle N, N_1 \rangle)(\varphi) = \mathbf{f}$ for some $J \in N_1$
	$\mathbf{f}$	iff $(J, \mathcal{M}, \langle N, N_1 \rangle)(\varphi) = \mathbf{t}$ for all $J \in N$
	$\mathbf{u}$	otherwise

An MKNF interpretation pair (M, N) is a *three-valued MKNF model* of an MKNF formula φ if $(M, N) \models \varphi$ and for all MKNF interpretation pairs (M', N') with $M \subseteq M'$ and $N \subseteq N'$, where at least one of the inclusions is proper and $M' = N'$ if $M = N$, $\exists I' \in M'$ such that $(I', \langle M', N' \rangle, \langle M, N \rangle)(\varphi) \neq \mathbf{t}$. As shown by Knorr et al. [8], an MKNF interpretation pair (M, M) that is a three-valued MKNF model of φ corresponds to a two-valued MKNF model M defined in [12].

MKNF interpretation pairs can be compared by an *order of knowledge*. Let (M_1, N_1) and (M_2, N_2) be MKNF interpretation pairs. $(M_1, N_1) \succeq_k (M_2, N_2)$ iff $M_1 \subseteq M_2$ and $N_1 \supseteq N_2$. A three-valued MKNF model (M, N) of an MKNF formula φ is called a *well-founded MKNF model* of φ if $(M_1, N_1) \succeq_k (M, N)$ for all three-valued MKNF models (M_1, N_1) of φ.

3.2 Hybrid MKNF Knowledge Bases

The critical issue of how to combine open and closed world reasoning is addressed in [12] by seamlessly integrating rules with DLs. A hybrid MKNF knowledge base

$\mathcal{K} = (\mathcal{O}, \mathcal{P})$ consists of a decidable description logic (DL) knowledge base $\mathcal{O}$, translatable into first-order logic and a rule base $\mathcal{P}$, which is a finite set of rules with modal atoms. The original work on hybrid MKNF knowledge bases [11,12] defines a two-valued semantics for such knowledge bases with disjunctive rules. In this paper, following [8], our focus is on nondisjunctive rules as presented in [11].

An MKNF rule (or simply a *rule*) r is of the form: $\mathbf{K}H \leftarrow \mathbf{K}A_1, \ldots,$ $\mathbf{K}A_m, \mathbf{not}\, B_1, \ldots, \mathbf{not}\, B_n$, where H, A_i, and B_j are function-free first-order atoms. Given a rule r, we let $hd(r) = \mathbf{K}H$, $bd^+(r) = \{\mathbf{K}A_i \mid i = 1..m\}$, and $bd^-(r) = \{B_i \mid i = 1..n\}$. A rule is *positive* if it contains no **not**-atoms. When all rules in $\mathcal{P}$ are positive, $\mathcal{K} = (\mathcal{O}, \mathcal{P})$ is called *positive*.

For the interpretation of a hybrid MKNF knowledge base $\mathcal{K} = (\mathcal{O}, \mathcal{P})$ in the logic of MKNF, a transformation $\pi(\mathcal{K}) = \mathbf{K}\pi(\mathcal{O}) \wedge \pi(\mathcal{P})$ is performed to transform $\mathcal{O}$ into a first-order formula and rules $r \in \mathcal{P}$ into a conjunction of first-order implications to make each of them coincide syntactically with an MKNF formula. More precisely,

$$\pi(r) = \forall \boldsymbol{x}: (\mathbf{K}H \subset \mathbf{K}A_1 \wedge \ldots \wedge \mathbf{K}A_m \wedge \mathbf{not}\, B_1 \wedge \ldots \wedge \mathbf{not}\, B_n)$$
$$\pi(\mathcal{P}) = \bigwedge_{r \in \mathcal{P}} \pi(r), \quad \pi(\mathcal{K}) = \mathbf{K}\pi(\mathcal{O}) \wedge \pi(\mathcal{P})$$

where $\boldsymbol{x}$ is the vector of free variables in r.

Under the additional assumption of DL-safety a first-order rule base is semantically equivalent to a finite ground rule base, in terms of two-valued MKNF models [12] as well as in terms of three-valued MKNF models [8]; hence decidability is guaranteed. In this paper, we assume that a given rule base is always DL-safe, and for convenience, when we write $\mathcal{P}$ we assume it is already grounded.

Given a hybrid MKNF knowledge base $\mathcal{K} = (\mathcal{O}, \mathcal{P})$, let $\mathsf{KA}(\mathcal{K})$ be the set of all (ground) **K**-atoms $\mathbf{K}\phi$ such that either $\mathbf{K}\phi$ occurs in $\mathcal{P}$ or $\mathbf{not}\,\phi$ occurs in $\mathcal{P}$. We generalize the notion of partition [8] from consistent pairs to all pairs: A *partition* of $\mathsf{KA}(\mathcal{K})$ is a pair (T, P) such that $T, P \subseteq \mathsf{KA}(\mathcal{K})$. A partition of the form (E, E) is said to be *exact*.

Intuitively, given a partition (T, P), T contains *true* modal **K**-atoms and P contains *possibly true* modal **K**-atoms. Thus, the complement of P is the set of *false* modal **K**-atoms and $P \backslash T$ the set of *undefined* modal **K**-atoms. If in addition we have the condition $T \subseteq P$, then these three sets are pairwise non-overlapping, hence (T, P) is consistent

Partitions are closely related to MKNF interpretation pairs. It is shown in [8,10] that an MKNF interpretation pair (M, N) induces a consistent partition (T, P) such that

1. $\mathbf{K}\xi \in T$ iff $\forall I \in M, (I, \langle M, N \rangle, \langle M, N \rangle)(\mathbf{K}\xi) = \mathbf{t}$,
2. $\mathbf{K}\xi \notin P$ iff $\forall I \in M, (I, \langle M, N \rangle, \langle M, N \rangle)(\mathbf{K}\xi) = \mathbf{f}$, and
3. $\mathbf{K}\xi \in P \backslash T$ iff $\forall I \in M, (I, \langle M, N \rangle, \langle M, N \rangle)(\mathbf{K}\xi) = \mathbf{u}$.

Given a set of first-order atoms S, we define the corresponding set of **K**-atoms as: $\mathbf{K}(S) = \{\mathbf{K}\phi \mid \phi \in S\}$.

Let S be a subset of $\mathsf{KA}(\mathcal{K})$. The *objective knowledge* of S relevant to $\mathcal{K}$ is the set of first-order formulas $\mathsf{OB}_{\mathcal{O},S} = \{\pi(\mathcal{O})\} \cup \{\xi \mid \mathbf{K}\xi \in S\}$.

Example 1. Consider a hybrid MKNF knowledge base $\mathcal{K} = (\mathcal{O}, \mathcal{P})$, where $\mathcal{O} = a \wedge (b \supset c) \wedge \neg f$ and $\mathcal{P}$ is

$$\mathbf{K}b \leftarrow \mathbf{K}a. \quad \mathbf{K}d \leftarrow \mathbf{K}c, \mathbf{not}\,e. \quad \mathbf{K}e \leftarrow \mathbf{not}\,d. \quad \mathbf{K}f \leftarrow \mathbf{not}\,b.$$

Reasoning with $\mathcal{K}$ can be seen as follows: Since $\mathbf{K}\mathcal{O}$ implies $\mathbf{K}a$, by the first rule we derive $\mathbf{K}b$; then due to $b \supset c$ in $\mathcal{O}$ we derive $\mathbf{K}c$. Thus its occurrence in the body of the second rule is true and can be ignored. For the $\mathbf{K}$-atoms $\mathbf{K}d$ and $\mathbf{K}e$ appearing in the two rules in the middle, without preferring one over the other, both can be undefined. Because both $\mathbf{not}\,b$ and $\mathbf{K}f$ are false (the latter is due to $\neg f$ in $\mathcal{O}$), the last rule is also satisfied. Now consider an MKNF interpretation pair $(M, N) = (\{I \mid I \models \mathcal{O} \wedge b\}, \{I \mid I \models \mathcal{O} \wedge b \wedge d \wedge e\})$, which corresponds to partition $(T, P) = (\{\mathbf{K}a, \mathbf{K}b, \mathbf{K}c\}, \{\mathbf{K}a, \mathbf{K}b, \mathbf{K}c, \mathbf{K}d, \mathbf{K}e\})$. For instance, we have that, for all $I \in M$, $(I, \langle M, N \rangle, \langle M, N \rangle)(\mathbf{K}a) = \mathbf{t}$ and $(I, \langle M, N \rangle, \langle M, N \rangle)(\mathbf{K}d) = \mathbf{u}$. The interpretation pair (M, N) is a three-valued MKNF model of $\mathcal{K}$; in fact, it is the well-founded MKNF model of $\mathcal{K}$. □

It is known that in general the well-founded MKNF model may not exist.

Example 2 [10]. Let us consider $\mathcal{K} = (\mathcal{O}, \mathcal{P})$, where $\mathcal{O} = (a \supset h) \wedge (b \supset \neg h)$ and $\mathcal{P}$ consists of

$$\mathbf{K}a \leftarrow \mathbf{not}\,b. \quad \mathbf{K}b \leftarrow \mathbf{not}\,a$$

Consider two partitions, $(\{\mathbf{K}a\}, \{\mathbf{K}a\})$ and $(\{\mathbf{K}b\}, \{\mathbf{K}b\})$. The corresponding MKNF interpretation pairs turn out to be two-valued MKNF models of $\mathcal{K}$. For example, for the former the interpretation pair is (M, M), where $M = \{\{a, h\}\}$. Since these two-valued MKNF models are not comparable w.r.t. undefinedness and there are no other three-valued MKNF models of $\mathcal{K}$, it follows that no well-founded MKNF model for $\mathcal{K}$ exists. □

4   Approximators for Hybrid MKNF KBs

In this section, we first show that the alternating fixpoint operator defined by Knorr et al. [8] can be recast as an approximator of AFT, and therefore can be applied to characterize all three-valued MKNF models automatically and naturally. We then study an important, technical issue that arises from treating AFT for the entire domain L^2 for arbitrary approximators (which may be non-symmetric). Finally in Subsect. 5, we formulate a richer approximator for hybrid MKNF knowledge bases and show that it can be applied to improve the computation of well-founded MKNF model.

Throughout this section, the underlying lattice is $(2^{\mathsf{KA}(\mathcal{K})}, \subseteq)$ and the induced bilattice is $(2^{\mathsf{KA}(\mathcal{K})})^2$.

We define an operator on $2^{\mathsf{KA}(\mathcal{K})}$, which is to be approximated by our approximators introduced shortly.

Definition 1. *Let $\mathcal{K} = (\mathcal{O}, \mathcal{P})$ be a hybrid MKNF knowledge base. Define an operator $\mathcal{T}_\mathcal{K}$ on $2^{\mathsf{KA}(\mathcal{K})}$ as follows:*

$$\mathcal{T}_\mathcal{K}(I) = \{\mathbf{K}a \in \mathsf{KA}(\mathcal{K}) \mid \mathsf{OB}_{\mathcal{O},I} \models a\} \cup$$
$$\{hd(r) \mid r \in \mathcal{P} : bd^+(r) \subseteq I, \mathbf{K}(bd^-(r)) \cap I = \emptyset\}$$

If $\mathcal{K}$ is a positive hybrid MKNF knowledge base, the operator $\mathcal{T}_\mathcal{K}$ is monotone and has a least fixpoint. If in addition $\mathcal{O}$ is an empty DL knowledge base, then $\mathcal{T}_\mathcal{K}$ is essentially the familiar *immediate consequence operator* of [16].

Knorr et al. [8] defined two kinds of transforms with consistent partitions. For the purpose of this paper, let us allow arbitrary partitions.

Definition 2. *Let $\mathcal{K} = (\mathcal{O}, \mathcal{P})$ be a hybrid MKNF knowledge base and $S \in 2^{\mathsf{KA}(\mathcal{K})}$. Define two forms of reduct:*

$$\mathcal{K}/S = \{\mathbf{K}a \leftarrow bd^+(r) \mid r \in \mathcal{P} : hd(r) = \mathbf{K}a, \mathbf{K}(bd^-(r)) \subseteq \mathsf{KA}(\mathcal{K}) \setminus S\}$$
$$\mathcal{K}//S = \{\mathbf{K}a \leftarrow bd^+(r) \mid r \in \mathcal{P} : hd(r) = \mathbf{K}a, \mathbf{K}(bd^-(r)) \subseteq \mathsf{KA}(\mathcal{K}) \setminus S,$$
$$\mathsf{OB}_{\mathcal{O},S} \not\models \neg a\}$$

We call $\mathcal{K}/S$ MKNF transform and $\mathcal{K}//S$ MKNF-coherent transform.

Since in both cases of $\mathcal{K}/S$ and $\mathcal{K}//S$ the resulting rule base is positive, a least fixpoint in each case exists. Let us define $\Gamma_\mathcal{K}(S) = lfp(\mathcal{T}_{\mathcal{K}/S})$ and $\Gamma'_\mathcal{K}(S) = lfp(\mathcal{T}_{\mathcal{K}//S})$. Then, we can construct two sequences $\mathbf{P}_i$ and $\mathbf{N}_i$,

$$\mathbf{P}_0 = \emptyset, \ldots, \mathbf{P}_{n+1} = \Gamma_\mathcal{K}(\mathbf{N}_n), \ldots, \mathbf{P}_\omega = \bigcup \mathbf{P}_i$$
$$\mathbf{N}_0 = \mathsf{KA}(\mathcal{K}), \ldots, \mathbf{N}_{n+1} = \Gamma'_\mathcal{K}(\mathbf{P}_n), \ldots, \mathbf{N}_\omega = \bigcap \mathbf{N}_i$$

Now let us place the construction above under AFT by formulating an approximator.

Definition 3. *Let $\mathcal{K} = (\mathcal{O}, \mathcal{P})$ be a hybrid MKNF knowledge base. Define an operator $\Phi_\mathcal{K}$ on $(2^{\mathsf{KA}(\mathcal{K})})^2$ as follows: $\Phi_\mathcal{K}(T, P) = (\Phi_\mathcal{K}(T, P)_1, \Phi_\mathcal{K}(T, P)_2)$, where*

$$\Phi_\mathcal{K}(T, P)_1 = \{\mathbf{K}a \in \mathsf{KA}(\mathcal{K}) \mid \mathsf{OB}_{\mathcal{O},T} \models a\} \cup$$
$$\{hd(r) \mid r \in \mathcal{P} : bd^+(r) \subseteq T, \mathbf{K}(bd^-(r)) \cap P = \emptyset\}$$
$$\Phi_\mathcal{K}(T, P)_2 = \{\mathbf{K}a \in \mathsf{KA}(\mathcal{K}) \mid \mathsf{OB}_{\mathcal{O},P} \models a\} \cup$$
$$\{hd(r) \mid r \in \mathcal{P} : hd(r) = \mathbf{K}a, \mathsf{OB}_{\mathcal{O},T} \not\models \neg a, bd^+(r) \subseteq P,$$
$$\mathbf{K}(bd^-(r)) \cap T = \emptyset\}$$

Intuitively, given a pair (T, P), the operator $\Phi_\mathcal{K}(\cdot, P)_1$, with P fixed, computes true $\mathbf{K}$-atoms w.r.t. (T, P) and operator $\Phi_\mathcal{K}(T, \cdot)_2$, with T fixed, computes the $\mathbf{K}$-atoms that are possibly true w.r.t. (T, P).

Notice that the least fixpoint of the operator $\Phi_\mathcal{K}(\cdot, P)_1$ corresponds to an element in the sequence $\mathbf{P}_i$, i.e., if P in $\Phi_\mathcal{K}(\cdot, P)_1$ is $\mathbf{N}_n$, then $lfp(\Phi_\mathcal{K}(\cdot, P)_1)$ is $\mathbf{P}_{n+1} = \Gamma_\mathcal{K}(\mathbf{N}_n)$. Similarly for operator $\Phi_\mathcal{K}(T, \cdot)_2$. In this way, the $\Phi_\mathcal{K}$ operator can be seen as a reformulation of the corresponding alternating fixpoint operator; namely, $\Phi_\mathcal{K}(\cdot, P)_1$ simulates operator $\mathcal{T}_{\mathcal{K}/P}$ and $\Phi_\mathcal{K}(T, \cdot)_2$ simulates operator $\mathcal{T}_{\mathcal{K}//T}$.

Proposition 1. $\Phi_{\mathcal{K}}$ is an approximator for $\mathcal{T}_{\mathcal{K}}$.

Proof. Let $(T_1, P_1) \subseteq_p (T_2, P_2)$. From $T_1 \subseteq T_2$ and $P_2 \subseteq P_1$, it is easy to verify that $\Phi_{\mathcal{K}}(T_1, P_1)_1 \subseteq \Phi_{\mathcal{K}}(T_2, P_2)_1$ and $\Phi_{\mathcal{K}}(T_2, P_2)_2 \subseteq \Phi_{\mathcal{K}}(T_1, P_1)_2$ (the latter is slightly more involved but still routine to confirm). Thus $\Phi_{\mathcal{K}}(T_1, P_1) \subseteq_p \Phi_{\mathcal{K}}(T_2, P_2)$. Furthermore, $\Phi_{\mathcal{K}}$ approximates $\mathcal{T}_{\mathcal{K}}$ since whenever $\Phi_{\mathcal{K}}(I, I)$ is consistent (even when $\mathrm{OB}_{\mathcal{O}, I}$ is unsatisfiable), $\Phi_{\mathcal{K}}(I, I) = (\mathcal{T}_{\mathcal{K}}(I), \mathcal{T}_{\mathcal{K}}(I))$. □

Next, we prove a property. Let us first explain the motivation. By going beyond symmetric approximators, one subtlety arises. Recall that for any approximator $A \in Appx(L^c)$ and any $(u, v) \in L^c$, the component operator $A(u, \cdot)_2$ is internal in $[u, \top]$, so that the least fixpoint is computed from the least element u in this domain. But when we expand the domain from $[u, \top]$ to L, we may compute with a different least fixpoint, if the operator A is not symmetric.

Example 3. Let $L = \{\bot, \top\}$ and A an identity function everywhere on L^2 except that $A(\bot, \top) = A(\bot, \bot) = (\top, \top)$. Clearly, when restricted to L^c, $A \in Appx(L^c)$, i.e., it is $\leq_p$-monotone, maps consistent pairs to consistent pairs, and approximates, e.g., the identify operator O on L. But it is not symmetric since $A(\bot, \top)_1 = \top$ and $A(\top, \bot)_2 = \bot$. Since $A \in Appx(L^c)$, according to consistent AFT [6], $A(\top, \cdot)_2$ is defined on $[\top, \top]$. Since $St_A(\top, \top) = (lfp(A(\cdot, \top)_1), lfp(A(\top, \cdot)_2)) = (\top, \top)$, it follows that $(\top, \top)$ is a stable fixpoint. Now consider a closely related approximator $A' \in Appx(L^2)$, where for all $x, y \in L$, $A'(\cdot, y)_1 = A(\cdot, y)_1$ and $A'(x, \cdot)$ behaves like $A(x, \cdot)$ everywhere except that it is defined on L. In this case, since $St_{A'}(\top, \top) = (lfp(A'(\cdot, \top)_1), lfp(A'(\top, \cdot)_2)) = (\top, \bot)$, $(\top, \top)$ is not a stable fixpoint of A'. This example is not a surprise since in general different domains may well lead to different least fixpoints.

Now consider another approximator $A'' \in Appx(L^2)$ such that A'' maps all pairs to $(\top, \top)$. It can be seen that A'' is $\leq_p$-monotone and $(\top, \top)$ is a stable fixpoint of A'' in both cases where $A''(\top, \cdot)_2$ is defined either as an operator on $[\top, \top]$ or an operator on L. That is, the least fixpoint of $A''(\top, \cdot)_2$ is coincidental on both domains.

As alluded to earlier, Denecker et al. [6] point out that it is possible to develop a generalization of AFT to L^2 without the symmetry assumption. We remarked that we first need to relax the definition of approximator for exact pairs. The above example indicates a second issue that needs to be addressed. To argue why such a generalization is not entirely trivial, let us digress briefly and discuss how such a generalization may be established.

In [7], a pair $(u, v) \in L^2$ is said to be *A-contracting* if $(u, v) \leq_p A(u, v)$.[2] A pair $(u, v) \in L^2$ is *A-prudent* if $u \leq lfp(A(\cdot, v)_1)$. Let us denote by L^{rp} the set of A-contracting and A-prudent pairs in L^2. We then need to show that, under the generalized definition of approximator (cf. Sect. 2), $\langle L^{rp}, \leq_p \rangle$ is a *chain-complete* poset that contains the least element $(\bot, \top)$.

---

[2] In [6], A-contracting pairs were called *A-reliable*.

Briefly, a *chain* in a poset $\langle L, \leq \rangle$ is a linearly ordered subset of L. A poset $\langle L, \leq \rangle$ is *chain-complete* if it contains a least element $\perp$ and if every chain $C \subseteq L$ has a least upper bound in L. A complete lattice is chain-complete, but the converse does not hold in general. However, the Knaster-Tarski fixpoint theory generalizes to chain-complete posets [6]: a monotone operator on a chain-complete poset possesses fixpoints and a least fixpoin.

We can show the following theorem.

Theorem 1. *Let $(L, \leq)$ be a complete lattice and A an approximator on L^2. Then, $\langle L^{rp}, \leq_p \rangle$ is a chain-complete poset that contains the least element $(\perp, \top)$.*

Now back to Example 3. For operator A, since $A(\perp, \top) = A(\perp, \perp) = (\top, \top)$, $A(\top, \perp) = (\top, \perp)$, $A(\top, \top) = (\top, \top)$, the pairs in L^{rp} are $(\perp, \top)$, $(\top, \perp)$ and $(\top, \top)$. Similarly for A'. The behaviors of A and A' still hold, i.e., $A(\top, \cdot)_2$ and $A'(\top, \cdot)_2$ have different least fixpoints, which are computed from different least elements of their respective domains. This shows that a straightforward generalization of AFT to L^2 may not even preserve consistent stable fixpoints.

Next, we show that this abnormal behavior does not happen to operator $\Phi_{\mathcal{K}}$.

Definition 4. *An approximator $A \in Appx(L^2)$ is said to be strong, if for each consistent stable fixpoint (u, v) of A, $lfp(A(u, \cdot)_2)$, where $A(u, \cdot)$ is defined on L, coincides with $lfp(A'(u, \cdot)_2)$ where $A'(u, \cdot)_2$ is the same operator as $A(u, \cdot)_2$ except that it is defined on $[u, \top]$.*

In other words, a strong approximator preserves consistent stable fixpoints. That is, for any $A \in Appx(L^2)$, if A^c, i.e., A restricted to L^c, happens to be a consistent approximator, then we want all stable fixpoints of A^c to be stable fixpoints of A.

Proposition 2. *$\Phi_{\mathcal{K}}$ is a strong approximator for $T_{\mathcal{K}}$.*

Proof. Let $\mathcal{K} = (\mathcal{O}, \mathcal{P})$. To show that $\Phi_{\mathcal{K}}$ is a strong approximator, assume (T, P) is a consistent stable fixpoint of $\Phi_{\mathcal{K}}$ such that $T \subseteq \Phi_{\mathcal{K}}(T, T)_2$. It is not difficult to show that $\Phi_{\mathcal{K}}(T, \cdot)_2$ is internal in $[T, \mathsf{KA}(\mathcal{K})]$. To be a strong approximator, $\Phi_{\mathcal{K}}(T, \cdot)_2$ must be coincidental on both domains $[T, \mathsf{KA}(\mathcal{K})]$ and $[\emptyset, \mathsf{KA}(\mathcal{K})]]$, which we show next. To be more precise, let us denote by $\Phi'_{\mathcal{K}}(T, \cdot)_2$ the same operator as $\Phi_{\mathcal{K}}(T, \cdot)_2$ except that it is defined on the interval $[T, \mathsf{KA}(\mathcal{K})]]$, i.e., $\Phi_{\mathcal{K}}(T, \cdot)_2$ is a monotone operator on $[T, \mathsf{KA}(\mathcal{K})]]$. Let $P' = lfp(\Phi'_{\mathcal{K}}(T, \cdot)_2)$. Since (T, P) is a consistent stable fixpoint of $\Phi_{\mathcal{K}}$, i.e., $T \subseteq P = lfp(\Phi_{\mathcal{K}}(T, \cdot)_2)$, at some point of iterated construction, T is already included in some intermediate set in the construction of $lfp(\Phi_{\mathcal{K}}(T, \cdot)_2)$. Then, by an induction on the parallel construction of the sequence generated from $lfp(\Phi'_{\mathcal{K}}(T, \cdot)_2)$ starting at T and the sequence from $lfp(\Phi_{\mathcal{K}}(T, \cdot)_2)$ starting at a point where T is already included, we can infer $P' \subseteq P$. For the other direction, it is easy to verify by definition that every intermediate set constructed for $lfp(\Phi'_{\mathcal{K}}(T, \cdot)_2)$ is a subset of P', and hence $P \subseteq P'$. Therefore, $P' = P$, and $\Phi_{\mathcal{K}}$ satisfies the condition of being strong. $\square$

Stable fixpoints of operator $\Phi_{\mathcal{K}}$ can be related to three-valued MKNF models in the following way.

Theorem 2. *Let $\mathcal{K} = (\mathcal{O}, \mathcal{P})$ be a hybrid MKNF knowledge base and (T, P) be a partition. Let further $(M, N) = (\{I \mid I \models \mathsf{OB}_{\mathcal{O},T}\}, \{I \mid I \models \mathsf{OB}_{\mathcal{O},P}\})$. Then, (M, N) is a three-valued MKNF model of $\mathcal{K}$ iff (T, P) is a consistent stable fixpoint of $\Phi_{\mathcal{K}}$ and $\mathsf{OB}_{\mathcal{O}, lfp(\Phi_{\mathcal{K}}(\cdot, T)_1)}$ is satisfiable.*

Note that in the formulation of approximator $\Phi_{\mathcal{K}}$, stable fixpoints are partitions that provide candidate interpretation pairs for three-valued MKNF models. That the extra condition $\mathsf{OB}_{\mathcal{O}, lfp(\Phi_{\mathcal{K}}(\cdot, T)_1)}$ is satisfiable means that, even if we make all non-true **K**-atoms false for the construction of the least fixpoint, it still does not cause contradiction with DL knowledge base $\mathcal{O}$. This provides a key insight in the semantics of hybrid MKNF knowledge bases.

Notice also that this theorem provides a naive method, based on guess-and-verify, to compute three-valued MKNF models of a given hybrid MKNF knowledge base $\mathcal{K}$ - guess a consistent partition (T, P) of $\mathsf{KA}(\mathcal{K})$ and check whether (T, P) is a stable fixpoint of $\Phi_{\mathcal{K}}$ and whether $\mathsf{OB}_{\mathcal{O}, lfp(\Phi_{\mathcal{K}}(\cdot, T)_1)}$ is satisfiable. Observe that the complexity of this checking is polynomial if the underlying DL is polynomial. Indeed, this theorem presents another piece of evidence that the data complexity of determining whether a three-valued MKNF model exists is NP-hard even when the underlying DL is polynomial. The claim was first proved in [10] by relating three-valued MKNF models with a notion called *stable partition*.

Proof (Sketch). ($\Leftarrow$) Let (T, P) be a consistent stable fixpoint of $\Phi_{\mathcal{K}}$ and $\mathsf{OB}_{\mathcal{O}, lfp(\Phi_{\mathcal{K}}(\cdot, T)_1)}$ satisfiable. As a consistent stable fixpoint of $\Phi_{\mathcal{K}}$, we have $T \subseteq P$ and $(T, P) = (lfp(\Phi_{\mathcal{K}}(\cdot, P)_1), lfp(\Phi_{\mathcal{K}}(T, \cdot)_2))$. Since $\mathsf{OB}_{\mathcal{O}, lfp(\Phi_{\mathcal{K}}(\cdot, T)_1)}$ is satisfiable and $lfp(\Phi_{\mathcal{K}}(\cdot, T)_1) \supseteq T$, $\mathsf{OB}_{\mathcal{O},T}$ is satisfiable. Next, by the definition of operator $\Phi_{\mathcal{K}}$, it can be verified that $lfp(\Phi_{\mathcal{K}}(\cdot, T)_1)$ and $lfp(\Phi_{\mathcal{K}}(T, \cdot)_2)$ coincide when $\mathsf{OB}_{\mathcal{O}, lfp(\Phi_{\mathcal{K}}(\cdot, T)_1)}$ is satisfiable, therefore $\mathsf{OB}_{\mathcal{O},P}$ is also satisfiable. It follows that the pair

$$(M, N) = (\{I \mid I \models \mathsf{OB}_{\mathcal{O},T}\}, \{I \mid I \models \mathsf{OB}_{\mathcal{O},P}\})$$

is an MKNF interpretation pair because $\emptyset \subset N \subseteq M$. Next, it is easy to show that there is a correspondence between (T, P) and (M, N): **K**$\xi \in T$ iff **K**ξ evaluates to **t** (w.r.t (M, N)); **K**$\xi \in P$ and **K**$\xi \notin T$ iff **K**ξ evaluates to **u**; otherwise, **K**ξ evaluates to **f**.

Then, we show that (M, N) is a three-valued MKNF model. Since $\mathsf{OB}_{\mathcal{O},T} = \{\pi(\mathcal{O})\} \cup \{\xi \mid \mathbf{K}\xi \in T\}$ and $\mathsf{OB}_{\mathcal{O},P} = \{\pi(\mathcal{O})\} \cup \{\xi \mid \mathbf{K}\xi \in P\}$, it follows $(M, N) \models \pi(\mathcal{O})$. For each $r \in \mathcal{P}$ with the form $\pi(r)$ and $\mathbf{K}\xi = hd(r)$, by definition of $\Phi_{\mathcal{K}}(T, P)_1$, the fixpoint construction confirms that if $body^+(r) \subseteq T$ and $\mathbf{K}(bd^-(r)) \cap P = \emptyset$, then $\mathbf{K}\xi \in T$; if $body^+(r) \subseteq P$, $\mathbf{K}(bd^-(r)) \cap T = \emptyset$ and $\mathsf{OB}_{\mathcal{O},T} \not\models \neg\xi$ then $\mathbf{K}\xi \in P$, therefore $(M, N) \models \pi(r)$. As this is applied to all rules in $\mathcal{P}$, we have $(M, N) \models \pi(\mathcal{K})$.

Next, if (M, N) is not a three-valued model of $\mathcal{K}$, then there exists pair (M', N'), with $M \subseteq M'$ and $N \subseteq N'$ where at least one inclusion is proper such that

$$(I, \langle M', N'\rangle, \langle M, N\rangle)(\pi(\mathcal{O}) \wedge \pi(\mathcal{P})) = \mathbf{t}$$

for any $I \in M'$. By Definition 15 of Knorr et al. [8], there is a partition (T', P') induced by (M', N'), surely $T' \subseteq T$ and $P' \subseteq P$, where at least one inclusion is proper. If we assume $T' \subsetneq T$, there exists $\mathbf{K}a$ such that $\mathbf{K}a \in T$ while $\mathbf{K}a \notin T'$. Then either $\mathsf{OB}_{\mathcal{O},T} \models a$ or $\exists r$ with $hd(r) = \mathbf{K}a$ such that $bd^+(r) \subseteq T, body^-(r) \cap P = \emptyset$ by definition of $\Phi_{\mathcal{K}}(T, P)_1$, in either case, one can check that $(M', N') \not\models \pi(\mathcal{K})$. Similarly, if we assume $P' \subsetneq P$, $(M', N') \not\models \pi(\mathcal{K})$. That leads to a contradiction, and thus (M, N) is a three-valued MKNF model.

($\Rightarrow$) First, as shown by Knorr et al. [8], given an MKNF interpretation (M', N'), there exists a partition (X, Y) induced by (M', N'), in the sense that $(M', N') = (\{I \mid I \models \mathsf{OB}_{\mathcal{O},X}\}, \{I \mid I \models \mathsf{OB}_{\mathcal{O},Y}\})$, such that $\mathbf{K}\xi \in X$ if $\mathbf{K}\xi$ evaluates to $\mathbf{t}$ w.r.t. (M', N'); $\mathbf{K}\xi \in Y$ if $\mathbf{K}\xi$ evaluates to $\mathbf{u}$ w.r.t. (M', N'); otherwise, $\mathbf{K}\xi$ evaluates to $\mathbf{f}$ w.r.t. (M', N').

When (M, N) is a three-valued model, the partition induced by (M, N) is just (T, P) such that $(M, N) = (\{I \mid I \models \mathsf{OB}_{\mathcal{O},T}\}, \{I \mid I \models \mathsf{OB}_{\mathcal{O},P}\})$. Since (M, N) is a three-valued model, (T, P) is consistent. $\mathsf{OB}_{\mathcal{O},lfp(\Phi_{\mathcal{K}}(\cdot,T)_1)}$ should also be satisfiable, as otherwise we derive $M = N = \emptyset$ by definition and thus (M, N) is not even an interpretation pair. We now show that (T, P) is a stable fixpoint of $\Phi_{\mathcal{K}}$, i.e., $T = lfp(\Phi_{\mathcal{K}}(\cdot, P)_1)$ and $P = lfp(\Phi_{\mathcal{K}}(T, \cdot)_2)$. First, we show that (T, P) is a fixpoint of $\Phi_{\mathcal{K}}$. By definition, we have $T \subseteq \Phi_{\mathcal{K}}(T, P)_1$. Now assume $T \subset \Phi_{\mathcal{K}}(T, P)_1$, then there exists a rule r with $hd(r) = \mathbf{K}a$ such that $\mathbf{K}a$ can only be derived from rules and $\mathbf{K}a \notin T$, then $\mathbf{K}a \in P \backslash T$, in this way $(M, N) \not\models \pi(r)$, contradicting to the three-valued model condition and thus $T = \Phi_{\mathcal{K}}(T, P)_1$. Similarly, we can show $P = \Phi_{\mathcal{K}}(T, P)_2$. If (T, P) is not a stable fixpoint, by constructing a pair (T', P) such that $T' = lfp(\Phi_{\mathcal{K}}(\cdot, P)_1)$, with $T' \subset T$, an MKNF interpretation pair (M', N) can be constructed with $M' = \{I \mid I \models \mathsf{OB}_{\mathcal{O},T'}\}$ and $M \subset M'$, it can be checked that $(I, \langle M', N \rangle, \langle M, N \rangle)(\pi(\mathcal{K})) = \mathbf{t}$, for any $I \in M'$, in this way (M, N) is not a three-valued model, a contradiction and therefore $T = lfp(\Phi_{\mathcal{K}}(\cdot, P)_1)$. Similarly, $P = lfp(\Phi_{\mathcal{K}}(T, \cdot)_2)$. Therefore, (T, P) is a stable fixpoint of $\Phi_{\mathcal{K}}$, and a consistent one. □

Example 4. We illustrate that inconsistent stable fixpoints are now possible. E.g., with $\mathcal{K}_1 = (\{\neg a\}, \{a \leftarrow \mathbf{not}\, b.\})$, we have the following sequence of stable revisions reaching the unique stable fixpoint of $\Phi_{\mathcal{K}_1}$:

$$(\emptyset, \{\mathbf{K}a, \mathbf{K}b\}) \Rightarrow (\emptyset, \emptyset) \Rightarrow (\{\mathbf{K}a, \mathbf{K}b\}, \emptyset)$$

As another example, let $\mathcal{K}_2 = (\{d\}, \mathcal{P})$ where $\mathcal{P} = \{\mathbf{K}a \leftarrow \mathbf{K}d, \mathbf{not}\, b., \mathbf{K}b \leftarrow \mathbf{not}\, a.\}$. There are four stable fixpoints, among which $(\{\mathbf{K}d\}, \{\mathbf{K}a, \mathbf{K}b, \mathbf{K}d\})$ is the least and $(\{\mathbf{K}d, \mathbf{K}a, \mathbf{K}b\}, \{\mathbf{K}d\})$ is an inconsistent one. Note that an inconsistent stable fixpoint may contain consistent information - this case in $\mathbf{K}d$.

5 A Richer Approximator for the Well-Founded Semantics

A question arises whether richer approximators for MKNF knowledge bases exist. For any two approximators A and B on L^2, A is *richer than* B (or *more precise than B*, in the terminology of [6]), denoted $B \leq_p A$, if for all $(x, y) \in L^2$, $B(x, y) \leq_p A(x, y)$.

There is a practical motivation for the question. Let (x, y) and (x', y') be the least fixpoints of B and A respectively. That A is richer than B means $(x, y) \leq_p (x', y')$. If A is strictly richer than B, and if (x', y') indeed corresponds to the well-founded MKNF model, then (x, y) cannot possibly correspond to the well-founded MKNF model. In this case, while (x', y') is iteratively computed for A, it can only be discovered for B by guess-and-verify, as indicated in Theorem 2. We now define such a richer approximator.

Definition 5. *Let* $\mathcal{K} = (\mathcal{O}, \mathcal{P})$ *be a hybrid MKNF knowledge base. Define operator* $\Psi_{\mathcal{K}}$ *on*$(2^{\mathsf{KA}(\mathcal{K})})^2$ *as: Given any pair* $(T, P) \in (2^{\mathsf{KA}(\mathcal{K})})^2$, *we define*

$$\Psi_{\mathcal{K}}(T, P) = (\Phi_{\mathcal{K}}(T, P)_1, \Psi_{\mathcal{K}}(T, P)_2)$$

where $\Psi_{\mathcal{K}}(T, P)_2$ *is defined by the following set*

$\{\mathbf{K}a \in \mathsf{KA}(\mathcal{K}) \mid \mathsf{OB}_{\mathcal{O}, P} \models a\} \cup$
$\{hd(r) \mid r \in \mathcal{P} : hd(r) = \mathbf{K}a, \mathsf{OB}_{\mathcal{O}, T} \not\models \neg a, bd^+(r) \subseteq P, \mathbf{K}(bd^-(r)) \cap T = \emptyset,$
$\quad \not\exists r' \in \mathcal{P} : hd(r') = \mathbf{K}b, \mathsf{OB}_{\mathcal{O}, T} \models \neg b, bd^+(r') \setminus \{\mathbf{K}a\} \subseteq T, \mathbf{K}(bd^-(r')) \cap P = \emptyset\}$

Operator $\Psi_{\mathcal{K}}$ differs from $\Phi_{\mathcal{K}}$ in the second component operator, with an extra condition for deriving $hd(r) = \mathbf{K}a$ (the last line in the definition above), which says that if for some rule r', whose objective head is already false and whose body excluding $\mathbf{K}a$ is already true, then, since the rule must be satisfied, $\mathbf{K}a$ must be false and thus should not be derived as possibly true. Notice that this is like embedding *unit propagation* into an approximator.

Proposition 3. $\Psi_{\mathcal{K}}$ *is a strong approximator of* $\mathcal{T}_{\mathcal{K}}$, *and* $\Phi_{\mathcal{K}} \leq_p \Psi_{\mathcal{K}}$.

Example 5. Let $\mathcal{K} = (\{\neg a\}, \mathcal{P})$, where $\mathcal{P}$ is

$$\mathbf{K}a \leftarrow \mathbf{K}b, not\, d. \quad \mathbf{K}b \leftarrow not\, c. \quad \mathbf{K}c \leftarrow not\, b.$$

Consider operator $\Phi_{\mathcal{K}}$ first. The least stable fixpoint of $\Phi_{\mathcal{K}}$ is $(T, P) = (\emptyset, \{\mathbf{K}b, \mathbf{K}c\})$. Since $\mathsf{OB}_{\mathcal{O}, lfp(\Phi_{\mathcal{K}}(\cdot, T)_1)}$ is unsatisfiable, according to Theorem 2, (T, P) does not correspond to a three-valued MKNF model. On the hand, the least stable fixpoint of $\Psi_{\mathcal{K}}$ is the exact partition $(\{\mathbf{K}c\}, \{\mathbf{K}c\})$, which corresponds to the well-founded MKNF model of $\mathcal{K}$, (M, M), where $M = \{\{c\}\}$.

Of course, the same idea can be applied to define an improved alternating fixpoint construction without placing it in the context of AFT. But a major advantage of the latter is that, due to the abstract power of AFT, the enhanced approximator characterizes all three-valued MKNF models, including (two-valued) MKNF models of [12].

Theorem 3. *Let* $\mathcal{K} = (\mathcal{O}, \mathcal{P})$ *be a hybrid MKNF knowledge base and* (T, P) *be a partition. Let further* $(M, N) = (\{I \mid I \models \mathsf{OB}_{\mathcal{O}, T}\}, \{I \mid I \models \mathsf{OB}_{\mathcal{O}, P}\})$. *Then,* (M, N) *is a three-valued MKNF model of* $\mathcal{K}$ *iff* (T, P) *is a consistent stable fixpoint of* $\Psi_{\mathcal{K}}$ *and* $\mathsf{OB}_{\mathcal{O}, lfp(\Psi_{\mathcal{K}}(\cdot, T)_1)}$ *is satisfiable.*

6　Related Work and Remarks

In [8], alternating fixpoint construction was defined to address the computation of the well-founded model, and in [10], the construction was related to a notion called *stable partition*. In this work, we characterize three-valued MKNF models in terms of stable fixpoints of two appropriate approximators; thanks to the power of abstraction of AFT, along the way we are able to improve the construction by Knorr et al.

The only other work that treats inconsistency explicitly is [2], where in case of inconsistency, instead of computing $(lfp(A(\cdot, v)_1), lfp(A(u, \cdot)_2))$ on the respective domains $[\perp, v]$ and $[u, \top]$, we compute $(lfp(A(\cdot, v)_1), A(u, v))$ because $lfp(A(u, \cdot)_2)$ may no longer be defined on $[u, \top]$. By computing $A(u, v)$ for the second component of the resulting pair, we may allow unfounded elements to be included as possibly true when inconsistency arises.

In [3], the authors remarked that inconsistencies may be derived by a set of *active integrity constraints* (AIGs). Since approximators are defined on L^2, inconsistencies arising in the context of AIGs can be captured. For instance, for the set of AICs, $\{\neg a \supset -a, a \supset +a\}$, assume $DB = \emptyset$, when starting with a pair $(\emptyset, \{+a\})$, it is mapped to the pair $(\{+a\}, \emptyset)$ by the approximator defined in [3]. It can be shown that this approximator is strong so that the abnormality that happened in Example 3 does not surface.

The desire to accommodate inconsistencies in AFT has been motivated in [5]. The precision order when applied to inconsistent pairs can be regarded as an order that measures the "degree of inconsistency", or "degree of doubt" [6]. If two inconsistent pairs satisfy $(x, y) \leq_p (x', y')$ where $y < x$ the latter can be viewed as at least as inconsistent as the former. In a more general context, the notion of inconsistency measures and the questions like "where is the inconsistency", "how severer it is", and how to make changes to an inconsistency theory have been the focus in some recent AI literature (see, e.g., [4]). A deeper understanding of inconsistencies in the context of AFT presents an interesting future direction.

References

1. Antić, C., Eiter, T., Fink, M.: HEX semantics via approximation fixpoint theory. In: Cabalar, P., Son, T.C. (eds.) LPNMR 2013. LNCS (LNAI), vol. 8148, pp. 102–115. Springer, Heidelberg (2013). https://doi.org/10.1007/978-3-642-40564-8_11
2. Bi, Y., You, J.-H., Feng, Z.: A generalization of approximation fixpoint theory and application. In: Kontchakov, R., Mugnier, M.-L. (eds.) RR 2014. LNCS, vol. 8741, pp. 45–59. Springer, Cham (2014). https://doi.org/10.1007/978-3-319-11113-1_4
3. Bogaerts, B., Cruz-Filipe, L.: Fixpoint semantics for active integrity constraints. Artif. Intell. **255**, 43–70 (2018)
4. De Bona, G., Hunter, A.: Localising iceberg inconsistencies. Artif. Intell. **246**, 118–151 (2017)
5. Denecker, M., Marek, V., Truszczyński, M.: Approximations, stable operators, well-founded fixpoints and applications in nonmonotonic reasoning. In: Minker, J. (ed.) Logic-based Artificial Intelligence, pp. 127–144. Springer, Boston (2000). https://doi.org/10.1007/978-1-4615-1567-8_6

6. Denecker, M., Marek, V.W., Truszczynski, M.: Ultimate approximation and its application in nonmonotonic knowledge representation systems. Inf. Comput. **192**(1), 84–121 (2004)

7. Denecker, M., Vennekens, J.: Well-founded semantics and the algebraic theory of non-monotone inductive definitions. In: Baral, C., Brewka, G., Schlipf, J. (eds.) LPNMR 2007. LNCS (LNAI), vol. 4483, pp. 84–96. Springer, Heidelberg (2007). https://doi.org/10.1007/978-3-540-72200-7_9

8. Knorr, M., Alferes, J.J., Hitzler, P.: Local closed world reasoning with description logics under the well-founded semantics. Artif. Intell. **175**(9–10), 1528–1554 (2011)

9. Lifschitz, V.: Nonmonotonic databases and epistemic queries. In: Proceedings of the 12th International Joint Conference on Artificial Intelligence, IJCAI 1991, pp. 381–386, Sydney, Australia (1991)

10. Liu, F., You, J.-H.: Three-valued semantics for hybrid MKNF knowledge bases revisited. Artif. Intell. **252**, 123–138 (2017)

11. Motik, B., Rosati, R.: A faithful integration of description logics with logic programming. In: Proceedings of the 19th International Joint Conference on Artificial Intelligence, IJCAI 2007, Hyderabad, India, pp. 477–482 (2007)

12. Motik, B., Rosati, R.: Reconciling description logics and rules. J. ACM **57**(5), 1–62 (2010)

13. Pelov, N., Denecker, M., Bruynooghe, M.: Well-founded and stable semantics of logic programs with aggregates. Theory Pract. Logic Program. **7**(3), 301–353 (2007)

14. Strass, H.: Approximating operators and semantics for abstract dialectical frameworks. Artif. Intell. **205**, 39–70 (2013)

15. Tarski, A.: A lattice-theoretical fixpoint theorem and its applications. Pac. J. Math. **5**(2), 285–309 (1955)

16. van Emden, M.H., Kowalski, R.A.: The semantics of predicate logic as a programming language. J. ACM **23**(4), 733–742 (1976)

17. Vennekens, J., Gilis, D., Denecker, M.: Splitting an operator: algebraic modularity results for logics with fixpoint semantics. ACM Trans. Comput. Logic **7**(4), 765–797 (2006)

Efficient TBox Reasoning with Value Restrictions—Introducing the $\mathcal{FL}_o wer$ Reasoner

Friedrich Michel, Anni-Yasmin Turhan$^{(\boxtimes)}$, and Benjamin Zarrieß

Institute for Theoretical Computer Science, Technische Universität Dresden,
Dresden, Germany
{friedrich.michel,anni-yasmin.turhan,benjamin.zarriess}@tu-dresden.de

Abstract. The Description Logic (DL) $\mathcal{FL}_0$ uses universal quantification, whereas its well-known counter-part $\mathcal{EL}$ uses the existential one. While for $\mathcal{EL}$ deciding subsumption in the presence of general TBoxes is tractable, this is no the case for $\mathcal{FL}_0$. We present a novel algorithm for solving the ExpTime-hard subsumption problem in $\mathcal{FL}_0$ w.r.t. general TBoxes, which is based on the computation of so-called *least functional models*. To build a such a model our algorithm treats TBox axioms as rules that are applied to objects of the interpretation domain. This algorithm is implemented in the $\mathcal{FL}_o wer$ reasoner, which uses a variant of the Rete pattern matching algorithm to find applicable rules. We present an evaluation of $\mathcal{FL}_o wer$ on a large set of TBoxes generated from real world ontologies. The experimental results indicate that our prototype implementation of the specialised technique for $\mathcal{FL}_0$ leads in most cases to a huge performance gain in comparison to the highly-optimised tableau reasoners.

1 Introduction

The Description Logic (DL) $\mathcal{FL}_0$ is a minimalistic DL, since it offers only the top concept, conjunction, and value restriction (universal quantification) as constructors for building complex concepts. It is the core part of one of the very first DLs called $\mathcal{FL}^-$ ($\mathcal{F}$rame $\mathcal{L}$anguage) introduced by Brachman and Levesque [8] for formalising frames. Unfortunately, in presence of a TBox value restrictions and conjunction have been identified as exactly those constructors that make the problem of deciding the subsumption relationship between two concepts hard [1, 15]. In particular, depending on the syntactical form of the TBox the complexity of deciding subsumption in $\mathcal{FL}_0$ takes a rollercoaster ride: it starts from PTIME with the empty TBox [8], jumps to CO-NP-completeness with acyclic TBoxes [15], then to PSPACE-completeness with cyclic definitions [12], culminates in ExpTime-completeness in presence of general TBoxes [1], and drops back to PTIME when restricted to Horn-TBoxes [13]. This is in sharp contrast to the robust behaviour of the popular DL $\mathcal{EL}$ that differs from $\mathcal{FL}_0$ only by using existential restrictions instead of value restrictions. In $\mathcal{EL}$, the complexity stays in PTIME even in the presence of general TBoxes [1].

© Springer Nature Switzerland AG 2019
P. Fodor et al. (Eds.): RuleML+RR 2019, LNCS 11784, pp. 128–143, 2019.
https://doi.org/10.1007/978-3-030-31095-0_9

In this paper we devise a novel algorithm for deciding subsumption w.r.t. general $\mathcal{FL}_0$-TBoxes, describe a first implementation of it in the new $\mathcal{FL}_o\textit{wer}$ reasoner and report on an evaluation of $\mathcal{FL}_o\textit{wer}$ on a large collection of ontologies. There are several reasons to study subsumption algorithms for $\mathcal{FL}_0$. First, for the Boolean-complete DL $\mathcal{ALC}$ subsumption w.r.t. TBoxes is still EXP-TIME-complete, thus its fragment $\mathcal{FL}_0$ hardly offers an optimal trade-off between worst-case complexity and expressiveness. Nevertheless, the volatile complexity of subsumption in $\mathcal{FL}_0$ raises the question of whether hard instances of the subsumption problem are even likely to occur in application ontologies.

Second, most state-of-the-art ontology reasoners are built and optimised for expressive DLs beyond $\mathcal{ALC}$, as for example FaCT++[1], HermiT[2] or Konclude[3]. Some DL reasoners such as Konclude and MORe [17] make use of specialised algorithms for certain language fragments as part of their overall reasoning algorithm. Thus an efficient subsumption algorithm for $\mathcal{FL}_0$ might be such a dedicated procedure that can augment general ontology reasoners.

Third, dedicated methods for standard and non-standard reasoning tasks in $\mathcal{FL}_0$ w.r.t. general TBoxes have been studied recently in [3–6]. Quite some of these useful inferences rely on subsumption tests as sub-procedures and $\mathcal{FL}_o\textit{wer}$ can supply a base for implementing these inferences.

$\mathcal{FL}_o\textit{wer}$'s subsumption algorithm for $\mathcal{FL}_0$ uses a characterisation of subsumption based on so-called tree-shaped *least functional models* [4]. For a given input concept and TBox, this algorithm generates a sufficiently large subtree of their least functional model by using the axioms from the TBox like rules to augment the tree. This process corresponds to deriving implicit consequences. To ensure termination it employs a blocking mechanism. We have implemented this algorithm in the new $\mathcal{FL}_o\textit{wer}$ reasoner.[4]

The key idea of the implementation is to use a variant of the well-known Rete algorithm for rule application [9] adapted to the case without negation: we translate the TBox to a Rete network and generate the tree representing the model of the TBox by propagating the nodes through the network. More precisely, the axioms in the TBox are applied as rules to nodes in the tree. For example, consider the following $\mathcal{FL}_0$ axiom:

$$\text{Animal} \sqcap \forall \text{eats.Plants} \sqsubseteq \text{Herbivore},$$

which essentially says that animals that eat only plants are herbivores. If a node in the current tree matches the left-hand side, which means it is labelled with the name Animal and its eats-child with Plants, then we add Herbivore to its label set. Since there are potentially many nodes and many axioms to consider, it is critical for performance reasons to avoid reiterating over all nodes and axioms after each change and so the key idea from the Rete algorithm is well-suited for our task. To support this claim we have conducted an evaluation on $\mathcal{FL}_o\textit{wer}$.

---

[1] owl.man.ac.uk/factplusplus.

[2] hermit-reasoner.com.

[3] konclude.com.

[4] https://github.com/attalos/fl0wer.

To create a large set of challenging $\mathcal{FL}_0$ ontologies we have transformed the OWL 2 EL ontologies of the OWL reasoner competition [16] into $\mathcal{FL}_0$ by replacing existential restrictions by value restrictions. It turns out that our reasoner $\mathcal{FL}_o wer$ is in many cases able to clearly outperform highly-optimised (hyper)tableaux reasoners like HermiT or JFact on reasoning tasks for this set of ontologies.

2   Preliminaries on $\mathcal{FL}_0$ and Least Functional Models

We define the DL $\mathcal{FL}_0$ and recall the characterisation of subsumption from [4] based on least functional models.

Syntax. Let $\mathsf{N_C}$ and $\mathsf{N_R}$ be disjoint *finite* sets of *concept names* and *role names*, respectively. An $\mathcal{FL}_0$ *-concept description* (*concept* for short) C is built according to the following syntax rule

$$C ::= A \mid \top \mid C \sqcap C \mid \forall r.C, \text{ where } A \in \mathsf{N_C}, r \in \mathsf{N_R}.$$

$\top$ is called the *top concept* and $\forall r.C$ is a *value restriction*. For nested value restrictions we use the following notation: let $\sigma = r_1 \cdots r_m \in \mathsf{N_R}^*$ for some $m \geq 0$ be a word over the alphabet $\mathsf{N_R}$ of role names. For some $A \in \mathsf{N_C}$ we write $\forall \sigma.A$ as an abbreviation of $\forall r_1. \cdots \forall r_m.A$. The empty word ϵ stands for A.

A *general concept inclusion* (GCI) is of the form $C \sqsubseteq D$, where C and D are concepts. A *TBox* is a finite set of GCIs.

Semantics. An interpretation $\mathcal{I}$ is a pair $\mathcal{I} = (\Delta^{\mathcal{I}}, \cdot^{\mathcal{I}})$, consisting of a non-empty set $\Delta^{\mathcal{I}}$ (*domain of $\mathcal{I}$*) and an *interpretation function* $\cdot^{\mathcal{I}}$ that maps every concept name $A \in \mathsf{N_C}$ to a subset of the domain $A^{\mathcal{I}} \subseteq \Delta^{\mathcal{I}}$ and every role name $r \in \mathsf{N_R}$ to a binary relation $r^{\mathcal{I}} \subseteq \Delta^{\mathcal{I}} \times \Delta^{\mathcal{I}}$. The interpretation function is extended to (complex) concepts as follows:

$$(C \sqcap D)^{\mathcal{I}} := C^{\mathcal{I}} \sqcap D^{\mathcal{I}}, \qquad (\top)^{\mathcal{I}} := \Delta^{\mathcal{I}}, \text{ and}$$
$$(\forall r.C)^{\mathcal{I}} := \{d \in \Delta^{\mathcal{I}} \mid \forall e \in \Delta^{\mathcal{I}}.(d, e) \in r^{\mathcal{I}} \longrightarrow e \in C^{\mathcal{I}}\}$$

A GCI $C \sqsubseteq D$ is *satisfied* in $\mathcal{I}$, denoted by $\mathcal{I} \models C \sqsubseteq D$, iff $C^{\mathcal{I}} \subseteq D^{\mathcal{I}}$. $\mathcal{I}$ is a *model* of a TBox $\mathcal{T}$, denoted by $\mathcal{I} \models \mathcal{T}$, iff $\mathcal{I}$ satisfies all GCIs in $\mathcal{T}$. The concept C is *subsumed by the concept D w.r.t. a TBox $\mathcal{T}$*, denoted by $C \sqsubseteq_{\mathcal{T}} D$, iff $C^{\mathcal{I}} \subseteq D^{\mathcal{I}}$ is satisfied in all models of $\mathcal{T}$.

Our focus is on a certain kind of tree-shaped interpretation. A *functional interpretation* is a tree with domain $\mathsf{N_R}^*$, where each element has exactly one child node for each role name.

Definition 1. *An interpretation $\mathcal{I} = (\Delta^{\mathcal{I}}, \cdot^{\mathcal{I}})$ is called a* functional interpretation *iff $\Delta^{\mathcal{I}} = \mathsf{N_R}^*$ and $r^{\mathcal{I}} = \{(\sigma, \sigma r) \mid \sigma \in \mathsf{N_R}^*\}$ for all $r \in \mathsf{N_R}$. A* functional interpretation $\mathcal{I}$ is called functional model *of a concept C w.r.t. TBox $\mathcal{T}$ iff $\mathcal{I} \models \mathcal{T}$ and $\epsilon \in C^{\mathcal{I}}$. For two functional interpretations $\mathcal{I}$ and $\mathcal{J}$ we write*

$$\mathcal{I} \subseteq \mathcal{J} \text{ iff } A^{\mathcal{I}} \subseteq A^{\mathcal{J}} \text{ for all } A \in \mathsf{N_C}.$$

The notion of a functional interpretation fixes the domain and the interpretation of role names. Thus, a functional interpretation is uniquely determined by the interpretation of concept names.

Definition 2. *Let C be a concept and $\mathcal{T}$ be a TBox. The interpretation $\mathcal{I}_{C,\mathcal{T}} = (N_R^*, \cdot^{\mathcal{I}_{C,\mathcal{T}}})$ is the functional interpretation satisfying*

$$A^{\mathcal{I}_{C,\mathcal{T}}} = \{\sigma \in N_R^* \mid C \sqsubseteq_{\mathcal{T}} \forall \sigma.A\} \text{ for all } A \in N_C.$$

We summarize some properties of $\mathcal{I}_{C,\mathcal{T}}$ which characterize $\mathcal{I}_{C,\mathcal{T}}$ as the *least* functional model of C w.r.t. $\mathcal{T}$.

Theorem 1 ([2,4]). *It holds that $\mathcal{I}_{C,\mathcal{T}}$ is a functional model of C w.r.t. $\mathcal{T}$. Furthermore, if $\mathcal{J}$ is a functional model of C w.r.t. $\mathcal{T}$, then $\mathcal{I}_{C,\mathcal{T}} \subseteq \mathcal{J}$.*

Subsumption can now be characterised as inclusion between least functional models. It follows that $C \sqsubseteq_{\mathcal{T}} D$ iff $\mathcal{I}_{C,\mathcal{T}} \subseteq \mathcal{I}_{D,\mathcal{T}}$.

3   Subsumption Algorithm for $\mathcal{FL}_0$ with General TBoxes

We define a decision procedure for subsumption of two concepts w.r.t. a TBox based on a finite representation of the least functional model obtained by "applying" GCIs. The algorithm expects input in a certain normal form. A concept is in *normal form* if it is either $\top$ or a conjunction of concept names and value restrictions of the form $\forall r.A$ with $r \in N_R$ and $A \in N_C$. A GCI $C \sqsubseteq D$ is in normal form if both C and D are in normal form and a TBox $\mathcal{T}$ is in normal form if all GCIs in $\mathcal{T}$ are in normal form. By using the equivalence $\forall r.(C \sqcap D) \equiv \forall r.C \sqcap \forall r.D$ and via "flattening" of nested value restrictions by introducing fresh concept names every TBox can be transformed in linear time into normal form.

In the remainder of this section $\mathcal{T}$ denotes a TBox in normal form. Given two concept names A and B mentioned in $\mathcal{T}$ we want to decide whether $A \sqsubseteq_{\mathcal{T}} B$ holds. The algorithm computes a finite subtree of the tree $\mathcal{I}_{A,\mathcal{T}}$ such that one can read off the named subsumers (concept names) of A w.r.t. $\mathcal{T}$ at the root. The structure that the algorithm operates on is a *partial functional interpretation*. It is the same as a functional interpretation but the domain is only a *finite* prefix-closed subset of N_R^*, that is, a finite tree.

Definition 3. *The interpretation $\mathcal{Y} = (\Delta^{\mathcal{Y}}, \cdot^{\mathcal{Y}})$ is a partial functional interpretation iff $\Delta^{\mathcal{Y}} \subset N_R^*$ is a finite prefix-closed set and for all $r \in N_R$ it holds that $r^{\mathcal{Y}} = \{(\sigma, \sigma r) \mid (\sigma, \sigma r) \in \Delta^{\mathcal{Y}} \times \Delta^{\mathcal{Y}}\}$.*

The algorithm for deciding $A \sqsubseteq_{\mathcal{T}} B$ generally proceeds as follows: it starts with a partial functional interpretation $\mathcal{Y}$ that just consists of

$$\Delta^{\mathcal{Y}} := \{\epsilon\} \qquad A^{\mathcal{Y}} := \{\epsilon\} \qquad B^{\mathcal{Y}} := \emptyset \text{ for all } B \in N_C \setminus \{A\}.$$

In each iteration a single element of the current tree and a single GCI from $\mathcal{T}$ is chosen such that the chosen element matches the left-hand side, but not the right-hand side of the GCI. Then minimal extensions of the tree are performed such that the element now matches also the right-hand side. The extension affects the domain and/or the interpretation of concept names. To ensure soundness we guarantee that for the tree $\mathcal{Y}$, the invariant $\mathcal{Y} \subseteq \mathcal{I}_{A,\mathcal{T}}$ is always satisfied. To ensure termination we distinguish blocked and non-blocked elements in the tree. The algorithm terminates if for all non-blocked elements and all GCIs, a match of the left-hand side implies a match of the right-hand side. For describing the procedure we have to define 1. what it means that a domain element of a partial interpretation matches a concept, 2. how to extend the tree to achieve a match with the right-hand side, and 3. how to distinguish blocked and non-blocked elements. For the first step we introduce the following auxiliary notions.

Definition 4. *Let* $\mathcal{Y} = (\Delta^{\mathcal{Y}}, \cdot^{\mathcal{Y}})$ *be a partial functional interpretation and* D *a concept in normal form. The set of elements in* $\Delta^{\mathcal{Y}}$ *that match* D, *denoted by* $\mathsf{match}(D, \mathcal{Y})$, *is defined inductively as follows:*

$$\mathsf{match}(\top, \mathcal{Y}) := \Delta^{\mathcal{Y}};$$
$$\mathsf{match}(A, \mathcal{Y}) := \{w \in \Delta^{\mathcal{Y}} \mid w \in A^{\mathcal{Y}}\} \text{ for all } A \in \mathsf{N_C};$$
$$\mathsf{match}(\forall r.A, \mathcal{Y}) := \{\sigma \in \Delta^{\mathcal{Y}} \mid \text{ there exists } \sigma r \in \Delta^{\mathcal{Y}} \text{ and } \sigma r \in A^{\mathcal{Y}}\}$$
$$\text{for all } r \in \mathsf{N_R} \text{ and } A \in \mathsf{N_C};$$
$$\mathsf{match}(C_1 \sqcap C_2, \mathcal{Y}) := \mathsf{match}(C_1, \mathcal{Y}) \cap \mathsf{match}(C_2, \mathcal{Y}).$$

Since $\mathcal{Y}$ is partial functional (i.e. has *at most* one child per node for each role name), it is easy to see that $\sigma \in \mathsf{match}(C, \mathcal{Y})$ implies $\sigma \in C^{\mathcal{Y}}$. The converse need not be true, as σ may have no r-child in $\Delta^{\mathcal{Y}}$. We define $\sigma \in \Delta^{\mathcal{Y}}$ *violates* $C \sqsubseteq D$ (in normal form) iff $\sigma \in \mathsf{match}(C, \mathcal{Y})$ and $\sigma \notin \mathsf{match}(D, \mathcal{Y})$. Given a TBox $\mathcal{T}$ in normal form and a partial functional interpretation $\mathcal{Y}$ we define the *set of all incomplete elements* as follows

$$\mathsf{ic}(\mathcal{Y}, \mathcal{T}) := \{\sigma \in \Delta^{\mathcal{Y}} \mid \text{ there is } E \sqsubseteq F \in \mathcal{T} \text{ such that } \sigma \text{ violates } E \sqsubseteq F\}.$$

Intuitively, the elements in $\mathsf{ic}(\mathcal{Y}, \mathcal{T})$ are those eligible for an extension of $\mathcal{Y}$ towards building a representation of the least functional model, while those in $\Delta^{\mathcal{Y}} \setminus \mathsf{ic}(\mathcal{Y}, \mathcal{T})$ are not. As an additional filter for extensions we define a blocking condition. First, we introduce the auxiliary notions for the blocking mechanism consisting of the standard notion of a prefix and proper prefix and a strict total order on $(\mathsf{N_R})^*$.

Definition 5. *Let* $\sigma, \rho \in \mathsf{N_R}^*$. *We write* $\rho \in \mathsf{prefix}(\sigma)$ *iff* $\sigma = \rho\hat{\sigma}$ *for some* $\hat{\sigma} \in \mathsf{N_R}^*$. *We write* $\rho \in \mathsf{pprefix}(\sigma)$ *iff* $\rho \in \mathsf{prefix}(\sigma)$ *and* $\rho \neq \sigma$ *to denote a proper prefix of* σ. *Let* $\mathsf{N_R} = \{r_1, \ldots, r_n\}$, $\sigma = r_{i_1} \cdots r_{i_k} \in \mathsf{N_R}^*$ *and* $\rho = r_{j_1} \cdots r_{j_\ell} \in \mathsf{N_R}^*$ *be two words with* $i_1, \ldots, i_k, j_1, \ldots, j_\ell \in \{1, \ldots, n\}$. *Let* $\prec_{\mathbb{N}}$ *denote the lexicographic order over tuples of natural numbers. We define*

$$\sigma \prec \rho \text{ iff } |\sigma| < |\rho| \text{ or if } k = \ell, \text{ then } (i_1, \ldots, i_k) \prec_{\mathbb{N}} (j_1, \ldots, j_\ell).$$

The blocking condition is defined by induction on $\prec$.

Definition 6. *Let* $\mathcal{Y} = (\Delta^{\mathcal{Y}}, \cdot^{\mathcal{Y}})$ *be a partial functional interpretation and let* $\sigma \in \Delta^{\mathcal{Y}}$. *By* $\mathcal{Y}(\sigma) := \{A \in \mathsf{N_C} \mid \sigma \in A^{\mathcal{Y}}\}$ *we denote the* label *of* σ *in* $\mathcal{Y}$. *The set of all blocked elements in* $\Delta^{\mathcal{Y}}$ *is defined as the smallest set satisfying all of the following conditions:*

1. ϵ *is not blocked.*
2. *Let* $\sigma \in \Delta^{\mathcal{Y}}$ *and assume for all* $\sigma' \in \Delta^{\mathcal{Y}}$ *with* $\sigma' \prec \sigma$ *the blocking status is already defined. The following conditions apply to* σ:
   (a) *If there exists* $\omega \in \Delta^{\mathcal{Y}}$ *with* $\omega \prec \sigma$ *such that* $\mathcal{Y}(\sigma) = \mathcal{Y}(\omega)$ *and* ω *is not blocked, then* σ *is blocked.*
   (b) *If there exists* $\rho \in \mathsf{pprefix}(\sigma)$ *such that* ρ *is blocked, then* σ *is blocked.*

The set of all non-blocked *elements in* $\mathcal{Y}$ *is denoted by* $\mathsf{nb}(\mathcal{Y})$.

Next, we define what an extension step is. Such a step expands a single non-blocked and incomplete element in a partial functional interpretation.

Definition 7. *Let* $\mathcal{Y}$ *and* $\mathcal{Z}$ *be two partial functional interpretations,* $\mathcal{T}$ *a TBox in normal form,* $m, n \geq 0$

$$\alpha = C \sqsubseteq (A_1 \sqcap \cdots \sqcap A_m \sqcap \forall r_1.B_1 \sqcap \cdots \forall r_n.B_n) \text{ and } \alpha \in \mathcal{T}$$

a GCI and $\sigma \in \mathsf{nb}(\mathcal{Y}) \cap \mathsf{ic}(\mathcal{Y}, \mathcal{T})$ *a non-blocked, incomplete element in* $\mathcal{Y}$ *violating* α. *Then* $\mathcal{Y}$ *expands* α *at* σ *to* $\mathcal{Z}$, *denoted* $\mathcal{Y} \vdash_{\alpha}^{\sigma} \mathcal{Z}$ *iff* $\mathcal{Z}$ *satisfies the conditions*

- $\Delta^{\mathcal{Z}} = \Delta^{\mathcal{Y}} \cup \{\sigma r_1, \ldots, \sigma r_n\}$;
- $A_i^{\mathcal{Z}} = A_i^{\mathcal{Y}} \cup \{\sigma\}$ *for all* $i = 1, \ldots, m$;
- $B_j^{\mathcal{Z}} = B_j^{\mathcal{Y}} \cup \{\sigma r_j\}$ *for all* $j = 1, \ldots, n$; *and*
- $Q^{\mathcal{Z}} = Q^{\mathcal{Y}}$ *for all* $Q \in \mathsf{N_C} \setminus (\{A_1, \ldots A_m, B_1, \ldots, B_n\})$.

A partial functional interpretation $\mathcal{Z}'$ *is a* $\mathcal{T}$-completion *of* $\mathcal{Y}$, *written as* $\mathcal{Y} \vdash_{\mathcal{T}} \mathcal{Z}'$, *iff there exists* $\alpha' \in \mathcal{T}$ *and* $\sigma' \in \mathsf{nb}(\mathcal{Y}) \cap \mathsf{ic}(\mathcal{Y}, \mathcal{T})$ *such that* $\mathcal{Y} \vdash_{\alpha'}^{\sigma'} \mathcal{Z}'$ *holds.*

For a given $\sigma \in \mathsf{nb}(\mathcal{Y}) \cap \mathsf{ic}(\mathcal{Y}, \mathcal{T})$ violating a GCI $\alpha \in \mathcal{T}$ there exists a unique $\mathcal{Z}$ with $\mathcal{Y} \vdash_{\alpha}^{\sigma} \mathcal{Z}$. The extension of $\mathcal{Y}$ leading to $\mathcal{Z}$ is minimal such that now σ matches the right-hand side of α in $\mathcal{Z}$. Depending on the choice of σ and the GCI there can be several $\mathcal{T}$-completions of $\mathcal{Y}$. Furthermore, it is guaranteed that either $\mathsf{nb}(\mathcal{Y}) \cap \mathsf{ic}(\mathcal{Y}, \mathcal{T}) = \emptyset$ or there exists a $\mathcal{T}$-completion of $\mathcal{Y}$.

Given the input $A, B \in \mathsf{N_C}$ and $\mathcal{T}$, the *algorithm* $\mathrm{SUBS}(A, B, \mathcal{T})$ for checking whether $A \sqsubseteq_{\mathcal{T}} B$ holds, computes a sequence of $\mathcal{T}$-completions until it reaches a partial functional interpretation where no non-blocked element violates any GCI. The algorithm starts with the following partial functional interpretation:

$$\Delta^{\mathcal{Y}_0} := \{\epsilon\}; \quad A^{\mathcal{Y}_0} := \{\epsilon\} \quad \text{and} \quad B^{\mathcal{Y}_0} := \emptyset \text{ for all } B \in \mathsf{N_C} \setminus \{A\}, \quad (1)$$

and computes a sequence

$$\mathcal{Y}_0 \vdash_{\mathcal{T}} \mathcal{Y}_1 \vdash_{\mathcal{T}} \cdots \mathcal{Y}_{(n-1)} \vdash_{\mathcal{T}} \mathcal{Y}_n$$

such that $\mathcal{Y}_n$ is complete in the sense that only blocked elements can remain incomplete, i.e. $\mathsf{nb}(\mathcal{Y}_n) \cap \mathsf{ic}(\mathcal{Y}_n, \mathcal{T}) = \emptyset$. It answers "yes" if $B \in \mathcal{Y}_n(\epsilon)$ (or equivalently $\epsilon \in B^{\mathcal{Y}_n}$) and "no" otherwise.

The choice of the next $\mathcal{T}$-completion is a *don't care nondeterministic choice*. For proving *soundness* it is sufficient to show that $\mathcal{Y}_i \subseteq \mathcal{I}_{A,\mathcal{T}}$ holds for each partial functional interpretation reachable from $\mathcal{Y}_0$ via $\vdash_{\mathcal{T}}$. By $\vdash_{\mathcal{T}}^{*}$ we denote the reflexive transitive closure of $\vdash_{\mathcal{T}}$.

Lemma 1. *Let $\mathcal{Y}_0$ be as in (1), $\mathcal{I}_{A,\mathcal{T}}$ the least functional model of A w.r.t. $\mathcal{T}$ and $\mathcal{Z}$ a partial functional interpretation satisfying $\mathcal{Y}_0 \vdash_{\mathcal{T}}^{*} \mathcal{Z}$. Then $\mathcal{Z} \subseteq \mathcal{I}_{A,\mathcal{T}}$.*

Proof (sketch). The proof is straightforward by induction on the length of the completion sequence.

As a consequence of this we get soundness of the overall procedure.

Lemma 2. *$Subs(A, B, \mathcal{T})$ is sound.*

Proof. If the answer is "yes", then $Subs(A, B, \mathcal{T})$ has obtained $\mathcal{Y}_n$ with $\mathcal{Y}_0 \vdash_{\mathcal{T}}^{*} \mathcal{Y}_n$ and $\epsilon \in B^{\mathcal{Y}_n}$. Since $\mathcal{Y}_n \subseteq \mathcal{I}_{A,\mathcal{T}}$, it follows that $\epsilon \in B^{\mathcal{I}_{A,\mathcal{T}}}$. From the definition of $\mathcal{I}_{A,\mathcal{T}}$ it follows that $A \sqsubseteq_{\mathcal{T}} B$ holds.

We prove that the algorithm terminates, which means that it always reaches a partial functional interpretation $\mathcal{Z}$, where the set $\mathsf{nb}(\mathcal{Z}, \mathcal{T}) \cap \mathsf{ic}(\mathcal{Z}, \mathcal{T})$ is empty. Using the blocking condition, the following lemma about the length of the elements in the tree follows immediately. The *length of an element* $\sigma \in \mathsf{N_R}^{*}$ is denoted by $|\sigma|$. We have $|\epsilon| = 0$ and $|\sigma' r| = |\sigma'| + 1$ with $r \in \mathsf{N_R}$.

Lemma 3. *Let $\mathcal{Z}$ be a partial functional interpretation such that $\mathcal{Y}_0 \vdash_{\mathcal{T}}^{*} \mathcal{Z}$. Then the set $\mathsf{nb}(\mathcal{Z})$ is prefix-closed and $\sigma \in \mathsf{nb}(\mathcal{Z})$ implies that $|\sigma| \leq 2^{|(\mathsf{N_C})|}$.*

Lemma 3 yields an upper bound on the depth of the trees that are the result of a sequence of $\mathcal{T}$-completions starting in the initial $\mathcal{Y}_0$ consisting only of ϵ labelled with A. The *depth* of a partial functional interpretation $\mathcal{Y} = (\Delta^{\mathcal{Y}}, \cdot^{\mathcal{Y}})$, denoted by $\mathsf{depth}(\mathcal{Y})$, is the maximum length defined by

$$\mathsf{depth}(\mathcal{Y}) := \max(\{|\sigma| \mid \sigma \in \Delta^{\mathcal{Y}}\}).$$

Lemma 4. *Let $\mathcal{Z}$ be a partial functional interpretation such that $\mathcal{Y}_0 \vdash_{\mathcal{T}}^{*} \mathcal{Z}$. It holds that*
$$\mathsf{depth}(\mathcal{Z}) \leq 2^{|(\mathsf{N_C})|} + 1.$$

The upper bound on the depth of the tree in a $\mathcal{T}$-completion sequence also yields an upper bound on its overall size. Furthermore, we observe that $\mathcal{Y} \vdash_{\mathcal{T}} \mathcal{Y}'$ implies that $\mathcal{Y} \subsetneq \mathcal{Y}'$, i.e. a $\mathcal{T}$-completion always adds something. At the same time, each label set can at most contain $|\mathsf{N_C}|$ many names. Thus, due to the depth bound and the upper bound on the label size there cannot be an infinite sequence of $\mathcal{T}$-completions. Hence, $Subs(A, B, \mathcal{T})$ always terminates.

Lemma 5. $Subs(A, B, \mathcal{T})$ *always terminates.*

After $Subs(A, B, \mathcal{T})$ terminates, it obtains a partial functional interpretation $\mathcal{Z}$ with $\mathsf{nb}(\mathcal{Z}) \cap \mathsf{ic}(\mathcal{Z}, \mathcal{T}) = \emptyset$. For the completeness proof we construct a model of $\mathcal{T}$ and A from $\mathcal{Z}$. In particular, the construction is based on the non-blocked elements in $\mathsf{nb}(\mathcal{Z})$. It might be the case that a non-blocked element has children (role successors) that are blocked. To handle this situation the following lemma is helpful. Recall that for an element $\sigma \in \Delta^{\mathcal{Z}}$ the *label set* $\mathcal{Z}(\sigma)$ is the set of all concept names that σ satisfies in $\mathcal{Z}$ (see Definition 6).

Lemma 6. *Let $\mathcal{Z}$ be a partial functional interpretation such that $\mathcal{Y}_0 \vdash_{\mathcal{T}}^* \mathcal{Z}$ and $\mathsf{nb}(\mathcal{Z}) \cap \mathsf{ic}(\mathcal{Z}, \mathcal{T}) = \emptyset$. For all $\sigma \in \mathsf{nb}(\mathcal{Z})$ and all $r \in \mathsf{N_R}$ it holds that*

$$\text{if } \sigma r \in \Delta^{\mathcal{Z}} \text{ then there exists } \rho \in \mathsf{nb}(\mathcal{Z}) \text{ such that } \mathcal{Z}(\sigma r) = \mathcal{Z}(\rho).$$

The construction of a model of $\mathcal{T}$ and A from the final partial functional interpretation $\mathcal{Z}$ obtained from the run of the algorithm is based on the label sets of the non-blocked elements. The lemma above guarantees that the label sets of blocked children of non-blocked elements can be found in $\mathsf{nb}(\mathcal{Z})$.

Definition 8. *Let $\mathcal{Z}$ be a partial functional interpretation such that $\mathcal{Y}_0 \vdash_{\mathcal{T}}^* \mathcal{Z}$ and $\mathsf{nb}(\mathcal{Z}) \cap \mathsf{ic}(\mathcal{Z}, \mathcal{T}) = \emptyset$. We define an interpretation $\mathsf{m}(\mathcal{Z})$ as follows:*

- $\Delta^{\mathsf{m}(\mathcal{Z})} := \{\mathcal{Z}(\sigma) \mid \sigma \in \mathsf{nb}(\mathcal{Z})\};$
- $Q^{\mathsf{m}(\mathcal{Y})} := \{X \in \Delta^{\mathsf{m}(\mathcal{Y})} \mid Q \in X\}$ *for all $Q \in \mathsf{N_C}$;*
- $r^{\mathsf{m}(\mathcal{Y})} := \{(\mathcal{Z}(\sigma), \mathcal{Z}(\sigma r)) \mid \sigma \in \mathsf{nb}(\mathcal{Z}), \sigma r \in \Delta^{\mathcal{Z}}\}$ *for all $r \in \mathsf{N_R}$.*

Lemma 6 ensures that the interpretation of role names is well-defined in $\mathsf{m}(\mathcal{Z})$. It easy to show that $\mathsf{m}(\mathcal{Z})$ is a model of $\mathcal{T}$ and A. As a consequence we get completeness of $Subs(A, B, \mathcal{T})$.

Theorem 2. $Subs(A, B, \mathcal{T})$ *is sound, complete and terminating.*

The algorithm $Subs(A, B, \mathcal{T})$ shares properties with the completion method for $\mathcal{EL}$ [1] as well as with tableau algorithms for expressive DLs [7]. Every single $\mathcal{T}$-completion step extends the label set of at least one node in the tree. Intuitively, adding the concept name C to the label set of domain element σ corresponds to deriving $A \sqsubseteq \forall \sigma.C$ as a consequence of $\mathcal{T}$. One single run of $Subs(A, B, \mathcal{T})$ not only decides whether $A \sqsubseteq B$ is entailed by $\mathcal{T}$ but computes *all* subsumers of A. This is similar to the $\mathcal{EL}$ completion method and other consequence-based calculi [18]. From tableau algorithms $Subs(A, B, \mathcal{T})$ inherits the blocking mechanism that ensures termination.

4   A Rete-Based Matching Algorithm for $\mathcal{FL}_0$-Concepts

Our implementation of $Subs(A, B, \mathcal{T})$ in $\mathcal{FL}_0wer$ employs a variant of the Rete algorithm [9] to obtain candidates for each $\mathcal{T}$-completion step. While the full

Rete algorithm admits also negation, we only need the positive part. In this section, we informally describe how this is realised—for full details see [14].

In order to compute a sequence of $\mathcal{T}$-completions $\mathcal{Y}_0 \vdash_\mathcal{T} \mathcal{Y}_1 \vdash_\mathcal{T} \mathcal{Y}_2 \vdash_\mathcal{T} \cdots$ starting from the tree $\mathcal{Y}_0$, one has to compute in each completion step i the pairs $(\sigma, C \sqsubseteq D) \in (\Delta^{\mathcal{Y}_i} \cap \mathsf{nb}(\mathcal{Y}_i)) \times \mathcal{T}$ such that σ matches C but not D in $\mathcal{Y}_i$. Thus, we can view a GCI $C \sqsubseteq D \in \mathcal{T}$ as a rule of the form

$$?\sigma \in \mathsf{match}(C, \mathcal{Y}_i) \rightarrow\ ?\sigma \in \mathsf{match}(D, \mathcal{Y}_i),$$

where $?\sigma$ ranges over the non-blocked domain elements of $\mathcal{Y}_i$. Since there is potentially a large number of elements in $\Delta^{\mathcal{Y}_i}$ that has to be matched with a large number of left-hand sides of GCIs (patterns) in the TBox in each step, we have chosen to implement this task using the Rete matching algorithm [9]. In each completion step the extension of the tree only affects a small number of elements: the matching element σ itself and/or its children. This makes the Rete-based algorithm particularly efficient in our setting because it stores matching information across completion steps to avoid reiterating over the whole set of pairs $(\Delta^{\mathcal{Y}_i} \cap \mathsf{nb}(\mathcal{Y}_i)) \times \mathcal{T}$ in each step. Only the elements with changes have to be rematched again in each step.

As a preprocessing phase $\mathcal{FL}_o wer$ compiles the TBox $\mathcal{T}$ in normal form into a *Rete network*. In general, the network consists of three kinds of nodes: a single root node, intermediate nodes and terminal nodes. A *terminal node* holds the right-hand side of a GCI that is ready to be applied to an element. The matching is done by passing so called tokens from the root node through the intermediate nodes to the terminal nodes. A *token* is a pair of the form $(\sigma, s) \in (\mathsf{N_R}^*, \mathsf{N_R} \cup \{\epsilon\})$. Intuitively, the token (σ, ϵ) is used to check whether σ matches the concept names at the top-level of a concept and a token of the form (σ, r) with $r \in \mathsf{N_R}$ is used to check whether σ matches value restrictions with role name r.

There are three types of *intermediate nodes* that process tokens arriving from predecessor nodes in the network:

- A *concept node* is labelled with a concept name $B \in \mathsf{N_C}$. It sends a token (σ, s) to all successor nodes iff $\sigma s \in B^{\mathcal{Y}_i}$.
- A *role node* is labelled with an $s \in \mathsf{N_R} \cup \{\epsilon\}$. An arriving token of the form (σ, s') is handled as follows. If $s \in \mathsf{N_R}$, then it sends (σ, s') to all successor nodes iff $s' = s$ and if $s = \epsilon$, then it sends the token $(\sigma s', \epsilon)$ to all successor nodes.
- An *inter-element node* is labelled with a tuple $(s_1, \ldots, s_m) \in (\mathsf{N_R} \cup \{\epsilon\})^m$. It stores all arriving tokens and sends a token (σ, ϵ) to its successor nodes once *all* tokens of the form $(\sigma, s_1), \ldots, (\sigma, s_m)$ have arrived at this node.

The overall network is structured in layers. The root node with no incoming edges is on top. To represent GCIs of the form $\top \sqsubseteq C$ the root node is connected to a terminal node with C and sends all tokens directly to this terminal node. All other successors of the root node are concept nodes. The root node takes an element of the form $\sigma = \rho r \in \mathsf{N_R}^*$ and sends the token (ρ, r) to all successor nodes. A successor of a concept node can only be another concept node or a role node. A role node leads directly to an inter-element node and inter-element nodes lead to terminal nodes.

Example 1. As an example consider a TBox that contains the following GCIs:

$$A_2 \sqcap A_4 \sqcap A_5 \sqcap \forall r_1.A_3 \sqcap \forall r_1.A_4 \sqcap \forall r_2.A_1 \sqsubseteq B_7,$$
$$\forall r_2.A_3 \sqcap \forall r_2.A_4 \sqsubseteq B_8,$$
$$\forall r_1.A_6 \sqsubseteq \forall r_1.B_9.$$

The corresponding Rete network is displayed in Fig. 1.

5 Implementation and Evaluation of $\mathcal{FL}_o wer$

The $\mathcal{FL}_o wer$ reasoner is implemented in Java and it takes as input a general $\mathcal{FL}_0$-TBox T in OWL format [10]. It implements three different reasoning tasks.

Subsumption: Given two OWL classes A and B decide whether $A \sqsubseteq_T B$ holds.

Subsumer set: Given an OWL class A compute all classes B in T for which $A \sqsubseteq_T B$ holds.

Classification: Decide for all pairs of named OWL classes A and B occurring in T whether $A \sqsubseteq_T B$ holds.

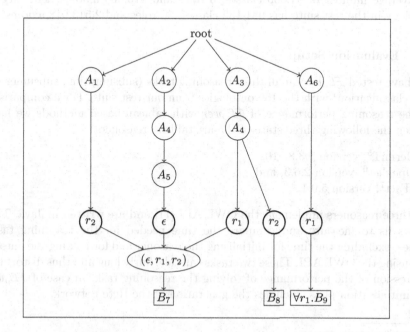

Fig. 1. Rete network for the TBox from Example 1.

To decide subsumption $\mathcal{FL}_o wer$ runs $Subs(A, B, \mathcal{T})$ but possibly stops already in case B occurs at the root of the tree. For computing the subsumer set of A a single complete run of $Subs(A, B, \mathcal{T})$ is sufficient, where the choice of B is irrelevant. All subsumers of A can be found at the root of the final tree. Classification is done by running $Subs(A, B, \mathcal{T})$ for each named class A in $\mathcal{T}$ separately (again B is irrelevant). The Rete network for $\mathcal{T}$ is created only once and is reused for the remaining runs of $Subs(A, *, \mathcal{T})$ multiple times in parallel. To this end the information that is stored on inter-element nodes and final nodes is stored in an external working memory.

5.1   Test Data

Since there are not enough $\mathcal{FL}_0$-ontologies around to yield a decently large test suite, we had to generate such ontologies. Our goal was to create ontologies, that are close to ontologies from applications rather than to construct purposefully complex ones. To create ontologies with a similar structure to application ontologies, we have chosen to use $\mathcal{FL}_0$-variants of the OWL 2 EL [10] ontologies of the OWL Reasoner Evaluation (ORE) competition benchmark [16]. For each ontology from that benchmark, we essentially flipped the quantifier, i.e., replaced $\exists$ by $\forall$. We have also dropped some axioms involving role inclusions and nominals that cannot be expressed in $\mathcal{FL}_0$. Additionally, ontologies containing fewer than 500 classes were discarded. This resulted in a test suite of 159 ontologies with an average number of 54.000 classes, 5 roles and 170.000 axioms. The largest ontology in the test suite has 981.152 classes, 50 roles and 2.513.918 axioms.

5.2   Evaluation Setup

We have tested $\mathcal{FL}_o wer$ on all three reasoning tasks (subsumption, subsumer set and classification) with the 159 ontologies from our test suite. For a comparison of the reasoning performance of $\mathcal{FL}_o wer$ with tableau-based methods we have chosen the following three state-of-the-art tableau reasoners

- HermiT[5], version 1.3.8.510,
- Openllet[6], version 2.6.3, and
- JFact[7], version 5.0.1.

All three reasoners implement the OWL API [11] and are written in Java. This allows us to measure and compare the time needed for the reasoning tasks alone—excluding the time for initialising the reasoner and for loading the ontologies using the OWL API. These two tasks can take fairly long and thus distort the impression of the performance of solving the reasoning task. In case of $\mathcal{FL}_o wer$ the initialisation phase includes the generation of the Rete network.

---

[5] hermit-reasoner.com.
[6] github.com/Galigator/openllet.
[7] jfact.sourceforge.net.

As a test system we have used an Intel Core i5 7600K with 4x 3.80 GHz and 16 GB of RAM. We have used a setup, where only the runtime of the reasoning task is measured and not the time of loading the ontology. Java was called with *-Xmx8g* to set the maximum allocation pool (heap) size to 8 GB. While this was sufficient for JFact, HermiT and $\mathcal{FL}_o wer$, Openllet ran into some *OutOfMemory* exceptions. Those runs were counted as unsuccessful for Openllet. For each reasoning task, the time out was set to 6 min.

5.3   Evaluation Results

We have obtained the following results for the three reasoning tasks.

Subsumption. Of the 159 ontologies with one subsumption call each for a randomly selected pair of classes, JFact was not able to decide the subsumption in 21 cases, Openllet in 8 cases, HermiT in 3 cases, while $\mathcal{FL}_o wer$ succeeded for all 159 ontologies. In Fig. 2 the runtime for the subsumption task is displayed in relation to the number of classes in the ontology. Note the use of a logarithmic scale in this and the following figures. It shows that $\mathcal{FL}_o wer$ performs best overall and is close to and often better than HermiT in terms of reasoning times.

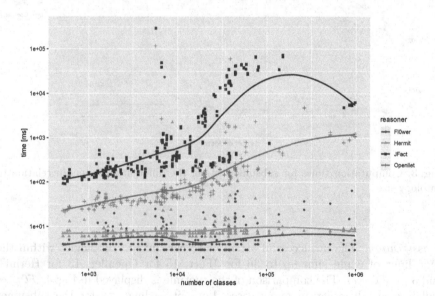

Fig. 2. Computation times for subsumption tests of the different reasoners in relation to ontology size.

Subsumer Set. The task was to compute the subsumer sets of a randomly selected class from each test ontology. This task was not solved within the given timeout of six minutes by JFact in 37 cases, Openllet in 15 cases, HermiT in 13 cases whereas $\mathcal{FL}_{o}wer$ computed all again. The comparison of the runtimes is displayed in Fig. 3. It shows that $\mathcal{FL}_{o}wer$ exhibit an almost linear behaviour and outperforms the other DL reasoners by at least one order of magnitude. Now, $\mathcal{FL}_{o}wer$ is especially efficient for this task as it computes the subsumer set of a given class within a single run, whereas the other reasoners need to perform a classification of the whole ontology (compare Figs. 3 and 4).

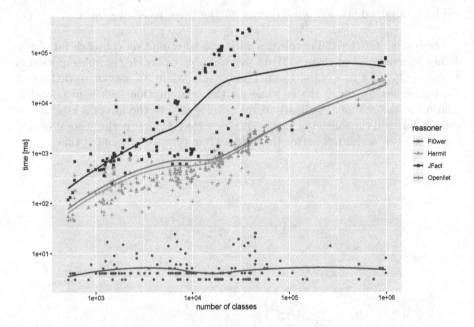

Fig. 3. Computation times for subsumer sets of the different reasoners in relation to ontology size.

Classification. The unsuccessful attempts to classify test ontologies within the time limit of 6 min sum up to 36 for JFact, 15 for Openllet, 13 for HermiT and 3 for $\mathcal{FL}_{o}wer$. The comparison of the runtime is displayed in Fig. 4. $\mathcal{FL}_{o}wer$ still has an advantage in many cases, but not as huge, as for the subsumer set computation. This seems bit surprising, since the classification method in $\mathcal{FL}_{o}wer$ is based on its fast subsumer set computation. We assume that this is mostly due to the fact that $\mathcal{FL}_{o}wer$ implements a naive classification algorithm that simply computes the subsumer set for each named class.

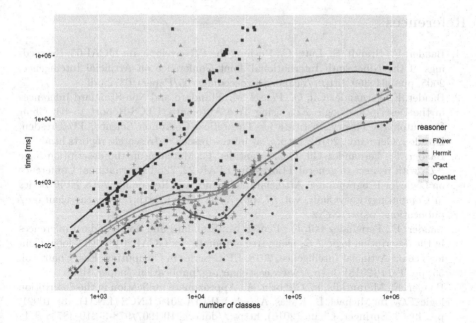

Fig. 4. Computation times for classification of the different reasoners in relation to ontology size.

6   Conclusions

We have presented a novel algorithm for deciding subsumption in the DL $\mathcal{FL}_0$ w.r.t. general TBoxes. The approach is to compute a (tree prefix of the) least functional model for the TBox and the potential subsumee. This approach is the basis for developing implementation friendly algorithms for other inferences in $\mathcal{FL}_0$. Furthermore, reasoners for expressive DLs often incorporate reasoners for special fragments and thus a dedicated reasoner for $\mathcal{FL}_0$ may be beneficial for them. Therefore our investigation presented here can contribute to develop a variety of DL reasoning systems.

Our new reasoner $\mathcal{FL}_o wer$ implements the algorithm for computing all subsumers using a variant of the Rete pattern matching algorithm. Our experimental results with our prototype showed that the specialised techniques lead in many cases to a huge performance gain in comparison to highly-optimised tableau reasoners that are designed for more expressive DLs. In particular $\mathcal{FL}_o wer$ is better in most cases than the other systems for testing a single subsumption and for computing classification. For the computation of all subsumers for a given concept $\mathcal{FL}_o wer$ truly outperforms the other DL reasoners by at least one order of magnitude. This is a remarkable results as it raises hopes that a naive method for classification can easily be sped up by simply using massively parallel hardware.

References

1. Baader, F., Brandt, S., Lutz, C.: Pushing the $\mathcal{EL}$ envelope. In: IJCAI-05, Proceedings of the Nineteenth International Joint Conference on Artificial Intelligence, 2005, pp. 364–369. http://ijcai.org/Proceedings/05/Papers/0372.pdf
2. Baader, F., Fernández Gil, O., Pensel, M.: Standard and Non-Standard Inferences in the Description Logic $\mathcal{FL}_0$ using Tree Automata. LTCS-Report 18–04, Chair for Automata Theory, Institute for Theoretical Computer Science, TU Dresden, Dresden, Germany (2018). http://lat.inf.tu-dresden.de/research/reports.html
3. Baader, F., Fernández Gil, O., Marantidis, P.: Matching in the description logic $\mathcal{FL}_0$ with respect to general TBoxes. In: LPAR-22, 22nd International Conference on Logic for Programming, Artificial Intelligence and Reasoning, 2018, EPiC Series in Computing, EasyChair, vol. 57, pp. 76–94 (2018). http://www.easychair.org/publications/paper/XrXz
4. Baader, F., Fernández Gil, O., Pensel, M.: Standard and non-standard inferences in the description logic $\mathcal{FL}_0$ using tree automata. In: GCAI-2018, 4th Global Conference on Artificial Intelligence, 2018, EPiC Series in Computing, EasyChair, vol. 55, pp. 1–14 (2018). http://www.easychair.org/publications/paper/H6d9
5. Baader, F., Marantidis, P., Okhotin, A.: Approximate unification in the description logic $\mathcal{FL}_0$. In: Michael, L., Kakas, A. (eds.) JELIA 2016. LNCS (LNAI), vol. 10021, pp. 49–63. Springer, Cham (2016). https://doi.org/10.1007/978-3-319-48758-8_4
6. Baader, F., Marantidis, P., Pensel, M.: The data complexity of answering instance queries in $\mathcal{FL}_0$. In: Companion of the The Web Conference 2018, WWW 2018, ACM, pp. 1603–1607 (2018). https://doi.org/10.1145/3184558.3191618
7. Baader, F., Sattler, U.: An overview of tableau algorithms for description logics. Studia Logica **69**(1), 5–40 (2001). https://doi.org/10.1023/A:1013882326814
8. Brachman, R.J., Levesque, H.J.: The tractability of subsumption in frame-based description languages. In: Proceedings of the National Conference on Artificial Intelligence, 1984, AAAI Press, pp. 34–37 (1984). http://www.aaai.org/Library/AAAI/1984/aaai84-036.php
9. Forgy, C.: Rete: a fast algorithm for the many patterns/many objects match problem. Artif. Intell. **19**(1), 17–37 (1982). https://doi.org/10.1016/0004-3702(82)90020-0
10. Cuenca Grau, B., Horrocks, I., Motik, B., Parsia, B., Patel-Schneider, P.F., Sattler, U.: OWL 2: the next step for OWL. J. Web Semant. **6**(4), 309–322 (2008)
11. Horridge, M., Bechhofer, S.: The OWL API: a java API for OWL ontologies. Seman. Web **2**(1), 11–21 (2011)
12. Kazakov, Y., de Nivelle, H.: Subsumption of concepts in $\mathcal{FL}_0$ for (cyclic) terminologies with respect to descriptive semantics is PSPACE-complete. In: Proceedings of the 2003 International Workshop on Description Logics (DL 2003), 2003, CEUR Workshop Proceedings. http://ceur-ws.org/Vol-81/kazakov.pdf
13. Krötzsch, M., Rudolph, S., Hitzler, P.: Complexity boundaries for Horn description logics. In: Proceedings of the Twenty-Second AAAI Conference on Artificial Intelligence, AAAI Press, pp. 452–457 (2007). http://www.aaai.org/Library/AAAI/2007/aaai07-071.php
14. Michel, F.: Entwurf und Implementierung eines Systems zur Entscheidung von Subsumption in der Beschreibungslogik $\mathcal{FL}_0$. Bachelor's thesis, TU Dresden (2017). (in German)
15. Nebel, B.: Terminological reasoning is inherently intractable. Artif. Intell. **43**(2), 235–249 (1990). https://doi.org/10.1016/0004-3702(90)90087-G

16. Parsia, B., Matentzoglu, N., Gonçalves, R.S., Glimm, B., Steigmiller, A.: The OWL reasoner evaluation (ORE) 2015 competition report. J. Autom. Reason. **59**(4), 455–482 (2017)
17. Armas Romero, A., Cuenca Grau, B., Horrocks, I.: MORe: modular combination of OWL reasoners for ontology classification. In: Cudré-Mauroux, P., et al. (eds.) ISWC 2012. LNCS, vol. 7649, pp. 1–16. Springer, Heidelberg (2012). https://doi.org/10.1007/978-3-642-35176-1_1
18. Simancik, F., Kazakov, Y., Horrocks, I.: Consequence-based reasoning beyond Horn ontologies. In: IJCAI 2011, Proceedings of the 22nd International Joint Conference on Artificial Intelligence, 2011, IJCAI/AAAI, pp. 1093–1098 (2011). https://doi.org/10.5591/978-1-57735-516-8/IJCAI11-187

Query Rewriting for DL Ontologies
Under the ICAR Semantics

Despoina Trivela[1]([⊠]), Giorgos Stoilos[2], and Vasilis Vassalos[1]

[1] Athens University of Economics and Business, Athens, Greece
despoina@aueb.gr
[2] Huawei Technologies, Research and Development, Edinburgh, UK

Abstract. In the current paper we propose a general framework for answering queries over inconsistent DL knowledge bases. The proposed framework considers the ICAR semantics and is based on a rewriting algorithm that can be applied over arbitrary DLs. Since the problem of ICAR-answering is known to be intractable for DLs other than DL-Lite, our algorithm may not terminate. However, we were able to describe sufficient termination conditions and to show that they are always satisfied for instance queries and TBoxes expressed in the semi-acyclic-$\mathcal{EL}_\perp$ as well as in DL-Lite$_{bool}$. Interestingly, recent results on UCQ-rewritability and existing techniques can be used within the proposed framework, to check if the conditions are satisfied for a given query and ontology expressed in a DL for which the problem is in general intractable.

1 Introduction

Query answering over data that are formally described using description logic (DL) knowledge bases (KBs) has received increasing attention in recent years. In this setting user queries are expressed in the ontology vocabulary and their answers reflect both the dataset (ABox) as well as the axioms specified in the ontology (TBox). The problem has been studied extensively for consistent datasets [11,16,21,24] as well as for datasets that contradict the ontological axioms [5,20,25,26].

In real-world applications data may often be inconsistent w.r.t. the ontology leading to logical contradictions. This is very likely in data integration applications where data originate from dispersed sources, or when data are generated automatically from an information extraction module. A straightforward approach to perform query answering would be to first remove the conflicting elements from the datasets. However, this is not always possible as the data may reside in distributed, or access restricted data sources, or be subject to frequent and diverse modifications. To address this issue, inconsistency-tolerant semantics have been proposed that describe which answers are meaningful to be returned in the presence of inconsistent data [9,18]. These semantics are typically based on the notion of the *repair*, that is a maximal (w.r.t. inclusion) consistent subset of the original dataset. Examples of such semantics are the

P. Fodor et al. (Eds.): RuleML+RR 2019, LNCS 11784, pp. 144–158, 2019.
https://doi.org/10.1007/978-3-030-31095-0_10

AR, IAR, ICAR semantics [18]. In the AR semantics queries are entailed from all the ABox repairs, while in the IAR semantics queries answers are obtained from the intersection of the ABox repairs. ICAR semantics consider data that are implied from the ontology and any consistent subset of the ABox, namely the *consistent logical consequences* of the TBox $\mathcal{T}$ and ABox $\mathcal{A}$, $clc(\mathcal{T}, \mathcal{A})$; the queries are entailed by the intersection of the $clc(\mathcal{T}, \mathcal{A})$ repairs.

As discussed in [18] query answering under the ICAR semantics come with a nice property that does not hold in the case of AR and IAR semantics. Consider two KBs $\mathcal{K} = (\mathcal{T}, \mathcal{A}), \mathcal{K}' = (\mathcal{T}, \mathcal{A}')$ that differ only in that $\mathcal{A}'$ includes some assertions that can be entailed from $\mathcal{T}$ and any consistent subset of $\mathcal{A}$. Differently from what one would expect, if we attempt to evaluate the same query over $\mathcal{K}$ and $\mathcal{K}'$ using either the AR or the IAR semantics, it is possible to obtain different results. This issue does not arise in the case of ICAR-answering because the logical implications of the ontology and consistent subsets of the ABox are taken into account to obtain the ICAR-answers. At the same time, in ICAR-answering it is possible to obtain answers from facts that can only be inferred from conflicting ABox assertions [4].

Regarding complexity both the IAR and ICAR semantics have shown to have nice computational properties since the query evaluation problem over ontologies expressed in the DL-Lite is in AC^0 w.r.t. data complexity [18,20], while it is in coNP for the AR semantics. However, computing answers using inconsistency-tolerant semantics has been proved quite difficult for DLs more expressive than DL-Lite. More precisely, Rosati [22] showed that the problem of IAR and ICAR-answering is at least coNP-hard w.r.t. data complexity for almost all well-known DLs from $\mathcal{EL}_\perp$ to $\mathcal{SHIQ}$. Moreover, in the $\mathcal{EL}_{\perp nr}$ fragment of $\mathcal{EL}_\perp$, where query answering is tractable for the IAR semantics, the problem remains in coNP for the ICAR semantics. Existing research results on consistent query answering over DL ontologies [5,13,20] focus on fragments of DL-Lite. In particular, the work presented in [13] employs the ICAR semantics and considers DL-Lite but in the OBDA setting, where the mappings relating the ontology terms to the data sources are also taken into account within query answering. In [26] a practical system was proposed for the ICAR and IAR semantics that computes upper approximations for DLs more expressive than DL-Lite. Moreover, in [25] an algorithm for IAR-answering was proposed that can handle arbitrary DLs but need not terminate. Despite the work presented in [26] the problem of designing practical ICAR-answering algorithms for expressive DLs is open.

In this work we study ICAR-answering over expressive DL ontologies. More precisely, we present a general framework for ICAR-answering that is based on query rewriting. Our rewriting algorithm has as a starting point the one presented in [19]. Given an input query and a DL ontology if our algorithm terminates, it computes a datalog program, extended with negation, that can be evaluated over the initial dataset to compute the ICAR-answers. Based on our analysis we extend previous results [25] to describe the conditions that ensure termination of our algorithm. The termination conditions are related to the notion of UCQ-rewritability that has been studied quite extensively in

DLs [1,6,15]. Consequently, we are able to provide positive results for instance queries and ontologies expressed in semi-acyclic-$\mathcal{EL}_\perp$ [7] and DL-Lite$_{bool}$, as well as UCQ-rewritable conjunctive queries over ontologies expressed in $\mathcal{EL}_{\perp nr}$. For DLs and queries for which the termination conditions are not generally satisfied, we exploit previous works [8,10,15] and provide an approach to check termination over a fixed input ontology and query. This allows us to design a framework for ICAR-answering over expressive DLs. Finally, we have conducted a preliminary evaluation of our approach. We obtained positive results showing that it is possible to perform ICAR-answering even in the case of DLs for which the problem is in general intractable.

2   Preliminaries

Description Logics. A DL *knowledge base* (KB) $\mathcal{K}$ consists of a TBox $\mathcal{T}$, and an ABox $\mathcal{A}$, $\mathcal{K} = \mathcal{T} \cup \mathcal{A}$. $\mathcal{T}$ and $\mathcal{A}$ are constructed from the countable and pairwise disjoint sets **C**, **R**, and **I** of atomic concepts (unary predicates), atomic roles (binary predicates), and individuals (constants). An ABox $\mathcal{A}$ is a finite set of *assertions* of the form $A(a)$ or $R(a,b)$ where $a, b \in$ **I**. A TBox $\mathcal{T}$ is a set of DL axioms. An $\mathcal{EL}_\perp$ concept is inductively defined by the syntax: $C := \top \mid \perp \mid A \mid C_1 \sqcap C_2 \mid \exists R.C$, where $A \in$ **C** and $R \in$ **R** and $C_{(i)}$ are $\mathcal{EL}_\perp$ concepts. An $\mathcal{EL}_\perp$ TBox $\mathcal{T}$ is a finite set of inclusions of the form $C_1 \sqsubseteq C_2$ with C_1, C_2 $\mathcal{EL}_\perp$ concepts. Inclusions of the form $C_1 \sqcap C_2 \sqsubseteq \perp$ (also written as $C_1 \sqsubseteq \neg C_2$) are called *negative* and the rest *positive*. DL-Lite$_R$ (or simply DL-Lite) restricts $\mathcal{EL}_\perp$ by allowing concepts of the form A, and $\exists R.\top$; R in DL-Lite can also be the *inverse* of a role of the form S^- and we can also have role inclusions of the form $S \sqsubseteq R$ or $S \sqsubseteq \neg R$ for S, R roles. An ABox $\mathcal{A}$ is *consistent* w.r.t. some TBox $\mathcal{T}$ if there exists a model for the KB $\mathcal{K} = \mathcal{T} \cup \mathcal{A}$; otherwise it is *inconsistent*. The semantics of DLs can be given by a well-known translation to First-Order Logic (FOL) [2]. Table 1 presents the translation of $\mathcal{EL}_\perp$ and DL-Lite axioms to first order clauses (inverse roles have been omitted). In the following we assume that the TBox axioms are translated into FOL.

Table 1. Translation of DL axioms into FOL

DL axiom	Clause
$B \sqsubseteq A$	$A(x) \leftarrow B(x)$
$A \sqcap B \sqsubseteq \perp$	$\perp \leftarrow A(x) \wedge B(x)$
$B_1 \sqcap B_2 \sqsubseteq A$	$A(x) \leftarrow B_1(x) \wedge B_2(x)$
$A \sqsubseteq \exists R.B$	$R(x, f(x)) \leftarrow A(x), B(f(x)) \leftarrow A(x)$
$\exists R \sqsubseteq A$	$A(x) \leftarrow R(x,y)$
$\exists R.B \sqsubseteq A$	$A(x) \leftarrow R(x,y) \wedge B(y)$
$P \sqsubseteq R$	$R(x,y) \leftarrow P(x,y)$
$P \sqsubseteq \neg R$	$\perp \leftarrow R(x,y) \wedge P(x,y)$

Datalog and Conjunctive Queries. A *disjunctive datalog clause* r (also called *rule*) is a function-free clause of the form $\forall \vec{x}, \vec{y}(\psi(\vec{x}) \leftarrow \phi(\vec{x}, \vec{y}))$ where $\phi(\vec{x}, \vec{y})$ is a conjunction of positive or negative atoms called the *body* of the clause and $\psi(\vec{x})$ is a disjunction of positive atoms called its *head*. For simplicity we will omit variable quantifiers and write $\psi(\vec{x}) \leftarrow \phi(\vec{x}, \vec{y})$. A *datalog clause* is a disjunctive datalog clause where the head contains a single atom. A *Horn-clause* is a datalog clause where the body contains only positive atoms. A *(disjunctive) datalog program* $\mathcal{P}$ is a finite set of (disjunctive) datalog clauses. We consider Herbrand models over all constants from $\mathcal{P}$. We say that a model M of $\mathcal{P}$ is *minimal* if there is no model M' of $\mathcal{P}$ such that M' is a subset of M. A positive ground atom $D(\vec{a})$ is entailed by $\mathcal{P}$ iff all minimal models of $\mathcal{P}$ contain $D(\vec{a})$; a negative ground atom $\neg D(\vec{a})$ is entailed by $\mathcal{P}$ iff $D(\vec{a})$ is not included in the minimal models of $\mathcal{P}$. The evaluation of $\mathcal{P}$ over an ABox $\mathcal{A}$ is the set of ground atoms entailed by the program $\mathcal{P} \cup \mathcal{A}$.

A *conjunctive query* (CQ) Q is a datalog clause with head predicate Q. The variables occurring in Q are called *answer variables*. A *boolean query* Q is a CQ with no answer variables. An instance query is a CQ of the form $Q(x) \leftarrow A(x)$ (we often simply write $A(x)$). A UCQ is a finite set of CQs. A tuple of constants $\vec{a}$ is a *certain answer* of Q over a KB $\mathcal{K} = \mathcal{T} \cup \mathcal{A}$ if the arity of $\vec{a}$ agrees with the arity of Q and $\mathcal{T} \cup \mathcal{A} \models Q(\vec{a})$, where $Q(\vec{a})$ denotes the Boolean query obtained by replacing the answer variables with $\vec{a}$. We use $\mathsf{cert}(Q, \mathcal{T} \cup \mathcal{A})$ to denote all certain answers of Q w.r.t. $\mathcal{K} = \mathcal{T} \cup \mathcal{A}$.

Definition 1. *Let $\mathcal{T}$ be a TBox and Q a CQ. A datalog-rewriting (or simply rewriting) of Q w.r.t. $\mathcal{T}$ is a datalog program $\mathcal{R}$ such that for any ABox $\mathcal{A}$ consistent w.r.t. $\mathcal{T}$ we have $\mathcal{T} \cup \mathcal{A} \models Q(\vec{a})$ iff $\mathcal{R} \cup \mathcal{A} \models Q(\vec{a})$, or in case Q is Boolean $\mathcal{T} \cup \mathcal{A} \models Q$ iff $\mathcal{R} \cup \mathcal{A} \models Q$. We say that a query Q is datalog-rewritable w.r.t. $\mathcal{T}$ if there exists a datalog-rewriting $\mathcal{R}$ of Q w.r.t. $\mathcal{T}$; if $\mathcal{R}$ is a UCQ, then Q is called UCQ-rewritable w.r.t. $\mathcal{T}$.*

Note that we will refer to a clause of the form $H(\vec{s}) \leftarrow \bigwedge_i \alpha_i \wedge \bigwedge_j \neg \mathcal{B}_j$, where α_i are positive atoms and $\mathcal{B}_j$ are conjunctions of positive atoms, as a datalog clause. Indeed such a clause is equisatisfiable to a datalog program that includes $H(\vec{s}) \leftarrow \bigwedge_i \alpha_i \wedge \bigwedge_j \neg \beta_j$ and $\beta_j \leftarrow \mathcal{B}_j$, for all j, where β_j are positive atoms.

Inconsistency-Tolerant Semantics. Definitions 2 and 3 recapitulate some of the notions used in the IAR and ICAR semantics [19]. Definition 4 formalises the notion of the rewriting under the IAR and ICAR semantics.

Definition 2. *Consider a TBox $\mathcal{T}$ and ABox $\mathcal{A}$ we define the consistent logical consequences of $\mathcal{T}, A$ as the set $clc(\mathcal{T}, \mathcal{A}) = \{a \mid \text{ some } S \subseteq \mathcal{A} \text{ exists s.t. } \mathcal{T} \cup S \models a \text{ and } S \text{ is consistent w.r.t. } \mathcal{T}\}$, where we use a to denote an assertion.*

Definition 3. *A repair of a set of assertions S w.r.t. a TBox $\mathcal{T}$ is any maximal (w.r.t. set inclusion) subset of S that is consistent w.r.t. $\mathcal{T}$.*

- We use $\mathcal{A}_{ir}$ to denote the intersection of all repairs of $\mathcal{A}$ w.r.t. $\mathcal{T}$. Let $\mathcal{Q}$ be a CQ and let $\mathcal{K} = \mathcal{T} \cup \mathcal{A}$ be a KB. A tuple of constants $\vec{a}$ is called an IAR-answer of $\mathcal{Q}$ over $\mathcal{K}$ if $\vec{a} \in \mathsf{cert}(\mathcal{Q}, \mathcal{T} \cup \mathcal{A}_{ir})$. We use $\mathsf{cert}_{ir}(\mathcal{Q}, \mathcal{T} \cup \mathcal{A})$ to denote the set of all IAR-answers of $\mathcal{Q}$ over $\mathcal{K}$ and we also write $\mathcal{T} \cup \mathcal{A} \models_{ir} \mathcal{Q}(\vec{a})$.
- We use $\mathcal{A}_{icar}$ to denote the intersection of all repairs of $clc(\mathcal{T}, \mathcal{A})$. Let $\mathcal{Q}$ be a CQ and let $\mathcal{K} = \mathcal{T} \cup \mathcal{A}$ be a KB. A tuple of constants $\vec{a}$ is called a ICAR-answer of $\mathcal{Q}$ over $\mathcal{K}$ if $\vec{a} \in \mathsf{cert}(\mathcal{Q}, \mathcal{T} \cup \mathcal{A}_{icar})$. We use $\mathsf{cert}_{icar}(\mathcal{Q}, \mathcal{T} \cup \mathcal{A})$ to denote the set of all ICAR-answers of $\mathcal{Q}$ over $\mathcal{K}$ and we also write $\mathcal{T} \cup \mathcal{A} \models_{icar} \mathcal{Q}(\vec{a})$.

The following example illustrates a nice property of the ICAR-semantics.

Example 1. Consider the TBox $\mathcal{T} = \{C(x) \leftarrow A(x), \bot \leftarrow A(x) \wedge B(x)\}$ and the ABox $\mathcal{A} = \{A(a), B(a)\}$. $\mathcal{A}$ has two repairs, that is $\{A(a)\}$ and $\{B(a)\}$, and hence their intersection is $\mathcal{A}_{ir} = \emptyset$. It is not hard to verify that $clc(\mathcal{T}, \mathcal{A})$ has two repairs, that is $\{A(a), C(a)\}$ and $\{B(a), C(a)\}$ and hence $\mathcal{A}_{icar} = \{C(a)\}$.

Next, consider the ABox $\mathcal{A}' = \mathcal{A} \cup \{C(a)\}$. Notice that $\mathcal{A}'$ differs from $\mathcal{A}$ only in that it contains $C(a)$ that can be entailed from $\mathcal{T}$ and the assertion $A(a)$ of $\mathcal{A}$. It holds that $\mathcal{A}'_{ir} = \{C(a)\}$ and $\mathcal{A}'_{icar} = \mathcal{A}_{icar}$. Interestingly, when we evaluate the same query $Q(x) \leftarrow C(x)$ over $\mathcal{T}, \mathcal{A}'$ and $\mathcal{T}, \mathcal{A}$ we yield different results in the case of the IAR-semantics but the same results for the ICAR-semantics.

Definition 4. *Given a TBox and a CQ $\mathcal{Q}$, an IAR-rewriting $\mathcal{R}^{ir}$ of $\mathcal{Q}$ w.r.t. $\mathcal{T}$ is a datalog program such that for every ABox $\mathcal{A}$ we have $\mathcal{T} \cup \mathcal{A} \models_{ir} \mathcal{Q}(\vec{a})$ iff $\mathcal{R}^{ir} \cup \mathcal{A} \models \mathcal{Q}(\vec{a})$. Similarly, for an ICAR-rewriting $\mathcal{R}^{icr}$ we have $\mathcal{T} \cup \mathcal{A} \models_{icr} \mathcal{Q}(\vec{a})$ iff $\mathcal{R}^{icr} \cup \mathcal{A} \models \mathcal{Q}(\vec{a})$.*

3   ICAR-rewriting over Expressive DLs

The problem of answering queries under the ICAR semantics was first investigated in [19]. Given an input TBox and query, the proposed algorithm computes a rewriting of the query w.r.t. the TBox that can be evaluated over any ABox to obtain the ICAR-answers. The rewriting technique proposed in [19] works for DL-Lite.

Example 2. Consider the DL-Lite TBox $\mathcal{T} = \{C(x) \leftarrow A(x), \bot \leftarrow A(x) \wedge B(x)\}$, the query $\mathcal{Q} = Q(x) \leftarrow C(x)$ and the ABox $\mathcal{A} = \{A(a), B(a)\}$ of Example 1.

In the first step, the algorithm in [19] computes the rewriting $\mathcal{R}$ of $\mathcal{Q}, \mathcal{T}$ under the standard semantics, $\mathcal{R} = \{Q(x) \leftarrow C(x), Q(x) \leftarrow A(x)\}$. Then, it extends the queries in $\mathcal{R}$ with the appropriate negative atoms, $\mathcal{R}' = \{Q(x) \leftarrow C(x), Q(x) \leftarrow A(x) \wedge \neg B(x)\}$. The negative atoms in $\mathcal{R}'$ guarantee that the evaluation of $\mathcal{R}'$ over $\mathcal{A}$ will only return answers from $\mathcal{A}_{ir}$. Indeed, atom $\neg B(x)$ prevents $Q(x) \leftarrow A(x) \wedge \neg B(x)$ from binding with $A(a)$ which is not included in $\mathcal{A}_{ir}$. At next step, the algorithm applies the rewriting procedure (under the standard semantics) once more on the elements of $\mathcal{R}'$ (only on the positive atoms) to obtain $\mathcal{R}'' = \mathcal{R}' \cup \{Q(x) \leftarrow A(x)\}$ that captures the assertions in $clc(\mathcal{T}, \mathcal{A})$. Indeed, when $Q(x) \leftarrow A(x)$ is evaluated over $\mathcal{A}$ we obtain the ICAR-answer $\{a\}$, $\mathsf{cert}(\mathcal{R}'', \mathcal{A}) = \mathsf{cert}_{icar}(\mathcal{Q}, \mathcal{T} \cup \mathcal{A})$. $\diamond$

A hybrid approach for ICAR-answering was presented in [26] that employs a rewriting, as well as an ABox saturation procedure. More precisely, given an input query $\mathcal{Q}$, a TBox $\mathcal{T}$, and an ABox $\mathcal{A}$, the algorithm in [26] exploits existing approaches [17,23] to compute the saturated ABox, that is the set of the assertions entailed from $\mathcal{A}$ and the axioms of $\mathcal{T}$ that can be translated into datalog. Then, it evaluates the IAR-rewriting of $\mathcal{Q}, \mathcal{T}$ over the saturated $\mathcal{A}$. The algorithm supports DL-Lite and it can be used to compute upper approximations of the ICAR-answers for more expressive DLs.

ICAR-answering over DL-Lite is FO-rewritable, and therefore in AC^0 in data complexity. However, it was shown that for more expressive DLs, consistent query answering under the ICAR is no longer tractable; actually, it is already coNP-hard in data complexity in $\mathcal{EL}_{\perp nr}$ [22]. Identifying DLs for which ICAR-answering is tractable is quite challenging. It was shown by Rosati [22] that tractability of IAR-answering does not imply tractability of ICAR-answering and the reason is the need to compute clc. Despite the theoretical studies over the ICAR semantics [19,22], there are no algorithms for ICAR-answering over expressive DLs.

In the following examples we attempt to employ the rewriting approach presented in [19] for an input TBox expressed in $\mathcal{EL}_\perp$.

Example 3. Consider the following $\mathcal{EL}_\perp$ TBox

$$\mathcal{T} = \{C(x) \leftarrow A(x) \wedge K(x)$$
$$K(x) \leftarrow B(x)$$
$$\perp \leftarrow A(x) \wedge B(x)\}$$

the query $\mathcal{Q} = Q(x) \leftarrow C(x)$ and the ABox $\mathcal{A} = \{A(a), B(a)\}$. It is not hard to verify that $clc(\mathcal{T}, \mathcal{A}) = \{A(a), B(a), K(a)\}$ and that $\mathcal{A}_{icar} = \{K(a)\}$. Hence, $\mathsf{cert}_{icar}(\mathcal{Q}, \mathcal{T} \cup \mathcal{A}) = \emptyset$.

Firstly, we compute the IAR-rewriting of $\mathcal{Q}, \mathcal{T}$. For this purpose, we apply the IAR-rewriting algorithm presented in [25] that takes as input an arbitrary DL TBox. We obtain the following IAR-rewriting:

$$\mathcal{R}^{ir} = \{Q(x) \leftarrow C(x) \tag{1}$$
$$Q(x) \leftarrow A(x) \wedge K(x) \wedge \neg(A(x) \wedge B(x)) \tag{2}$$
$$Q(x) \leftarrow A(x) \wedge B(x) \wedge \neg(A(x) \wedge B(x))\} \tag{3}$$

Next, in the same spirit as in [19], we apply a rewriting procedure on the elements of $\mathcal{R}^{ir}$ ignoring their negative part (we omit clause (3)):

$$(1) \rightsquigarrow \qquad Q(x) \leftarrow A(x) \wedge K(x) \tag{4}$$
$$Q(x) \leftarrow A(x) \wedge B(x) \tag{5}$$
$$(2) \rightsquigarrow Q(x) \leftarrow A(x) \wedge B(x) \wedge \neg(A(x) \wedge B(x)) \tag{6}$$

Finally, we construct the set $\mathcal{R}' = \mathcal{R}^{ir} \cup \{(4), (5)\}$.

Notice that when $Q(x) \leftarrow A(x) \wedge B(x)$ of $\mathcal{R}'$ is evaluated over $\mathcal{A}$ we obtain $Q(a)$, but $\{a\}$ is not in $\text{cert}_{\text{icar}}(\mathcal{Q}, \mathcal{T}, \mathcal{A})$; hence $\mathcal{R}'$ is not an ICAR-rewriting.

In order to fix this issue, one could check if the clause $\perp \leftarrow A(x) \wedge B(x)$ is entailed from $\mathcal{T}$ to decide whether $Q(x) \leftarrow A(x) \wedge B(x)$ is included in the ICAR-rewriting. In particular, since $\mathcal{T} \models \perp \leftarrow A(x) \wedge B(x)$, any set of the form $\{A(a), B(a)\}$ is inconsistent w.r.t. $\mathcal{T}$, and hence it cannot be used to infer an assertion included in $clc(\mathcal{T}, \mathcal{A})$. Consequently, the clause $Q(x) \leftarrow A(x) \wedge B(x)$ that bounds to assertions of the form $\{A(a), B(a)\}$ cannot be used to yield an ICAR-answer. In the same vein, the clause $Q(x) \leftarrow A(x) \wedge K(x)$ should be included in the output ICAR-rewriting since it holds that $\mathcal{T} \not\models \perp \leftarrow A(x) \wedge K(x)$. By eliminating (5) from $\mathcal{R}'$ we obtain the ICAR-rewriting $\mathcal{R}^{\text{icr}} = \mathcal{R}^{\text{ir}} \cup \{(4)\}$. $\Diamond$

Example 4. Consider the following $\mathcal{EL}_\perp$ TBox

$$\mathcal{T} = \{A(x) \leftarrow R(x, y) \wedge K(y) \tag{7}$$
$$A(x) \leftarrow R(x, y) \wedge A(y) \tag{8}$$
$$\perp \leftarrow K(x) \wedge R(x, y)\} \tag{9}$$

query $Q(x) \leftarrow A(x)$ and ABox $\mathcal{A} = \{R(a, b), K(b), R(b, a)\}$.

Similarly to the previous example, we compute the IAR-rewriting of $\mathcal{Q}$ w.r.t. $\mathcal{T}$ by applying the calculus of [25]. We obtain the following rewriting:

$$\mathcal{R}^{\text{ir}} = \{Q(x) \leftarrow A(x) \tag{10}$$
$$A(x) \leftarrow R(x, y) \wedge K(y) \wedge \neg(R(x, y) \wedge K(x)) \wedge \neg(R(y, z) \wedge K(y)) \tag{11}$$
$$A(x) \leftarrow R(x, y) \wedge A(y) \wedge \neg(R(x, y) \wedge K(x))\} \tag{12}$$

Next, we apply a rewriting procedure on positive parts of the clauses in $\mathcal{R}^{\text{ir}}$. Since we cannot further rewrite clause (11) we rewrite (10), (12) and obtain:

$$(10), (12) \rightsquigarrow A(x) \leftarrow R(x, y) \wedge K(y) \qquad (7)$$
$$A(x) \leftarrow R(x, y) \wedge A(y) \qquad (8)$$

In line with Example 4 notice that for the clauses (7), (8) in $\mathcal{R}'$ it holds that $\mathcal{T} \not\models \perp \leftarrow R(x, y) \wedge K(y)$, $\mathcal{T} \not\models \perp \leftarrow R(x, y) \wedge A(y)$. However, $\mathcal{R}' = \mathcal{R}^{\text{ir}} \cup \{(7), (8)\}$ is not an ICAR-rewriting. Indeed, when we evaluate (7) over $\mathcal{A}$ we obtain $A(a)$ and because of (8) we derive $A(b)$. However, b is not an ICAR-answer since $A(b) \notin clc(\mathcal{T}, \mathcal{A})$. This is because to derive $A(b)$ we have used $\{R(a, b), K(b), R(b, a)\}$ which is inconsistent w.r.t. $\mathcal{T}$. $\Diamond$

As illustrated in Example 4 in order to introduce a recursive clause of the form $A(x) \leftarrow R(x, y) \wedge A(y)$ in the ICAR-rewriting it is not sufficient to examine if $\mathcal{T} \not\models \perp \leftarrow R(x, y) \wedge A(y)$; since the concept A participates in a recursion, there is an infinite number of negative clauses for which we should examine if they are entailed from $\mathcal{T}$. Intuitively, this is the reason for the co-NP data complexity [22] of the ICAR-answering problem (already for the DL $\mathcal{EL}_{\perp nr}$ for which IAR-answering is in P): if the input query contains concepts involved in some recursion (such as concept A in our example), then the number of ABox assertions that can be used to infer an assertion in $clc(\mathcal{T}, \mathcal{A})$ is unbounded.

4 An ICAR-rewriting Algorithm

Based on the ideas discussed in Sect. 3 we propose an algorithm for ICAR-rewriting over a TBox expressed in an arbitrary DL. Definition 5 describes the notion of the negative closure that was used in [25] to obtain the IAR-rewriting. Intuitively, the negative closure $\mathcal{T}_{cn}$ of a TBox $\mathcal{T}$ is a finite set of negative clauses that can capture the negative clauses entailed from $\mathcal{T}$. We use $\mathcal{T}_{cn}$ to examine if the condition described in Example 3 holds.

Definition 5. *A negative closure of a TBox $\mathcal{T}$, denoted by $\mathcal{T}_{cn}$, is a finite set of negative clauses such that $\mathcal{T} \models \bot \leftarrow \bigwedge \beta_i$ iff some $\bot \leftarrow \bigwedge \alpha_i$ in $\mathcal{T}_{cn}$ exists with $\bot \leftarrow \bigwedge \alpha_i \models \bot \leftarrow \bigwedge \beta_i$.*

Algorithm 1. ICAR-Rewriting

Input: a CQ $\mathcal{Q}$ and a $\mathcal{L}$-TBox $\mathcal{T}$
1: Compute a negative closure $\mathcal{T}_{cn}$ of $\mathcal{T}$
2: Compute the IAR-rewriting $\mathcal{R}^{ir}$ of $\mathcal{Q}$ w.r.t. $\mathcal{T}$.
3: $\mathcal{R}^{icr} := \mathcal{R}^{ir}$
4: **for** $H(\bar{s}) \leftarrow \bigwedge_i \alpha_i \wedge \bigwedge_j \neg\beta_j \in \mathcal{R}^{ir}$ **do**
5:     Compute a UCQ-rewriting $\mathcal{R}_\alpha$ of $Q(\bar{s}) \leftarrow \bigwedge_i \alpha_i$ w.r.t. $\mathcal{T}$
6:     **for** each $Q(\bar{s}) \leftarrow \bigwedge_i \alpha'_i \in \mathcal{R}_\alpha$ **do**
7:         **if** for every clause $\mathcal{C} \in \mathcal{T}_{cn}$ it holds $\mathcal{C} \not\models \bot \leftarrow \bigwedge_i \alpha'_i$ **then**
8:             $\mathcal{R}^{icr} = \mathcal{R}^{icr} \cup \{H(\bar{s}) \leftarrow \bigwedge_i \alpha'_i \wedge \bigwedge_j \neg\beta_j\}$
9:         **end if**
10:     **end for**
11: **end for**
12: **return** $\mathcal{R}^{icr}$

Algorithm 1 computes the IAR-rewriting $\mathcal{R}^{ir}$ (line (2)) by applying the procedure presented in [25]. Then, in order to build the set $\mathcal{R}^{icr}$, it applies a rewriting procedure on the elements of $\mathcal{R}^{ir}$ by neglecting their negative part. More precisely, for the positive body atoms α_i of every element in $\mathcal{R}^{ir}$, it constructs the UCQ-rewriting of the query $Q(\bar{s}) \leftarrow \bigwedge_i \alpha_i$ (lines (4)–(5)). Condition in line (7) is necessary, as explained in Sect. 3, in order to only retrieve facts from $clc(\mathcal{T}, \mathcal{A})$ when evaluating $\mathcal{R}^{icr}$ over $\mathcal{A}$.

Example 5. Consider the following $\mathcal{EL}_\bot$ TBox $\mathcal{T}$ and query $Q(x) \leftarrow A(x)$.

$$\mathcal{T} = \{A(x) \leftarrow R(x,y) \wedge B(y) \tag{13}$$
$$B(x) \leftarrow C(x) \tag{14}$$
$$\bot \leftarrow B(x) \wedge K(x)\} \tag{15}$$

In the first step, Algorithm 1 computes the IAR-rewriting of $\mathcal{Q}$:

$$\mathcal{R}^{ir} = \{Q(x) \leftarrow A(x) \tag{16}$$
$$A(x) \leftarrow R(x,y) \wedge B(y) \wedge \neg(B(y) \wedge K(y)) \tag{17}$$
$$A(x) \leftarrow R(x,y) \wedge C(y) \wedge \neg(C(y) \wedge K(y))\} \tag{18}$$

Next, by considering the clauses (16), (17), (18) in $\mathcal{R}^{ir}$ it computes the UCQ-rewriting of $Q(x) \leftarrow A(x)$, $Q(x) \leftarrow R(x,y) \wedge B(y)$, $Q(x) \leftarrow R(x,y) \wedge C(y)$:

$$(16) \rightsquigarrow Q(x) \leftarrow R(x,y) \wedge B(y) \tag{19}$$

$$Q(x) \leftarrow R(x,y) \wedge C(y) \tag{20}$$

$$17 \rightsquigarrow A(x) \leftarrow R(x,y) \wedge C(y) \wedge \neg(B(y) \wedge K(y)) \tag{21}$$

The negative closure of $\mathcal{T}$ is $\mathcal{T}_{cn} = \{\bot \leftarrow B(x) \wedge K(x), \bot \leftarrow C(x) \wedge K(x)\}$. Therefore, the clauses $\bot \leftarrow R(x,y) \wedge B(y)$, and $\bot \leftarrow R(x,y) \wedge C(y)$ are not entailed by $\mathcal{T}_{cn}$ and condition in line (7) is satisfied. Finally, Algorithm 1 outputs $\mathcal{R}^{icr} = \mathcal{R}^{ir} \cup \{(19), (20), (21)\}$ that is an ICAR-rewriting of Q w.r.t. $\mathcal{T}$.   ◇

Theorem 1. *Let $\mathcal{T}$ be a DL TBox and let Q be a CQ. If Q is UCQ-rewritable w.r.t. $\mathcal{T}$ and there exists a negative closure $\mathcal{T}_{cn}$ of $\mathcal{T}$, then Algorithm 1 terminates and computes the ICAR-rewriting of Q w.r.t. $\mathcal{T}$.*

Proof. (sketch) In [25] it was shown that if there exists a negative closure of $\mathcal{T}$, and Q is datalog rewritable, then there always exists an IAR-rewriting $\mathcal{R}^{ir}$ of Q w.r.t. $\mathcal{T}$ (Theorem 6). Therefore, if there exists a negative closure of $\mathcal{T}$ and a UCQ-rewriting $\mathcal{R}$ of Q w.r.t. $\mathcal{T}$, then there also exists an IAR-rewriting of Q w.r.t. $\mathcal{T}$, and Algorithm 1 terminates. To prove correctness of Algorithm 1 we first show that $\mathcal{R}^{icr} \cup \mathcal{A} \models Q(\vec{a})$ iff $\mathcal{R}^{ir} \cup clc(\mathcal{T}, \mathcal{A}) \models Q(\vec{a})$. By definition of $\mathcal{R}^{ir}$ it holds that $\mathcal{R}^{ir} \cup clc(\mathcal{T}, \mathcal{A}) \models Q(\vec{a})$ iff $\mathcal{R} \cup clc(\mathcal{T}, \mathcal{A}) \models_{ir} Q(\vec{a})$, where $\mathcal{R}$ is the rewriting of Q w.r.t. $\mathcal{T}$ under the standard semantics. Finally, by definition of $\mathcal{A}_{icar}$ we conclude that $\mathcal{R}^{icr} \cup \mathcal{A} \models Q(\vec{a})$ iff $\mathcal{R} \cup \mathcal{A} \models_{icr} Q(\vec{a})$.   □

5   Positive Results for ICAR-Answering

In this section we exploit recent results on UCQ-rewritability of queries over a range of DLs along with the ICAR-rewriting approach of Sect. 4, to provide positive results for ICAR-answering over DLs that do not fall into the DL-Lite fragment.

As it can be seen from the previous section Algorithm 1 need not terminate. There two reasons for non-termination: 1. non-existence of a negative closure and 2. non-UCQ-rewritability of the input query.

Example 6. Consider the following $\mathcal{EL}_\bot$ TBox $\mathcal{T}$

$$\mathcal{T} = B(x) \leftarrow R(x,y) \wedge B(y) \tag{22}$$

$$\bot \leftarrow B(x) \wedge K(x)\} \tag{23}$$

and the instance query $Q(x) \leftarrow K(x)$.

Assume we attempt to compute $\mathcal{T}_{cn}$. We resolve (23) with (22) to obtain $\bot \leftarrow R(x,y) \wedge B(y) \wedge K(x)$; this clause can then be resolved with (22) to derive the clause $\bot \leftarrow R(x,y) \wedge R(y,z) \wedge B(z) \wedge K(x)$. Since none of the resolvents entails the other, by definition of $\mathcal{T}_{cn}$ it must contain both. It can be seen that once we could generate an infinite number of clauses of all of which must belong to $\mathcal{T}_{cn}$. $\diamond$

Example 6 illustrates the case where there is no negative closure for a given $\mathcal{EL}_{\bot}$ TBox $\mathcal{T}$. Intuitively, the non-existence of $\mathcal{T}_{cn}$ is related to concepts in the negative clauses of $\mathcal{T}$ that participate in some recursion (like concept B). Such recursions do not occur in DL-Lite but can be met already in $\mathcal{EL}_{\bot}$ causing a blow-up in data complexity from P to coNP [22]. In [25] a condition was presented that can be used to check if there exists a negative closure for a given TBox expressed in a Horn-DL. The following lemma describes this condition.

Lemma 1. *Let $\mathcal{T}$ be a $\mathcal{L}$ TBox where $\mathcal{L}$ is a Horn-DL. Let the set of concepts $\mathcal{S} = \{A_i(x) \mid \bot \leftarrow A_1(x) \wedge \ldots \wedge A_m(x) \in \mathcal{T}\}$. If every instance query $Q(x) \leftarrow A_i(x)$ in $\mathcal{S}$ is UCQ-rewritable w.r.t. $\mathcal{T}$ and consistent ABoxes, then there exists a negative closure $\mathcal{T}_{cn}$ of $\mathcal{T}$.*

In [22] the $\mathcal{EL}_{\bot nr}$ fragment of $\mathcal{EL}_{\bot}$ was studied for which IAR-answering remains tractable. Intuitively, in $\mathcal{EL}_{\bot nr}$ concepts that appear in negative clauses are not involved in recursions. Moreover, in [25] it was shown that there always exists a negative closure for an $\mathcal{EL}_{\bot nr}$ TBox. However, in the case of ICAR semantics query answering remains intractable for the DL $\mathcal{EL}_{\bot nr}$. Roughly, this is because the input query may contain concepts involved in some recursion and hence there is no limit in the number of ABox assertions that can be used to infer assertions in $clc(\mathcal{T}, \mathcal{A})$ (see Example 4). Although the problem of ICAR-answering of CQs over $\mathcal{EL}_{\bot nr}$ TBoxes is in general intractable, for a given UCQ-rewritable CQ we obtain the following result.

Theorem 2. *Let $\mathcal{T}$ be a $\mathcal{EL}_{\bot nr}$ TBox and let Q be a CQ that is UCQ-rewritable. Then, on input $\mathcal{T}$ and Q Algorithm 1 terminates and computes an ICAR-rewriting of Q w.r.t. $\mathcal{T}$.*

The authors in [7] showed that instance queries over semi-acyclic-$\mathcal{EL}_{\bot}$ TBoxes are always UCQ-rewritable. Moreover, in [25] it was shown that there always exists a negative closure for a semi-acyclic-$\mathcal{EL}_{\bot}$ TBox. Theorem 3 follows.

Theorem 3. *Let $\mathcal{T}$ be a semi-acyclic-$\mathcal{EL}_{\bot}$ TBox and let Q be an instance query. Then, on input $\mathcal{T}$ and Q, Algorithm 1 terminates and computes an ICAR-rewriting of Q w.r.t. $\mathcal{T}$.*

In [12] a goal-oriented procedure was presented that computes a datalog rewriting of a given DL-Lite$_{bool}$ TBox. By exploiting these results in [25] it was shown that there always exists a negative closure for a DL-Lite$_{bool}$ TBox. Moreover, instance query answering in DL-Lite$_{bool}$ is known to be UCQ-rewritable [1]. Therefore, we can obtain the following positive result on ICAR-rewritability for the non-Horn DL-Lite$_{bool}$.

Theorem 4. *Let $\mathcal{T}$ be a DL-Lite*$_{bool}$ *TBox and let $\mathcal{Q}$ be an instance query. Then, on input $\mathcal{T}, \mathcal{Q}$ Algorithm 1 terminates and computes an ICAR-rewriting of $\mathcal{Q}$ w.r.t. $\mathcal{T}$ that is a datalog program.*

To check if a given query is UCQ-rewritable we can exploit results in UCQ-rewritability of queries over DLs that are not always UCQ-rewritable [6,8,15]. The authors in [8] study UCQ-rewritability of a given instance query over Horn-DLs, like $\mathcal{EL}_\perp$, $\mathcal{ELI}_\perp$ and Horn-$\mathcal{SHIF}$. These results were used to design a practical algorithm for checking UCQ-rewritability of instance queries [15]. Subsequently, the system of [15] was extended to support rooted CQs [14].

Therefore, given a TBox $\mathcal{T}$ expressed in a Horn-DL one can decide on the existence of a negative closure by using the system of [15] to check UCQ-rewritability of all relevant instance queries described in Lemma 1. Moreover, the system of [15] can be used to check UCQ-rewritability of the input query $\mathcal{Q}$. If the conditions of Theorem 1 are satisfied, Algorithm 1 can be used to obtain an ICAR-rewriting of $\mathcal{Q}, \mathcal{T}$.

6   Evaluation

We have created a prototype system to perform a preliminary experimental evaluation of the proposed framework. Our system is based on the implementation of Algorithm 1. At first step, the UCQ-rewritability of the input query $\mathcal{Q}$ is examined by using the system Grind [14]. Next, our system uses the framework implemented in [25] to decide if there exists a negative closure of the input TBox $\mathcal{T}$ and if so, to compute an IAR-rewriting of $\mathcal{Q}$ and $\mathcal{T}$. If $\mathcal{Q}$ is not UCQ-rewritable, or if a negative closure cannot be computed for $\mathcal{T}$, then the system reports that it cannot output an ICAR-rewriting. Otherwise, it proceeds in computing the ICAR-rewriting $\mathcal{R}^{icr}$ as described in lines 3–10. The whole system currently supports ontologies expressed in $\mathcal{EL}_\perp$ which is the DL supported by Grind.

To generate our experimental setting we examined the ontologies ENVO, FBbi, MOHSE, NBO, Not-Galen that were used in [14] to evaluate Grind. We did not consider SO, as it was reported in [25] that a negative closure cannot be constructed for this ontology; this is because concept engineered_region(x) participates in a recursion that causes Lemma 1 to fail. The ontologies ENVO and FBbi include negative axioms. For the rest ontologies, that is MOHSE and Not-Galen, we manually added negative axioms. For this purpose we tried to use concepts that appear in different levels in the concept hierarchy, so that these affect large or small parts of the ontology. Each ontology used in [14] came with 10 handcrafted queries. Among them we used only those queries that include concepts involved in some negative axiom and for which the Grind system reported they are UCQ-rewritable. We have also manually constructed test queries that each one of them contains at least one body atom that uses a concept or role involved in a negative axiom. More precisely, for an axiom of the form $B \sqsubseteq \neg C$ we have constructed queries $Q(x) \leftarrow A(x)$ and $Q(x) \leftarrow D(x)$ such that $\mathcal{T} \models A \sqsubseteq B$ and $\mathcal{T} \models B \sqsubseteq D$. Overall, for each ontology we used 10 test queries that satisfy the UCQ-rewritability condition.

Table 2. Results for computation of ICAR-rewritings.

	$t_\mathcal{R}$	$|\mathcal{R}^{icr}|$	q^-	max	avg
envo	1 889	44 623	96%	7	6.5
	1 025	21 171	96%	7	6.4
	1 140	21 170	96%	7	6.5
	1 091	21 927	96%	14	6.7
	1314	22 694	96%	7	6.5
	129	75	86%	4	4.0
	1 441	21 932	97%	7	6.5
	1 126	21 170	96%	7	6.5
	2 538	21 170	96%	7	6.5
	2 385	21 934	96%	7	6.5
FBbi	194	285	7%	7	4.2
	148	406	5%	7	3.2
	85	13	100%	13	4.1
	88	29	100%	4	2.8
	51	307	7%	7	4.2
	72	9	100%	10	2.8
	90	555	54%	7	3.5
	76	295	51%	1	1.0
	56	280	8%	7	4.2
	90	547	53%	7	3.9

	$t_\mathcal{R}$	$|\mathcal{R}^{icr}|$	q^-	max	avg
MOHSE	17 037	98 119	96%	50	1
	32 213	101 646	96%	50	1
	15 651	101 620	96%	2	1.1
	1 049	3 511	92%	2	1.1
	1 701	31	80%	50	50
	1 923	3	66%	2	2.0
	1 269	43	72%	50	50.0
	1 314	43	51%	50	50
	2 9375	98115	96%	50	21.5
	34 318	98115	96%	50	21
Not-Galen	178 351	82 885	80%	6	1.0
	175 581	82 885	80%	6	1.6
	172 970	82 886	80%	6	1.57
	36	33	36%	6	1.7
	176 622	83 884	80%	6	1.0
	176 102	82 886	80%	6	1.0
	171 392	82 885	80%	5	1.0
	32	41	51%	1	1.0
	132 558	82 885	80%	6	1.0
	170 489	82 885	80%	6	1.0

Our results are depicted in Table 2. Columns $t_\mathcal{R}$, $|\mathcal{R}^{ir}|$ and q^- present the time to obtain the ICAR-rewriting in ms, the number of clauses in the output rewriting, and the percentage of the clauses in the output that contain a negative part. Finally columns max and avg present the maximum and average number of negative atoms in the elements of the rewriting. In most cases the ICAR-rewriting was obtained within a few seconds. In contrast, in the case of Not-Galen the time to compute the rewriting was up to 3 min. This is because, the rewriting procedure (described in line 5, Algorithm 1) was applied on every element of the IAR-rewriting which was quite large for almost all test queries. One could avoid several calls of the rewriting procedure and exploit the rewritings that have already been constructed during the IAR-rewriting process. For example, for an input TBox $\mathcal{T} = \{A(x) \leftarrow A_1(x), \bot \leftarrow A(x) \wedge B(x)\}$ and query $\mathcal{Q} = Q(x) \leftarrow A(x) \wedge C(x)$ the IAR-rewriting is of the form $\mathcal{R}^{ir} = \{Q(x) \leftarrow A(x) \wedge C(x) \wedge \neg(A(x) \wedge B(x)) \wedge \neg(A_1(x) \wedge B(x)), Q(x) \leftarrow A_1(x) \wedge C(x) \wedge \neg(A_1(x) \wedge B(x))\}$ and to obtain $\mathcal{R}^{ir}$ the standard rewriting $\mathcal{R} = \{Q(x) \leftarrow A(x) \wedge C(x), Q(x) \leftarrow A_1(x) \wedge C(x)\}$ must be computed. Therefore, to construct the ICAR-rewriting one could make use of $\mathcal{R}$ instead of applying anew rewriting procedure on every element of $\mathcal{R}^{ir}$. Our implementation does not involve such optimisations however we feel that they could reduce the rewriting times.

Regarding the size of the output rewriting, one could design optimisations to eliminate the redundant elements from the output, $\mathcal{R}^{icr}$. For example, the clause $Q(x) \leftarrow A_1(x) \wedge C(x) \wedge \neg(A(x) \wedge B(x)) \wedge \neg(A_1(x) \wedge B(x))$ is subsumed by

$Q(x) \leftarrow A_1(x) \wedge C(x) \wedge \neg(A_1(x) \wedge B(x))$ and hence the former can be discarded from $\mathcal{R}^{\text{icr}}$. Further work is required in that respect to reduce the size of $\mathcal{R}^{\text{icr}}$.

Finally, the number of negative conjuncts in the elements of the rewriting was quite small (up to 50). Note that the evaluation in [26] showed that triple-store systems can handle a large number of negative atoms (even more than one hundred). In conclusion, in most cases we have been able to obtain within reasonable time an ICAR-rewriting for our test ontologies and queries.

Overall, our evaluation results show that we were able, in most cases, to compute an ICAR-rewriting for the given TBoxes for which the problem is in general intractable. Moreover, computing the ICAR-rewriting can be done relatively efficiently and the number of negative atoms added in the clauses was usually quite small.

7   Conclusions

In the current work we have studied query answering over knowledge bases where the dataset is inconsistent with respect to the ontology. We have extended our previous work [25] on IAR semantics to provide an algorithm that computes the rewriting of the given query and arbitrary DL TBox for the ICAR semantics. Provided that ICAR-answering for expressive DLs is intractable, our algorithm may not terminate; if it terminates, it outputs an ICAR-rewriting, that is a datalog program with negation that can be evaluated over the initial ABox. Next, we studied the reasons for non-termination of our algorithm and developed conditions that ensure its termination. We showed that these conditions hold for semi-acyclic-$\mathcal{EL}_\perp$ and DL-Lite$_{\text{bool}}$. Interestingly, we can exploit recent results and use practical systems that have already been developed, to check termination of our algorithm for arbitrary fixed Horn-DL ontologies. Our experiments provided encouraging results as in almost all cases we were able to compute an ICAR-rewriting in reasonable time.

Overall, we have provided an approach for ICAR-answering over expressive DL-ontologies for which the problem is known to be intractable. To our knowledge this is the first attempt in the context of consistent query answering to produce rewritings for arbitrary DLs. It is left open for investigation if these rewritings could be used in other settings, e.g. to repair datasets, or to obtain positive results in the OBDA setting with GAV mappings [3]. Further experimental evaluation to examine whether the conditions apply in practice is left for future work.

Acknowledgements. Research supported by EU's Horizon 2020 research and innovation programme, grant agreement No. 720270 (HBP SGA1) and by the Research Centre of the Athens University of Economics and Business, in the framework of "Research Funding at AUEB for Excellence and Extroversion" (Action 2, 2017–2018). Part of this work was conducted when Giorgos Stoilos was working at AUEB.

References

1. Artale, A., Calvanese, D., Kontchakov, R., Zakharyaschev, M.: The DL-Lite family and relations. J. Artif. Intell. Res. **36**, 1–69 (2009)
2. Baader, F., McGuinness, D.L., Nardi, D., Patel-Schneider, P.F.: The Description Logic Handbook: Theory, Implementation and Applications. Cambridge University Press, Cambridge (2002)
3. Bienvenu, M.: Inconsistency-tolerant ontology-based data access revisited: taking mappings into account. In: Proceedings of the Twenty-Seventh International Joint Conference on Artificial Intelligence, IJCAI 2018, 13–19 July 2018, Stockholm, Sweden, pp. 1721–1729 (2018)
4. Bienvenu, M., Bourgaux, C.: Inconsistency-tolerant querying of description logic knowledge bases. In: Reasoning Web: Logical Foundation of Knowledge Graph Construction and Query Answering - 12th International Summer School 2016, Aberdeen, UK, 5–9 September 2016, Tutorial Lectures, pp. 156–202 (2016)
5. Bienvenu, M., Bourgaux, C., Goasdoué, F.: Querying inconsistent description logic knowledge bases under preferred repair semantics. In: AAAI, pp. 996–1002 (2014)
6. Bienvenu, M., Hansen, P., Lutz, C., Wolter, F.: First order-rewritability and containment of conjunctive queries in horn description logics. In IJCAI: International Joint Conference on Artificial Intelligence (2016)
7. Bienvenu, M., Lutz, C., Wolter, F.: Deciding FO-rewritability in $\mathcal{EL}$. In: Proceedings of the Twenty-Fifth International Workshop on Description Logics (2012)
8. Bienvenu, M., Lutz, C., Wolter, F.: First-order rewritability of atomic queries in Horn description logics. In: Proceeding of the Twenty-Third International Joint Conference on Artificial Intelligence (2013)
9. Bienvenu, M., Rosati, R.: Tractable approximations of consistent query answering for robust ontology-based data access. In: IJCAI 2013, Proceedings of the 23rd International Joint Conference on Artificial Intelligence, Beijing, China, 3–9 August 2013, pp. 775–781 (2013)
10. Bienvenu, M., Cate, B.T., Lutz, C., Wolter, F.: Ontology-based data access: a study through disjunctive datalog, CSP, and MMSNP. ACM Trans. Database Syst. **39**(4), 33:1–33:44 (2014)
11. Calvanese, D., De Giacomo, G., Lembo, D., Lenzerini, M., Rosati, R.: Tractable reasoning and efficient query answering in description logics: the DL-Lite family. J. Autom. Reasoning **39**(3), 385–429 (2007)
12. Grau, B.C., Motik, B., Stoilos, G., Horrocks, I.: Computing datalog rewritings beyond Horn ontologies. In: Proceedings of the Twenty-Third International Joint Conference on Artificial Intelligence (2013)
13. De Giacomo, G., Lembo, D., Oriol, X., Savo, D.F., Teniente, E.: Practical update management in ontology-based data access. In: d'Amato, C., et al. (eds.) ISWC 2017. LNCS, vol. 10587, pp. 225–242. Springer, Cham (2017). https://doi.org/10.1007/978-3-319-68288-4_14
14. Hansen, P., Lutz, C.: Computing FO-rewritings in EL in practice: from atomic to conjunctive queries. In: The Semantic Web - ISWC 2017–16th International Semantic Web Conference, Vienna, Austria, 21–25 October 2017, Proceedings, Part I, pp. 347–363 (2017)
15. Hansen, P., Lutz, C., Seylan, I., Wolter, F.: Efficient query rewriting in the description logic $\mathcal{EL}$ and beyond. In: Proceedings of the Twenty-Fourth International Joint Conference on Artificial Intelligence, pp. 3034–3040 (2015)

16. Kikot, S., Kontchakov, R., Zakharyaschev, M.: Conjunctive query answering with OWL 2 QL. In: Proceedings of the Thirteenth International Conference on Principles of Knowledge Representation and Reasoning (2012)

17. Kiryakov, A., Bishoa, B., Ognyanoff, D., Peikov, I., Tashev, Z., Velkov, R.: The features of BigOWLIM that enabled the BBCs world cup website. In: Workshop on Semantic Data Management (SemData), pp. 13–17 (2010)

18. Lembo, D., Lenzerini, M., Rosati, R., Ruzzi, M., Savo, D.F.: Inconsistency-tolerant semantics for description logics. In: Hitzler, P., Lukasiewicz, T. (eds.) RR 2010. LNCS, vol. 6333, pp. 103–117. Springer, Heidelberg (2010). https://doi.org/10.1007/978-3-642-15918-3_9

19. Lembo, D., Lenzerini, M., Rosati, R., Ruzzi, M., Savo, D.F.: Query rewriting for inconsistent DL-Lite ontologies. In: Rudolph, S., Gutierrez, C. (eds.) RR 2011. LNCS, vol. 6902, pp. 155–169. Springer, Heidelberg (2011). https://doi.org/10.1007/978-3-642-23580-1_12

20. Lembo, D., Lenzerini, M., Rosati, R., Ruzzi, M., Savo, D.F.: Inconsistency-tolerant query answering in ontology-based data access. J. Web Semant. **33**, 3–29 (2015)

21. Pérez-Urbina, H., Motik, B., Horrocks, I.: Tractable query answering and rewriting under description logic constraints. J. Appl. Logic **8**(2), 186–209 (2010)

22. Rosati, R.: On the complexity of dealing with inconsistency in description logic ontologies. In: Proceedings of the Twenty-Second International Joint Conference on Artificial Intelligence, pp. 1057–1062 (2011)

23. Stoilos, G., Grau, B.C., Motik, B., Horrocks, I.: Repairing ontologies for incomplete reasoners. In: Proceedings of the 10th International Semantic Web Conference, Bonn, Germany, pp. 681–696 (2011)

24. Trivela, D., Stoilos, G., Chortaras, A., Stamou, G.: Optimising resolution-based rewriting algorithms for OWL ontologies. J. Web Semant. **33**, 30–49 (2015)

25. Trivela, D., Stoilos, G., Vassalos, V.: A framework and positive results for IAR-answering. In: Proceedings of the Thirty-Second AAAI Conference on Artificial Intelligence, New Orleans, Louisiana, USA, 2–7 February 2018 (2018)

26. Tsalapati, E., Stoilos, G., Stamou, G.B., Koletsos, G.: Efficient query answering over expressive inconsistent description logics. In: Proceedings of the Twenty-Fifth International Joint Conference on Artificial Intelligence, pp. 1279–1285 (2016)

Technical Communication Papers

Complementing Logical Reasoning
with Sub-symbolic Commonsense

Federico Bianchi[1,2(✉)], Matteo Palmonari[1], Pascal Hitzler[2,3],
and Luciano Serafini[4]

[1] University of Milan-Bicocca, Milan, Italy
{federico.bianchi,matteo.palmonari}@unimib.it
[2] Wright State University, Dayton, OH, USA
pascal.hitzler@wright.edu
[3] Kansas State University, Manhattan, KS, USA
[4] Fondazione Bruno Kessler, Trento, Italy
serafini@fbk.com

Abstract. Neuro-symbolic integration is a current field of investigation in which symbolic approaches are combined with deep learning ones. In this work we start from simple non-relational knowledge that can be extracted from text by considering the co-occurrence of entities inside textual corpora; we show that we can easily integrate this knowledge with Logic Tensor Networks (LTNs), a neuro-symbolic model. Using LTNs it is possible to integrate axioms and facts with commonsense knowledge represented in a sub-symbolic form in one single model performing well in reasoning tasks. In spite of some current limitations, we show that results are promising.

1 Introduction

Neuro-symbolic integration models [11,12] aim at combining properties of symbolic reasoning and neural networks, to account both for data-driven learning and high-level reasoning, two tightly related aspects of human cognition. Additional advantages of this combination can be found in a higher explainability of learned knowledge and in the capability of softening some aspects of crisp logic-based reasoning approaches. This integration is also connected to the combination of sub-symbolic perception with high-level reasoning, a critical task in artificial intelligence [19].

Logic Tensor Networks (LTNs) [9,24] are an example of a neuro-symbolic model that embeds first-order fuzzy logic in a vector space. In LTNs logic constants are represented as vectors and n-ary predicates are n-ary functions whose values are real numbers in the range $[0, 1]$. A neural network for each predicate learns both the representation of logic constants and the weights that characterize the n-ary function. Learning is based on a set of axioms.

On the other hand, computational linguistics has developed distributional models of language that have been found cognitively plausible at a large extent

P. Fodor et al. (Eds.): RuleML+RR 2019, LNCS 11784, pp. 161–170, 2019.
https://doi.org/10.1007/978-3-030-31095-0_11

by psychologists [18]. We believe that these models, once adapted to be easily integrated with existing logical frameworks that combine learning and reasoning, can provide an account for commonsense knowledge and structured inferences that go beyond crisp reasoning approaches.

In this paper, we focus on the integration of two aspects of knowledge: (i) *sub-symbolic* common sense knowledge [17] that accounts for some kind of intuitive understanding of the world [7] and (ii) axiomatic knowledge has the one found in knowledge bases that accounts for structured inference. As an example of how this combination might work, imagine an agent that has access to the following set of axioms $\{species(cat), mammal(tiger), bird(penguin), \forall x(mammal(x) \rightarrow animal(x))\}$, we refer to the first three as instantiated atoms or facts and to the latter one as universally quantified formula; this axiomatic knowledge is not enough to infer $mammal(cat)$. However, if the agent knows that cats and tigers are similar to each other and both are dissimilar to penguins, she might infer that cats are mammals too (i.e., $mammal(cat)$). Once the latter instantiated atom has been inferred, the agent can make use of the axiom $\forall x(mammal(x) \rightarrow animal(x))$ to infer that cats are also animals (i.e., $animal(cat)$), bridging the gap with more complex inferences. We believe that combining these two worlds would bring great benefits in reasoning approaches since one requires the help of the other.

We present a first approach towards this direction that feeds Entity Embedding (EEs) generated using distributional semantics, i.e., vector-based representations of entities generated from text using Word2Vec [3], to a knowledge base represented in LTNs. This EEs encode the similarity between entities based on the principle that entities that share more contexts within a text corpus are more similar to each other. Distributional semantics has been found to provide representations that are strongly correlated with associative learning [18]; we thus refer to these representations as *sub-symbolic* commonsense. While in LTNs, neural representations of axioms are usually learned only from a partial (structured) knowledge base, EEs are used here as representations for the LTNs constants. In this way, LTNs will only need to learn the representation of the predicate network. Moreover, with the use of pre-trained representations, we can make inferences on entities that do not occur in the knowledge base as long as we have a sub-symbolic commonsense representation of those entities. Figure 1 shows the elements of our model that combines logical reasoning and sub-symbolic commonsense knowledge.

In once sentence, the major contribution of this work is to show that combining commonsense knowledge under the form of text-based entity embeddings with LTNs is not only simple, but it is also promising. Our experiments explore a limited part of a knowledge base but results show that the model is flexible and can be useful under different settings and use-cases.

The paper is organized as follows: in Sect. 2 we summarize related work and in Sect. 3 we outline the two main components of our model, namely the embeddings and LTNs and we show how we can combine the strengths of both; in Sect. 4 we develop and experiment comparing our model with some baselines; eventually we end the paper in Sect. 5 with conclusions and future work.

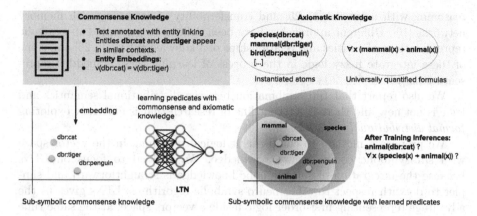

Fig. 1. We learn embeddings from text and we use LTNs to learn how to represent predicates with the network. "dbr:" stands for the DBpedia Knowledge Graph namespace.

2 Related Work

While recently deep learning [13] has shown great capabilities in many different tasks its techniques are still limited when they have to take into account the same reasoning and knowledge transformation capabilities that symbolic approaches show. However, symbolic artificial intelligence is constrained by computational limits and knowledge acquisition bottlenecks. The neuro-symbolic field was introduced to address the limits of both approaches by at the same time taking advantage of the capabilities of each of them. In this section, we present recent works from both the symbolic/statistical relational learning and the neuro-symbolic fields.

Symbolic and Statistical Relational Learning Approaches. Different symbolic/statistical relational learning approaches have been devised to treat inference; Recently, ProbLog [8] has been proposed as a probabilistic logic programming language that can be used to combine probability and logical inference, allowing the user to treat both probabilistic uncertainty and classical inference. Another approach to inference is the one represented by Probabilistic Soft Logic (PSL) [1] that is a statistical relational learning model that comes from the family of Markov Logic Networks (MLNs) [20].

Neuro-symbolic Approaches. We refer to recent surveys for discussions of different neuro-symbolic approaches proposed in the literature [11,12] while hereby we cite some examples of relevant approaches. DeepProbLog [19] is, for example, a "deep" extension of ProbLog [8], that show that it is possible to combine the power of deep nets with the expressive capabilities of logical reasoning. On the other hand, the Neural Theorem Prover (NTP) was introduced as an extension of the Prolog language that supports soft unification rules with the use of similarity between embedded representations [23]. Other approaches address reasoning with networks [16,25] and transferability of reasoning with memory networks [10]. Different approaches have been defined to generate sub-symbolic representations of entities and relationships of a knowledge base [5,26] and some of these integrate fuzzy logic in the process of learning these embedded representations [14].

We also report that the combination between distributional semantics and logic is not new and there is a recent line of research that is currently exploring *formal distributional semantics* [4].

We propose a method to complement logical reasoning in the vector space with sub-symbolic commonsense knowledge. We decided to focus on LTNs because the integration of sub-symbolic knowledge is straightforward and simpler to do with respect to other neuro-symbolic algorithms: LTNs gives us the advantage representing first-order logic inside a vectors space; at the same time we use entity embeddings to represent commonsense knowledge as the starting vector space on which LTNs learn to do reasoning.

3   Logical Reasoning with Sub-symbolic Commonsense

3.1   Logical Reasoning with Logic Tensor Networks

LTNs [9,24] use first-order fuzzy logic and represent terms, functions, and predicates in a vector space. Connectives are interpreted as binary operations over real numbers in $[0,1]$. For example, *t-norms* are used in place of the conjunction from classical logic (e.g., the t-norm can be interpreted as the min between two truth values). The action of representing elements of the logic language as elements in the vector space is referred to as grounding.

In LTNs, constants are grounded to vectors in $\mathbb{R}^n$ and predicates are grounded to neural network operations which output values in $[0,1]$. The neural network learns to define the truthness of an atom $P(c_1,\ldots,c_n)$ as a function of the grounding of the terms $c_1,\ldots,c_n$ [24]. For a predicate of arity m and for which $\mathbf{v}_1,\ldots,\mathbf{v}_m \in \mathbb{R}^n$ are the groundings of m terms, the grounding of the predicate is defined as $\mathcal{G}(P)(\mathbf{v}) = \sigma(u_P^T(tanh(\mathbf{v}^T W_P^{[1:k]}\mathbf{v} + V_P\mathbf{v} + B_P)))$ where $\mathbf{v} = \langle \mathbf{v}_1,\ldots,\mathbf{v}_m \rangle$ represents the concatenation between the vectors $\mathbf{v}_i$, σ is the sigmoid function, W, V, B and u are parameters to be learned by the network while k is layer size of the tensor.

LTNs reduce the learning problem to a maximum satisfiability problem: the task is to find groundings for terms and predicates that maximize the satisfiability of the formulas in the knowledge base. For example, for a grounded formula like *mammal(cat)*, the network updates the representation of the predicate *mammal* (i.e., the parameters in the tensor layer) and the representation of *cat* (i.e., its vector) in such a way that the degree of truth of an instantiated atom is closer to 1. Optimization also works in place of quantified formulas (e.g., $\forall x(mammal(x) \rightarrow animal(x))$); In fact, the universally quantified formulas are computed by using an aggregation operation [24] defined over a subset of the domain space $\mathbb{R}^n$. LTNs can be be used to do after-training reasoning over combinations of axioms on which it was not trained on (e.g., ask the truth value of queries like $\forall x(\neg mammal(x) \rightarrow species(x))$; this property allows us to explore the learned knowledge base with different combinations of predicates.

3.2 Sub-symbolic Commonsense with Entity Embeddings

Sub-symbolic representations of knowledge are popular in the deep learning community. In the NLP area, pre-trained representation of words (aka, word embeddings [21]) based on in-text co-occurrence are used frequently to enhance the performance on several tasks. In the same way, embeddings of entities and relationships that come from a knowledge base are becoming widely used in several contexts [5].

We use text-based embeddings of entities [3]. This approach is grounded in distributional semantics, originally introduced for words: similar words appear in similar context share similar meanings [15]; the same is true for entities [3], with two main advantages: (i) entities identifiers are not ambiguous and (ii) entities identifiers are interpreted as logical constants. The original work [3] presented also embeddings of ontological types, we ignore this component in our work since in this work we interpret entity types as unary predicates in LTNs. Starting from a text T, containing a sequence of words $w_1, \ldots, w_n$ we use entity linking tools [22] to find entities and to generate an annotated text that contains sequences of entity identifiers $e_1, \ldots, e_m$. The word2vec algorithm [21] is used to learn an embedding function ϕ based on the co-occurrence of entity mentions in the text $\phi(e_i) = \mathbf{e_i}$. Word2vec lets the user decide the dimension of the embedding and a window size to define the width of the context for each entity.

3.3 Combining Sub-symbolic Commonsense and Logical Reasoning

In our *commonsense* vector space, logical constants are represented by a vector and thus we can use LTNs to learn representations for the predicates over the *commonsense* vector space. Thus, we used the entity embeddings $\mathbf{e_1}, \ldots, \mathbf{e_m}$ as vectors to feed to LTNs. The truth value computed by LTNs is function not only of the parameters of the networks but also of the text-based pre-trained representations. While LTNs generally needs to learn the representation of vector from scratch in our setting are already learned (sub-symbolic commonsense) and do not need any more training.

Figure 1 shows a summary of the components of our model. We generate distributional embeddings from text and then we use axiomatic knowledge to learn the representations of predicates. After training we can use the model to reason over new axioms. A good way of understanding how LTNs work is to consider the learned predicate network as an area for which vectors have a truth value of 1 in certain locations that decreases to values close to 0 when vectors are distant from those locations.

4   Experiments

The main motivation that guides this experimental section is to show the capabilities of a model that combines logical reasoning and sub-symbolic commonsense knowledge like the one defined in the previous section.

We use 100 dimensional DBpedia entity embeddings [3]; these embeddings are generated first by using an automatic annotator (DBpedia Spotlight) and then by using word2vec (Skip-gram) [21]. LTNs were initialized with $k = 20$ and we used the fuzzy Lukasiewicz t-norm as in [24]. Code, data and architectures to replicate our experiments are available online[1].

4.1   Reference Knowledge Base

We create three small knowledge completion tasks for our experiments that are based on a common reference knowledge base introduced in this paper[2]. Our knowledge base is based on DBpedia and contains: a set of predicates P (e.g., *mammal*); a set of constants C (e.g., dbr:cat[3]); a set I of instantiated atoms, i.e., facts such as (e.g., *mammal*(dbr:cat)); a set Q of universally quantified formulas that represent the dependency in the DBpedia ontology (e.g., $\forall x\ mammal(x) \rightarrow animal(x)$); the set I^Q of formulas closed under the application of standard FOL inference to the previous set I (e.g., *animal*(dbr:cat)); a set of negated I^N instantiated atoms that are derived as follows: all the instantiated atoms built with predicates in P that are not in I^Q and I (e.g., $\neg fungus$(dbr:cat)). The reference knowledge base D is $I \cup I^Q \cup I^N$.

We first of all extract entities (C) from DBpedia and its ontology of the following classes (note that some classes are much less represented than others): Mammal (0.38%), Fungus (0.17%), Bacteria (0.03%), Plant (0.42%). We add the universally quantified formulas Q to derive inferences for predicates Animal, Eukaryote and Species for each atom, and apply this axiom to generate the set of instantiated axioms I^Q. Finally, we also generate all the negative instantiated atoms in I^N (e.g., $\neg fungus$(dbr:cat)). Considering positive and negative instantiated axioms this reference knowledge base contains 35,133 elements. We test the following three tasks by splitting the reference knowledge base D in training and testing:

- **D1. Objective:** evaluate the performance of the algorithms in a task in which only positive atoms are given, not all the atoms can be influenced from the axioms. As training, we have 1,400 positive atoms and we ask the models to

---

[1] https://github.com/vinid/logical_commonsense.

[2] Other knowledge base exist but some are too big to be explored [5] and others can be completed with simple axioms [6].

[3] We are aware that in some cases there is a subtle difference between what can be considered an instance and what is instead a type; *cat* can be for example the type of all the instances of cats. Since this generally depends on the granularity of the knowledge base we think that this does not affect the general applicability of the proposed experiment.

find all the other 7,077 atoms related to the entities found in the 1,400 atoms. For example, models have to infer atoms about the instance "dbr:cat" even if the only know that "dbr:cat" is a Species.

- **D2. Objective:** evaluate the performance of the algorithms in a task in which both positive and negative atoms are given; each entity in the training set appears also in the test set. As training, we have 7,026 atoms both positive and negatives and as in D1 we ask the models to find other 20,890 atoms (positive and negative).
- **D3. Objective:** evaluate the performance of the algorithms in a task in which both positive and negative atoms are given, but the test set will also contain atoms of entities not present in the training set: The models will need to rely on the sub-symbolic commonsense vectors. As training, we have 1,756 atoms and the models are now asked to infer the value of 33,377 atoms (positive and negative).

Domain Theory. We define a set of universally quantified axioms to be used by the models that contains 22 axioms. The complete list is available online and we hereby show some of them.

- $\forall x(plant(x) \rightarrow eukaryote(x))$
- $\forall x(mammal(x) \rightarrow animal(x))$
- $\forall x(plant(x) \rightarrow \neg mammal(x))$
- $\forall x(fungus(x) \rightarrow \neg animal(x))$

Note that this set is different from the set Q: the models will not know for example that $\forall(x : animal(x) \rightarrow eukaryote(x))$.

Baseline. We will compare the LTN_{EE} model (trained over atoms and universally quantified formulas to reach 0.99 satisfiability over the input knowledge base) with the following competitors.

- Simple LTNs model not initialized with pre-trained embeddings. We use this model to show that the use of pre-trained representation is useful.
- Probabilistic Soft Logic [1], the main competitor for the symbolic field that will be trained on both atoms and universally quantified formulas. We use the tool provided by the original authors with default parameters[4].
- Deep Neural Network initialized with EEs trained to assign 0 or 1 to instantiated atoms (note that we cannot use universally quantified formulas here), we explored several architectures often obtaining similar results. The DNN referenced in the results embeds the pre-trained representations of entities and a one-hot representation of predicates in 20 dimensions, concatenate them and apply another transformation to 1 dimension plus a sigmoid as non-linearity. Validation is done on 20% of the input data. Note that DNN cannot make use of the axioms of the domain theory.

---

[4] https://psl.linqs.org/.

4.2 Results

Table 1 shows the results of the different models over the different settings with the F1 measure for each predicate. In the following sections we discuss the result on each dataset.

Experiments on D1. In this setting we compare LTN_{EE} with LTN and PSL. We cannot use DNN because we are using only positive atoms. We also report that a simple rule-based model that uses axioms to complete the knowledge base would be able to infer only 45% of the axioms (with a 100% precision). The LTN_{EE} approach is the best performing one. The comparison between pure LTN and PSL suggests that while the latter performs better their difference is not high in this setting.

Experiments on D2. In this experiment each entity for which we require to find other instantiated atoms appear at least one time in the training set; this allows us to use PSL as a baseline in this setting. PSL performance is more or less similar to the one shown for the D1 dataset, but the performance between LTN_{EE} and DNN is comparable. In this settings the domain theory does not seem to provide increases in performance, but we remark that LTNs provide a model that can be queried after training.

Experiments on D3. From this experiment it is clear that LTN_{EE} generalizes slightly better than the competitor, and this could be due to both the domain theory and the fact that LTN_{EE} trains each predicate as a separate tensor layer. While the F1 score for many classes are comparable, the ones for Fungus and Bacteria reached a lower score than in the previous experiment: this might be due to the fact that the representation of elements of the class Fungus are similar to those of the class Plant while the Bacteria class as only a few instances in this experiment. Even if DNN and LTN_{EE} performances are similar (as expected, since both are neural models), we stress the fact that LTN_{EE} can be used for after-training logical inferences and this is a key aspect.

4.3 Examples and Limits

After training we can evaluate the truthfulness of axioms for which it was not specifically trained on. Table 2 reports some examples. We also explored the possibilities given by a more complex example that contains KG triples with facts *nationality(Person, Country)*, *bornIn(Person, City)* and *locatedIn(City, Country)* with 200 training examples (for which we also defined some simple axioms like $\forall x, \forall y, \forall z(bornIn(x, y) \land locatedIn(y, z) \rightarrow nationality(x, z))$ during training, but not the ones we show in the Table). It is interesting how LTNs can learn to reason on non-trivial axiomatic properties like the fact that being born in New York makes one American. The small experiment with KG triples is limited by the fact that the current implementation of LTNs suffers from heavy computational requirements in the presence of predicates in combination with quantifiers [2]. While the use of quantifiers extends the expressive power of the model it certainly downgrades the efficiency.

Table 1. $F1$ score per tested class.

D1	A_{F1}	F_{F1}	M_{F1}	P_{F1}	B_{F1}	E_{F1}	S_{F1}
LTN_{EE}	0.81	0.74	0.84	0.66	0.52	0.97	1.00
LTN	0.40	0.14	0.12	0.10	0.03	0.93	1.00
PSL	0.54	0.19	0.15	0.14	0.07	0.93	1.00
D2	A_{F1}	F_{F1}	M_{F1}	P_{F1}	B_{F1}	E_{F1}	S_{F1}
LTN_{EE}	0.91	**0.86**	0.91	0.86	0.63	**0.99**	1.00
DNN	**0.93**	0.82	**0.93**	0.87	0.54	**0.99**	1.00
PSL	0.56	0.20	0.20	0.17	0.10	0.88	0.98
D3	A_{F1}	F_{F1}	M_{F1}	P_{F1}	B_{F1}	E_{F1}	S_{F1}
LTN_{EE}	0.88	0.80	0.89	0.82	0.60	0.99	1.00
DNN	0.87	0.64	0.85	0.77	0.47	0.98	1.00

Table 2. The truth values of novel axioms.

Axiom	Truth
$\forall x(species(x) \to animal(x))$	0
$\forall x(eukaryote(x) \to \neg bacteria(x))$	0.73
$\exists x(eukaryote(x) \land \neg plant(x))$	1
$\forall x, y, z(nationality(x, y) \land locatedIn(y, z) \to bornIn(x, z))$	0.33
$\exists x(nationality(x, Canada) \land bornIn(x, Montreal))$	1
$\forall x(bornIn(x, New\ York) \to nationality(x, United\ States))$	0.88

5 Conclusions

In this paper, we have shown that the combination of sub-symoblic commonsense representations, under the form of entity embeddings generated from text, and logical reasoning in vector spaces is flexible and can be used to solve completion tasks. Since LTNs are based on Neural Networks, they reach similar results while also achieving high explainability due to the fact that they ground first-order logic. The real advantage comes from the fact that LTNs allow us to get the best of both the symbolic and connective worlds and to easily integrate additional knowledge like sub-symbolic commonsense knowledge. Despite the limitations and the simple experimental setting, the preliminary results show that the approach is promising. The key point of this paper is that with the combined model we can inject domain knowledge in a network (using LTNs) and at the same time use pre-trained representations. Our futures steps include improving LTNs training to treat bigger knowledge bases [5], introducing commmonsense knowledge within other frameworks and testing natural language inference tasks.

Acknowledgment. We gratefully acknowledge the support of NVIDIA Corporation with the donation of the Titan Xp GPU used for this research.

References

1. Bach, S.H., Broecheler, M., Huang, B., Getoor, L.: Hinge-loss markov random fields and probabilistic soft logic. J. Mach. Learn. Res. **18**, 1–67 (2017)
2. Bianchi, F., Hitzler, P.: On the capabilities of logic tensor networks for deductive reasoning. In: AAAI Spring Symposium: Combining Machine Learning with Knowledge Engineering (2019)
3. Bianchi, F., Palmonari, M., Nozza, D.: Towards encoding time in text-based entity embeddings. In: Vrandečić, D., et al. (eds.) ISWC 2018. LNCS, vol. 11136, pp. 56–71. Springer, Cham (2018). https://doi.org/10.1007/978-3-030-00671-6_4
4. Boleda, G., Herbelot, A.: Formal distributional semantics: introduction to the special issue. Comput. Linguist. **42**(4), 619–635 (2016)
5. Bordes, A., Usunier, N., Garcia-Duran, A., Weston, J., Yakhnenko, O.: Translating embeddings for modeling multi-relational data. In: NIPS, pp. 2787–2795 (2013)

6. Bouchard, G., Singh, S., Trouillon, T.: On approximate reasoning capabilities of low-rank vector spaces. In: Integrating Symbolic and Neural Approaches, AAAI Spring Syposium on Knowledge Representation and Reasoning (KRR) (2015)
7. Chudnoff, E.: Intuitive knowledge. Philos. Stud. **162**(2), 359–378 (2013)
8. De Raedt, L., Kimmig, A.: Probabilistic (logic) programming concepts. Mach. Learn. **100**(1), 5–47 (2015)
9. Donadello, I., Serafini, L., d'Avila Garcez, A.: Logic tensor networks for semantic image interpretation. In: IJCAI, pp. 1596–1602 (2017)
10. Ebrahimi, M., Sarker, M.K., Bianchi, F., Xie, N., Doran, D., Hitzler, P.: Reasoning over RDF knowledge bases using deep learning. arXiv preprint arXiv:1811.04132 (2018)
11. Garcez, A.d., Gori, M., Lamb, L.C., Serafini, L., Spranger, M., Tran, S.N.: Neural-symbolic computing: An effective methodology for principled integration of machine learning and reasoning. arXiv preprint arXiv:1905.06088 (2019)
12. Garcez, A.S., Lamb, L.C., Gabbay, D.M.: Neural-symbolic cognitive reasoning. Springer, Heidelberg (2008). https://doi.org/10.1007/978-3-540-73246-4
13. Goodfellow, I., Bengio, Y., Courville, A.: Deep Learning. MIT Press, Cambridge (2016)
14. Guo, S., Wang, Q., Wang, L., Wang, B., Guo, L.: Jointly embedding knowledge graphs and logical rules. In: EMNLP, pp. 192–202 (2016)
15. Harris, Z.S.: Distributional structure. Word **10**(2–3), 146–162 (1954)
16. Hohenecker, P., Lukasiewicz, T.: Deep learning for ontology reasoning. arXiv preprint arXiv:1705.10342 (2017)
17. Kuipers, B.: On representing commonsense knowledge. In: Findler, N.V. (ed.) Associative Networks, pp. 393–408. Elsevier, New York (1979)
18. Lenci, A.: Distributional semantics in linguistic and cognitive research. Ital. J. Linguist. **20**(1), 1–31 (2008)
19. Manhaeve, R., Dumančić, S., Kimmig, A., Demeester, T., De Raedt, L.: Deepproblog: neural probabilistic logic programming. arXiv preprint arXiv:1805.10872 (2018)
20. Meza-Ruiz, I., Riedel, S.: Jointly identifying predicates, arguments and senses using markov logic. In: NAACL, pp. 155–163. ACL (2009)
21. Mikolov, T., Sutskever, I., Chen, K., Corrado, G.S., Dean, J.: Distributed representations of words and phrases and their compositionality. In: NIPS, pp. 3111–3119 (2013)
22. Rizzo, G., Troncy, R.: NERD: a framework for unifying named entity recognition and disambiguation extraction tools. In: EACL, pp. 73–76. ACL (2012)
23. Rocktäschel, T., Riedel, S.: End-to-end differentiable proving. In: NIPS, pp. 3788–3800 (2017)
24. Serafini, L., d'Avila Garcez, A.S.: Learning and reasoning with logic tensor networks. In: Adorni, G., Cagnoni, S., Gori, M., Maratea, M. (eds.) AI*IA 2016. LNCS (LNAI), vol. 10037, pp. 334–348. Springer, Cham (2016). https://doi.org/10.1007/978-3-319-49130-1_25
25. Socher, R., Chen, D., Manning, C.D., Ng, A.: Reasoning with neural tensor networks for knowledge base completion. In: Advances in neural information processing systems, pp. 926–934 (2013)
26. Trouillon, T., Welbl, J., Riedel, S., Gaussier, É., Bouchard, G.: Complex embeddings for simple link prediction. In: ICML, pp. 2071–2080 (2016)

Adding Constraint Tables to the DMN Standard: Preliminary Results

Marjolein Deryck[(✉)], Bram Aerts, and Joost Vennekens

Department of Computer Science, KU Leuven, Campus De Nayer,
Sint-Katelijne-Waver, Belgium
{marjolein.deryck,b.aerts,joost.vennekens}@kuleuven.be

Abstract. The DMN standard allows users to build declarative models of their decision knowledge. The standard aims at being simple enough to allow business users to construct these models themselves, without help from IT staff. To this end, it combines simple decision tables with a clear visual notation. However, for real-life applications, DMN sometimes proves too restrictive. In this paper, we develop an extension to DMN's decision table notation, which allows more knowledge to be expressed, while retaining the simplicity of DMN. We demonstrate our new notation on a real-life case study on product design.

Keywords: Decision Model and Notation · First Order Logic · Constraint modelling

1 Introduction

Recently, the Object Management Group (OMG) has developed a new standard, the *Decision Model and Notation* [1], as a declarative representation for decision knowledge. It states explicitly that *[t]he primary goal of DMN is to provide a common notation that is readily understandable by all business users[...]* [1, p.13].

In a recent project [2], we used DMN to model the decision process followed by product engineers to design a specific kind of industrial component to match customer requirements. Here, DMN's ability to be understood by "business users" (in this case, the product engineers) was a key advantage. The multinational company with which we collaborate did not have a standardized product design process. Therefore, a significant knowledge elicitation effort was required, in which engineers from all over the world were brought together in order to define a single design process, that could then be partially automated into a decision support system. In these knowledge elicitation workshops, it was key to make use of a formal notation that could be understood by both our knowledge experts and the company's domain experts. This helped to avoid misunderstandings, ensured smooth communication, and allowed certain well-delineated parts of the

This work is supported by the Flemish Agency for Innovation and Entrepreneurship, TETRA HBC.2017.0039 and R&D project HBC.2017.0417.

M. Deryck and B. Aerts—Joint primary author.

© Springer Nature Switzerland AG 2019
P. Fodor et al. (Eds.): RuleML+RR 2019, LNCS 11784, pp. 171–179, 2019.
https://doi.org/10.1007/978-3-030-31095-0_12

decision model to be assigned as "homework" to specific experts. Moreover, the fact that the product engineers not only understood but even helped to build the formal model is also important for its maintenance, for which the engineers themselves will mainly be responsible.

DMN achieves its readability by offering a visual notation (the *Decision Requirements Diagram* or *DRD*) to decompose a big decision into smaller sub-decisions, a visual *decision table* notation to model individual decisions, and the intuitive *S-FEEL language* that can be used inside the decision tables.

However, the simplicity of DMN comes at a cost. While we found that this format elegantly handled the majority of the decision processes of our case study, a limited number of key decisions simply did not fit into the framework. In this paper, we propose an extension to the DMN standard, which aims to make it better suited for complex real-life situations. In [3], it was shows that DMN can be seen as an intuitive notation for certain formulas in First-Order Logic (FO). Following this approach, we will define the newly introduced constraint tables by means of a transformation to FO. The main advantage of this approach is that we can then feed this representation into a model generation system for FO, such as IDP [4], MiniSat [5] or z3 [6], and thereby immediately obtain an implementation of our extension.

In Sect. 2, we describe the business case that we will use as running example. We explain the difficulties we faced when formalizing this case in DMN (Sect. 3), followed by proposing constraint tables as an extension to DMN in Sect. 4. Then, we discuss the semantics of DMN decision tables and the newly introduced constraint tables (Sect. 5). Section 6 discusses how constraint tables facilitates handling the running example. Related work is presented in the Sect. 7. Finally, conclusions and future work are discussed in Sect. 8.

2   Running Example: Product Design

We will apply DMN and our extension to a real-life use case at a company that manufactures highly specialized products to order. This use case is discussed extensively in [2]. In this section, we recall a simplified version.

The task of the company's product engineers is to design a product that consists of two mandatory components: a body and a spring. In addition, there may be a third component, called a wiper. The engineers are given a number of specifications: the desired dimension of the product, the temperature range and pressure in which it should function.

They need to determine: 1. Which type of body (closed or open) to use; 2. What material to use for it; 3. Whether to use a normal or a thick spring.

They have to make these choices in such a way that: 1. The materials used can cope with the given temperature range; 2. When the materials shrink due to cold, the spring should prevent the component from falling out of the cavity in which it is placed; 3. A wiper is included if the component is to operate in a dirty environment. The materials have different costs, with cheaper materials typically being weaker than more expensive ones. In general, the engineers look for the cheapest design that will work.

3   DMN Case

We present a DMN model that is based on the typical decision procedure followed by the engineers. As we will show, parts of this procedure fit naturally into DMN, while others require cumbersome work-arounds. Figure 1 shows our DRD, which captures the general structure of the decision logic. It starts from customer requirements at the bottom and has the decisions that must be made at the top, with certain subdecisions in between. Each rectangle corresponds to a decision table, which can be found in Fig. 2. The different components of such a table are identified in Fig. 3. The (dark) green headers represent input columns, while the (light) blue headers are output columns. As can be seen in these figures, all of our tables use the *U(nique)* hit policy, which means that the conditions on the input columns in different rows must be mutually exclusive.

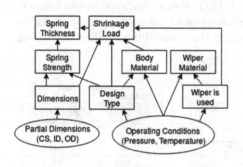

Fig. 1. DRD model of application.

To design a component, the engineers typically proceed as follows:

1. There are 3 relevant dimensions *: the outer diameter (OD) of the component, its inner diameter (ID) and its cross-section (CS). The customer provides 2 of these and the engineer computes the third ($CS = (OD - ID)/2$).* The corresponding DMN table consists of three different rows (one per dimension that might be missing), that all contain essentially the same information.

2. An initial design type is chosen, depending on the required operating conditions. This decision fits well within the mold of a DMN table.

3. Based on the temperature restrictions and design type, the "best" material is selected for the body. When representing this decision in a DMN table, we lose the distinction between physical constraints (i.e., the fact that some materials simply cannot cope with certain temperatures) and preferences (i.e., the fact that some material is not chosen because a cheaper material is available). Because this distinction is lost, is not clear how to update this table when, e.g., the price of a certain material drops: there may be cases in which this material was originally not chosen because it was too expensive, but we cannot discern them from cases where the material was not chosen because it would melt.

4. The expected shrinkage is computed, based on the design type, dimensions, temperature and selected materials. While we omit the details, the formula for this calculation can easily be placed in the "Shrinkage Load" table.

5. Based on the selected design type, the spring strength of a standard spring is computed. If this is enough to cope with the expected shrinkage, the standard spring is selected. Otherwise, the engineers switch to a thick spring and recompute the spring strength of the design. Since DMN does not allow to "recompute" a value, the "Spring Strength" table computes both the spring strength that the design would have with a standard spring and that with a thick spring.

The "Spring Thickness" table then uses these two values to decide which spring type to use. Because of this, the DRD has an edge *from* "Spring Strength" *to* "Spring Thickness". This is counterintuitive, because in reality it is the thickness of the spring that determines the design's spring strength, not the other way around.

6. *If, for an open design, even the thick spring does not provide enough spring strength, the engineers' final option is to switch to a closed design type, but since this is an initial choice upon which all further choices are based, this means that the engineers essentially start over the entire process from scratch.* If the thick spring's strength does not suffice, our DMN table will fail to assign a value to "Spring Thickness" (i.e., it will be null). This will alert the user that something is wrong, but the backtracking step of redoing the entire design process with a closed design is not possible in DMN. For this reason, we have implemented a workaround in which the "Design Type" table is overly cautious: in low temperatures, it will always choose a closed design, even in cases when an open design could suffice. This is sub-optimal, but avoids the possibility of failing the spring thickness check at the end.

Dimensions

U	OD	ID	CS	OD	ID	CS
1	null	-	-	ID+2*CS		
2	-	null	-		OD-2*CS	
3	-	-	null			(OD-ID)/2

Body Material

U	Tempe-rature	Design Type	Body Material
...	...	...	...
5	[-50, 80]	Closed	M1
7	(80, 150]	-	M3
...	...	...	...

Spring Thickness

U	Shrinkage Load	Spring Thickness
1	<= Standard Spring Strength	Standard
2	(Standard Spring Strength, Thick Spring Strength]	Thick

Design Type

U	Pressure	Tempe-rature	Design Type
1	<= 100	-	Closed
2	(100,150]	<= -50	Closed
3	(100,150]	> -50	Open
4	> 150	-	Open

Spring Strength

U	Design Type	Standard Spring Strength	Thick Spring Strength
1	Closed	15	20
2	Open	7	11

Shrinkage Load

U	Design Type	Shrinkage Load
1	...	f(Cross Section, Material of Body, Temperature)

Fig. 2. Decision tables for component design (Color figure online)

Fig. 3. The constitutive elements of the "Wiper is used" decision table

4   Constraint Tables

In this section, we extend DMN with *constraint tables*. This will allow a more direct representation of the constraints that are relevant for constructing a design and will thereby avoid the issues highlighted in the previous section.

A DMN decision table uniquely defines the value of its outputs. This is due to two properties. First, while cells in an input column may contain different kinds of S-FEEL expression, only single values can be used in output columns. Therefore, if a row matches the input, the corresponding single value must be assigned to the output. Second, DMN allows to designate a default value for each output column: if no row matches the input, then the output takes on this default value. Alternatively, when there is no default, the output is assigned the special value *null*, which is typically taken to indicate an error in the specification.

Our new constraint tables change both of these properties. First, we allow the same S-FEEL expressions that can be used in input columns to appear in output columns as well. For instance, the *Material of Body* table in Fig. 4 states that if a closed design is used, material M2 cannot be used, without specifying which of the other materials should be used in its place. Second, the rows of a constraint table are viewed as logical *implications*, in the sense that if the conditions on the inputs are satisfied, then the conditions on the outputs must also be satisfied. This means that if, for instance, none of the rows are applicable, the outputs can take on an arbitrary value, as opposed to being forced to *null* (in constraint tables, no default values can be assigned).

We introduce a new hit policy to identify constraint tables. We call this the *Every* hit policy, denoted as E^*, because it expresses that every one of these implications must be satisfied. Consider, for intance, the *Design Type* table in Fig. 4. Regardless of hit policy, this table states that if the pressure exceeds 150, the design type must be open. The effect of its E^* hit policy is seen when the pressure *does not* exceed 150. In this case, our constraint table imposes *no* restriction on the design: an open and a closed design are both possible. The *Material of Body* table combines the $E*$ hit policy with the ability to use S-FEEL expressions in the output. It states that if the design is closed, material $M2$ cannot be used for the body. Again, if an open design is used, no constraints are imposed on the body material.

5   The Semantics of DMN and Constraint Tables

Calvanese et al. express the formal semantics of decision tables in First-Order Logic [3]. We will use this as a starting point for the semantics of constraint tables, so we repeat some of this formalization here. For reasons of space, we restrict attention to the fragment of DMN used in our running example.

Each column in a decision table corresponds to an FO constant, that is mentioned in the column's header. Each cell in the column represents a condition that this constant may satisfy. Such a condition Q is transformed into an FO formula $\Phi_Q(x)$ in one free variable x. The easiest case is when the condition consists of a term T (a term is either a constant of an n-ary function applied to n terms), which is short for the equality "$= T$". In this case, $\Phi_Q(x)$ is the formula $x = T$. If Q is an interval $[i, j]$, then $\Phi_Q(x)$ is $x \geq i \wedge x \leq j$. If Q is a list $(C_1, \ldots, C_n)$ of constants, then Φ_Q is $x = C_1 \vee \ldots \vee x = C_n$. Other kinds of intervals and comparisons are defined in a similar way. Table 1 summarizes a number of the possibilities.

Table 1. The FO translation Φ_Q of different S-FEEL conditions Q.

Q	–	T	$not(T)$	$\leq T$	$(i, j]$	Q_1, Q_2
$\Phi_Q(x)$	true	$x = T$	$\neg \Phi_T$	$x \leq T$	$x > i \wedge x \leq j$	$x = \Phi_{Q_1} \vee x = \Phi_{Q_2}$

A row in a decision table corresponds to the conjunction of all these condition formulas, applied to their respective column headers. For instance, the second row of the "Design Type" table corresponds to:

$Pressure > 100 \wedge Pressure \leq 150 \wedge Temp \geq -50 \wedge DesignType = Closed.$

An entire table then corresponds to the disjunction of all its rows, e.g.:

$(Pressure > 100 \wedge Pressure \leq 150 \wedge Temp \geq -50 \wedge DesignType = Closed)$
$\vee (Pressure \leq 100 \wedge true \wedge DesignType = Closed) \vee \ldots$

In constraint tables, we represent each condition Q in a table row by the same formula Φ_Q, as shown in Table 1. The difference with decision tables lies in how we combine these individual formulas Φ_Q. In a constraint table, a row no longer corresponds to a conjunction, but to an implication. To be more concrete, if the first m columns of the table are inputs and the next $n - m$ columns are output, a row $r = (Q_1, \ldots, Q_m, Q_{m+1}, Q_n)$ corresponds to the *implicative formula* $\Phi_{\vec{H}}(r)$:

$$\bigwedge_{i=0}^{m} \Phi_{Q_i} \Rightarrow \bigwedge_{i=0}^{n} \Phi_{Q_{m+i}}.$$

For instance, the first row of the "Material of Body" table corresponds to the formulas $DesignType = Closed \Rightarrow MaterialOfBody \neq M2$. The semantics of constraint table T is then defined as the conjunction $\bigwedge_{r \in R} \Phi_{\vec{H}}(r)$.

6   Discussion

We discuss some features of our constraint table representation of the running example in more detail.

Dimensions. DMN tools use decision tables in a strictly feed-forward way, deriving outputs from inputs. By contrast, when we feed the constraint tables into a FO model generation tool, this can use the constraint in different ways. For instance, it can use the *Dimensions* table to derive eithe CS, OD or ID from the other two.

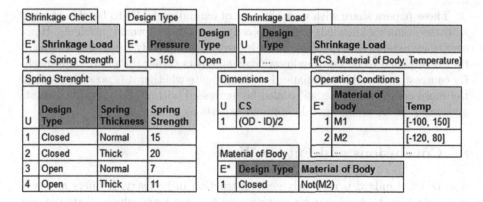

Fig. 4. Combination of constraint and decision tables

Design Type. The only physical design type constraint is that a closed design is not possible for pressures larger than 150. Nevertheless, the corresponding table in our original DMN representation (Fig. 3) also contained the temperature as an input column. This was to avoid running into cases in which the shrinkage cannot be addressed by simply adding a thick spring to an open design. By contrast, our constraint representation states directly that a design type must be chosen in accordance with the constraints expressed in the *Design Type* table and the *Shrinkage Load* table, avoiding the need for the original work-around.

Materials. Each material has a minimum/maximum temperature, defined in the *Operating Conditions* table. It states that, if a material is used, the operating temperature should be within the allowed temperature range of that material.

7   Related Work

The limitations of DMN decision tables have been recognized before. In addition to the S-FEEL language that we have used in this paper, the DMN standard itself also puts forward the more general FEEL language to allow much more complex expressions. However, FEEL is a full programming language with its own syntax, and hence not really suitable for direct use by domain experts [1].

Also a number of DMN tools provide ways to deal with the limitations. For instance, OpenRules allows to insert Java-snippets to express complex parts of the logic. The advantage of this approach is that it still keeps the overall idea of decision tables that can be maintained by business experts, while allowing an IT-expert to code specific complex parts of the decision logic. In addition to allowing imperative code snippets, several approaches also allow DMN to be extended by more declarative representations. For instance, the aforementioned OpenRules also offers an interface to a constraint solver [7], while [8] allows DMN to be enriched with domain knowledge expressed in Description Logic.

These papers share with ours the goal of extending DMN to be able to cope with the complex knowledge that typically arises in real-world problems. However, they extend DMN's decision tables with a completely separate formalism, that is intended to be used by knowledge engineers rather than DMN approach. By contrast, our goal in this paper is to investigate whether it is possible to allow the more complex knowledge to also be expressed within DMN's table format, allowing it to be maintained by domain experts as well.

8   Conclusions and Future Work

The DMN standard is currently gaining traction in industry. It offers an easy-to-use decision table format for representing decision logic, which allows domain experts to write and maintain the models, without requiring interventions by IT staff. Despite DMN's increasing popularity, it often still lacks the expressivity necessary to tackle real-world problems. In this paper, we have given an example of such a real-world problem, we have presented a DMN model for it, and discussed why this model is not really satisfactory.

Our main contribution is the extension of DMN with constraint tables. Crucially, constraint tables allow complex knowledge to be expressed, while still retaining the appealing visual and "syntax-less" representation that has made DMN itself a success. While extending DMN in this way may make the language harder to learn, we believe that domain experts will still be able to cope with constraint tables in DMN. While perhaps building a model from scratch would be more of a challenge, we are confident that, at least, if a model has initially been built in collaboration with a knowledge engineer, the domain experts will be able to maintain it without further help. Indeed, when looking at the constraint representation of our running example, we believe that the meaning of each of these tables will be clear enough to domain experts to make this feasible.

Following [3], we have defined the semantics of constraint tables by a transformation to first-order logic. The resulting formulas can be fed into a solver, providing a working implementation. Currently, we still perform this transformation by hand, but we are working on a fully automatic implementation.

References

1. OMG: Decision Model and Notation 1.2 (2018)
2. Aerts, B., Vennekens, J.: Application of logic-based methods to machine component design. In: Technical Communications of the 34th International Conference on Logic Programming (ICLP 2018) (2018)
3. Calvanese, D., Dumas, M., Laurson, Ü., Maggi, F.M., Montali, M., Teinemaa, I.: Semantics and analysis of DMN decision tables. In: La Rosa, M., Loos, P., Pastor, O. (eds.) BPM 2016. LNCS, vol. 9850, pp. 217–233. Springer, Cham (2016). https://doi.org/10.1007/978-3-319-45348-4_13
4. De Cat, B., Bogaerts, B., Bruynooghe, M., Janssens, G., Denecker, M.: Predicate logic as a modeling language: the IDP system, pp. 279–323 (2018)

5. Eén, N., Sörensson, N.: An extensible SAT-solver. In: Giunchiglia, E., Tacchella, A. (eds.) SAT 2003. LNCS, vol. 2919, pp. 502–518. Springer, Heidelberg (2004). https://doi.org/10.1007/978-3-540-24605-3_37

6. de Moura, L., Bjørner, N.: Z3: an efficient SMT solver. In: Ramakrishnan, C.R., Rehof, J. (eds.) TACAS 2008. LNCS, vol. 4963, pp. 337–340. Springer, Heidelberg (2008). https://doi.org/10.1007/978-3-540-78800-3_24

7. Feldman, J.: Representing and solving rule-based decision models with constraint solvers. In: Olken, F., Palmirani, M., Sottara, D. (eds.) RuleML 2011. LNCS, vol. 7018, pp. 208–221. Springer, Heidelberg (2011). https://doi.org/10.1007/978-3-642-24908-2_23

8. Calvanese, D., Dumas, M., Maggi, F.M., Montali, M.: Semantic DMN: formalizing decision models with domain knowledge. In: Costantini, S., Franconi, E., Van Woensel, W., Kontchakov, R., Sadri, F., Roman, D. (eds.) RuleML+RR 2017. LNCS, vol. 10364, pp. 70–86. Springer, Cham (2017). https://doi.org/10.1007/978-3-319-61252-2_6

Detecting "Slippery Slope" and Other Argumentative Stances of Opposition Using Tree Kernels in Monologic Discourse

Davide Liga(✉) ⓘ and Monica Palmirani(✉) ⓘ

CIRSFID, Alma Mater Studiorum - University of Bologna, Bologna, Italy
{davide.liga2,monica.palmirani}@unibo.it

Abstract. The aim of this study is to propose an innovative methodology to classify argumentative stances in a monologic argumentative context. Particularly, the proposed approach shows that Tree Kernels can be used in combination with traditional textual vectorization to discriminate between different stances of opposition without the need of extracting highly engineered features. This can be useful in many Argument Mining sub-tasks. In particular, this work explores the possibility of classifying opposition stances by training multiple classifiers to reach different degrees of granularity. Noticeably, discriminating support and opposition stances can be particularly useful when trying to detect Argument Schemes, one of the most challenging sub-task in the Argument Mining pipeline. In this sense, the approach can be also considered as an attempt to classify stances of opposition that are related to specific Argument Schemes.

Keywords: Argument Mining · Tree Kernels · Argument Schemes

1 Introduction

In many legal systems, there is an obligation to open a public review on the bill during the legislative process or on technical-administrative guidelines. In the information society, the attitude to open web portals for collecting opinions and comments from citizens is very frequent and also social media have recently been used to support participation. One of the main problems of this approach is to lose the argumentative threads of posts and to have, conversely, a flat chat flow. It is extremely difficult for the decision maker to recompose a discussion with hundreds of posts, or to extract a useful map of pros and cons from the debate. Moreover, it is difficult to recognize arguments and counter-arguments, or fallacies like "Slippery Slope" that produces polarization and emphasizes the discussion. This paper presents a method which is based on Argument Schemes and uses a tree kernel approach for detecting "Slippery Slope" and other argumentative stances of opposition. A use case in legal domain was considered: a corpus of monologic texts gathered from the website of Nevada Legislature,

© Springer Nature Switzerland AG 2019
P. Fodor et al. (Eds.): RuleML+RR 2019, LNCS 11784, pp. 180–189, 2019.
https://doi.org/10.1007/978-3-030-31095-0_13

specifically, from the opinions against the Senate Bill 165, which aims to regulate Euthanasia. The paper is organized as follows: Sect. 2 introduces the main idea of the solution and the methodology; Sect. 3 reports the state of the art and related works; Sect. 4 describes the corpus and the annotation; Sect. 5 exposes the experiment; Sect. 6 reports the results; the Sect. 7 presents conclusions and future works.

2 Methodology

2.1 The Argument Mining Pipeline

The main target of Argument Mining (AM) is extracting argumentative units, and their relations, from discourse [2,12]. A major characteristic of AM is its multidisciplinary nature, which physiologically fosters cooperation among different fields.

The reason why AM is prone to be multidisciplinary is that it is a combination of multifaceted problems. For the same reason, AM is often described as a pipeline (with much research focused on one or more of the involved steps).

For the purposes of this paper, we will refer to the two-step pipeline proposed by Cabrio and Villata [2], where the first step is the identification of arguments and the second step is the prediction of argument relations.

There can be a further step to be undertaken in an ideal AM pipeline, just after having detected the argumentative units and their relations (which include not only premises and conclusions but also heterogeneous relations such as support and attack). This step is that of fitting the "map" of the argumentative relations into a suitable Argument Scheme (e.g., argument from Example, "Slippery Slope" argument, argument from Expert Opinion).

As argued in this paper, a key step towards the achievement of this complex AM sub-task can be the creation of classifiers able to detect argumentative units that can be specific of an Argument Scheme.

The present work describes a solution for a classification problem. In a nutshell, the described approach uses Tree Kernels (TKs, described in [15]) to classify stances of opposition. Some of the classes of the classification discussed in this work are markedly related to specific Argument Schemes, which means that this classification solution can be exploited as a way to detect Argument Schemes, a highly complex AM sub-task. Particularly, the proposed methodology aims to detect the famous "Slippery slope" argument and other kind of argumentative oppositions, in a monologic context.

2.2 Tree Kernels Methods

Kernel machines are a well-known typology of classifiers, which also includes support-vector machine (SVM). In general, a kernel can be considered as a *similarity measure* capable to generating an implicit mapping of the inputs of a vector space $\mathcal{X}$ into a high-dimensional space $\mathcal{V}$. In other words, a kernel can be represented as an implicit mapping $\varphi : \mathcal{X} \to \mathcal{V}$.

The kernel function $k(\mathbf{x}, \mathbf{x}')$ (where $\mathbf{x}$ and $\mathbf{x}'$ belong to the input space $\mathcal{X}$) can be represented as an inner product in a high-dimensional space $\mathcal{V}$ and can be written as follows:

$$k(\mathbf{x}, \mathbf{x}') = \langle \varphi(\mathbf{x}), \varphi(\mathbf{x}') \rangle_{\mathcal{V}} \tag{1}$$

Where $\langle ., . \rangle_{\mathcal{V}}$ must be considered an inner product. Given a training dataset composed of n examples $\{(\mathbf{x}_i, y_i)\}_{i=1}^{n}$, where $y \in \{c_1, c_2\}$ with c_1 and c_2 being the two classes of a binary classification, the final classifier $\hat{y}$ can be calculated in the following way:

$$\hat{y} = \sum_{i=1}^{n} w_i y_i k(\mathbf{x}_i, \mathbf{x}') \tag{2}$$

Where w_i are the weights learned by the trained algorithm. Finally, exploiting what is described in Eq. 1, the Eq. 2 becomes:

$$\hat{y} = \sum_{i=1}^{n} w_i y_i \varphi(\mathbf{x}).\varphi(\mathbf{x}') \tag{3}$$

As far as TKs are concerned, they are a particular group of kernel functions specifically designed to operate on tree-structured data. In other words, a TK can be considered a *similarity measure* able to evaluate the differences between two trees.

Importantly, before selecting the appropriate TK function, there are two important steps to follow. The first step is to select the type of tree representation. For example, in this work, sentences have been converted into a particular kind of tree-structured representation called Grammatical Relation Centered Tree (GRCT), which involves PoS-Tag units and lexical terms. A description of various kind of tree representations can be found in Croce et al. [3]. The second step is to choose what type of substructures will be involved in the calculations. In fact, since TKs calculate the similarities of tree structures by watching at their fragments, it is crucial to establish what kind of substructures must be considered. In this work, the above-mentioned GRCT structures have been divided into Partial Trees (PTs) fragments, where each node is composed of any possible sub-tree, partial or not. Noticeably, this kind of substructures are able to provide a high generalization. The resulting TK function is called Partial Tree Kernel (PTK) and can be described as follows [15]:

$$K(T_1, T_2) = \sum_{n_1 \in N_{T_1}} \sum_{n_2 \in N_{T_2}} \Delta(n_1, n_2) \tag{4}$$

The above equation describes the kernel which calculates the similarity between the trees T_1 and T_2, where N_{T_1} and N_{T_2} are their respective sets of nodes and $\Delta(n_1, n_2)$ is the number of common fragments in nodes n_1 and n_2. More information about fragments of trees can be found in Moschitti [15] and Nguyen et al. [17].

The reason for using Tree Kernels is that they can be able to classify tree-structured data (in this case, tree-structured sentences), without the need of extracting highly engineered features. This is possible because Tree Kernels are able to measure the similarity among tree-sentences by watching at the fragments of their tree-representations. Intuitively, tree portions can be thought as "features" in a high dimensional space.

3   Related Works

The aim of this work is to classify argumentative opposition and facilitate Argument Scheme detection. Currently, only a few studies contribute to this part of the AM pipeline. Feng and Hirst [4], for instance, achieved an accuracy ranging from 63 to 91% in one-against-others classification and 80–94% in pairwise classification using a complex pipeline of classifiers. Lawrence and Reed [8] deployed highly engineered features to achieve F-scores ranging from 0.78 to 0.91. The present study, however, is an attempt to perform a simpler task of classification avoiding the use of highly-engineered features while keeping a high level of generalization. In fact, the present methodology shows that Tree Kernels can be used not only to classify argumentative stances, but also to facilitate Argument Scheme detection, without requiring highly-engineered features and keeping a high degree of generalization.

TKs have already been used successfully in several NLP-related tasks (e.g., question answering [6], metaphor identification [7], semantic role labelling [16]). However, the domain of AM has often preferred methodologies which resort to highly engineered feature sets, while the applications of TKs have been relatively limited. In spite of this, the results of these applications have been strongly encouraging, showing the ability of TKs to perform well. Rooney et al. [18] is one of the first studies that used TKs (in their study, they employed TKs and Part-of-Speech tags sequences). In 2015, Lippi and Torroni [12] suggested that TKs could be used for the detection of arguments. An year after, they presented MARGOT, a web application tool for the automatic extraction of arguments from text [13]. Importantly, TKs have been used in a wide range of domains. For instance, important results have been presented in the legal domain [10,11], while Mayer et al. [14] used TKs to analyze Clinical Trials.

The present study is among the first ones that use TKs to both classify argumentative evidences (*premises*) and to facilitate Argument Schemes detection. This approach is the continuation of a previous work (currently under publication [9]), which aimed at discriminating between different kinds of argumentative support (supporting evidences). These two works are an attempt to find a working methodology to discriminate among stances of support and stances of opposition by using Tree Kernels. Being able to classify different kinds of support and opposition is a crucial aspect when dealing with the classification of Argument Schemes.

4 Corpus and Annotation

The analyzed sentences have been gathered from public available data. A group of 638 sentences has been extracted and annotated from the "Opinion Poll" section of the official website of Nevada Legislature. More specifically, from the opinions against the Senate Bill 165. Clearly, being informal texts, the sentences are sometimes incomplete or segmented and mistakes are frequent, which makes the annotation task particularly complex.

Following an empirical analysis, we tried to select groups of sentences which could represent different types of reason for the opposition stance. Watching at those reasons and at their similarities, we selected those groups of reasons which had common characteristics at different levels of granularity. After this preliminary empirical analysis, each sentence of the corpus has been annotated by hand following the classes listed in Table 1.

This annotation scheme is designed to achieve different degrees of granularity of classification by training multiple classifiers and grouping some of the classes in *superclasses*, as described in Table 2. The classes PERSONAL EXPERIENCE and NOT PERSONAL EXPERIENCE have not been used yet, but they could give a contribution as soon as the process of annotation will be completed. Also the distinction between JUDGEMENTS SIMPLE and JUDGEMENT MORAL has not been exploited yet.

Table 1. The annotation classes with some examples.

Classes	Examples
SLIPPERY SLOPE	- *This would turn physicians into legal murderers*
JUDGEMENT SIMPLE	- *This bill is terrible*
JUDGEMENT MORAL	- *This bill is an affront to human dignity*
MORAL ASSUMPTIONS	- *Only God should decide when a person is supposed to die*
	- *Being a Christian, I cannot accept this bill*
	- *This is totally against the Hippocratic Oath!*
STUDY STATISTICS	- *Our country already experienced 20% increase of suicide rate*
ANECDOTAL	- *The bible says that this is wrong*
(PERSONAL EXPERIENCE)	- *My husband struggled a lot of years and [...]*
(NOT PERSONAL EXPERIENCE)	- *In Oregon this bill created the chaos*
OTHER/NONE	All the sentences that does not belong to the above classes

Even if the process of annotation is not yet completed, we can empirically state that these are some of the most frequent classes that characterize the comments against the Bill 165. Those comments which do not give any clue or explanation for the opposition (e.g. exhortations like "Please, vote no!") have

Table 2. The granularity levels and the grouping options.

Granularity 1	Granularity 2	Granularity 3	Granularity 4
SLIPPERY SLOPE	SLIPPERY SLOPE	SLIPPERY SLOPE	SLIPPERY SLOPE
OTHER/NONE	TESTIMONY	TESTIMONY	ANECDOTAL
			STUDY STATISTICS
	OTHER/NONE	JUDGEMENTS MORAL	JUDGEMENT(sim.+mor.)
			MORAL ASSUMPTIONS
		OTHER/NONE	OTHER/NONE

been considered in the class OTHER/NONE. The reason for this choice is that we aim to find out how debating people explain their opposition in a monologic environment. The focus of this annotation is *why* people are expressing a stance of opposition.

5   The Experiment

The annotation process, which gathered 638 sentences so far, is still ongoing under the supervision of experts of domain. The number of sentences grouped by class is described in Table 3.

Table 3. Number of sentences depending on class and granularity.

Classes	Gr.4	Gr.3	Gr.2	Gr.1
SLIPPERY SLOPE		82		
STUDY STATISTICS	26			
ANECDOTAL *(PERSONAL EXPERIENCE)* *(NOT PERSONAL EXPERIENCE)*	107	133		556
JUDGEMENT SIMPLE JUDGEMENT MORAL	54	140	423	
MORAL ASSUMPTIONS	86			
OTHER/NONE	283			

After having extracted the sentences, a Grammatical Relation Centered Tree (GRCT) representation was created for each sentence of the corpus. Furthermore, a TFIDF (term frequency-inverse document frequency) vectorization has been applied. In this regard, we tried three different TFIDF vectorizations considering monograms, 2-grams and 3-grams, in order to assess the effects of n-grams on the results.

In other words, the sentences of the corpus were converted into two kinds of "representation", with each labelled example having both a Grammatical Relation Centered Tree and a TFIDF BoW/n-grams representation (which, in

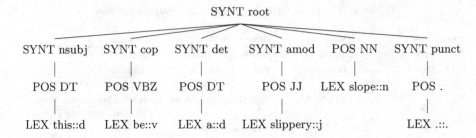

Fig. 1. The GCRT representation for the sentence *"This is a slippery slope."*

turn, can consider monograms, 2-grams and 3-grams). Figure 1 shows the GCRT representation for the sentence *"This is a slippery slope."*.

For each level of granularity, a classifier has been trained on the three different TFIDF vectorizations (monograms, 2-grams, and 3-grams), which resulted in twelve possible combinations.

All these classifiers were trained on the GRCT and TFIDF representations by using KeLP [5]. This operation was performed by randomly dividing the corpus of 638 sentences into a test set of 191 sentences and a training set of 446 sentences and by using a One-vs-All classification, which is one of the most common approach for multi-class problems. Noticeably, KeLP allows to combine multiple kernel functions. In this work, the classification algorithm was built as a combination of a Linear Kernel and a Partial Tree Kernel (PTK) [15], with the first kernel related to the TFIDF vectors and the second kernel related to the GRCT representations. More details on kernel combinations can be found in Shawe-Taylor and Cristianini [19].

6   Results

The scores of all the classifiers can be seen in Table 4, grouped by granularity. Also, the mean F1 scores of a stratified baseline were added. Given the unbalanced distribution of classes, a stratified baseline was preferred to others, because it reflects the natural distribution of classes in the training set.

Overall, when trying to achieve a deeper granularity, the mean F1 scores of the classifiers decrease. More precisely, the Mean F1 scores ranges from 0.76 to 0.81 at granularity 1, from 0.76 to 0.78 at granularity 2, from 0.70 to 0.71 for granularity 3, from 0.53 to 0.58 for granularity 4.

The classifiers showing best performances are probably those of granularity 2 and 3, since they are the most balanced in terms of number of instances. Noticeably, monograms show better performances at granularity 1 and 2, while 2-grams and 3-grams outperform monograms at granularity 4.

While the mean F1 scores of the baseline are low especially at higher degrees of granularity (showing that the classification attempted in this study is not trivial), all the other classifications outperformed the stratified baseline, showing a good ability of the proposed classifiers to solve the classification problem.

Table 4. The F1 scores of the classifiers grouped by granularity (P = Precision, R = Recall, F1 = F1 score). Close to the class name, the number of instances is specified. SS = SLIPPERY SLOPE, O = OTHER, T = TESTIMONY, JM = JUDGEMENTS AND MORAL, ST = STUDY STATISTICS, A = ANECDOTAL, MA = MORAL ASSUMPTIONS, J = JUDGEMENTS.

Classes	TK+Monograms			TK+2-grams			TK+3-grams			Stratified baseline
	P	R	F1	P	R	F1	P	R	F1	
				Granularity 1						
SS (82)	0.75	0.60	**0.67**	0.79	0.44	0.56	0.79	0.44	0.56	
O (556)	0.94	0.97	**0.96**	0.92	0.98	0.95	0.92	0.98	0.95	
Mean F1		**0.81**			0.76			0.76		0.54
				Granularity 2						
SS (82)	0.76	0.64	**0.70**	0.79	0.60	0.68	0.79	0.60	0.68	
T (133)	0.68	0.79	**0.73**	0.67	0.71	0.69	0.67	0.71	0.69	
O (423)	0.92	0.90	**0.91**	0.89	0.92	0.90	0.89	0.92	0.90	
Mean F1		**0.78**			0.76			0.76		0.31
				Granularity 3						
SS (82)	0.76	0.64	**0.70**	0.71	0.60	0.65	0.78	0.56	0.65	
T (133)	0.67	0.85	**0.75**	0.65	0.82	0.73	0.67	0.82	0.74	
JM (140)	0.66	0.49	0.56	0.72	0.55	0.63	0.75	0.57	**0.65**	
O (283)	0.75	0.81	0.78	0.77	0.82	0.80	0.77	0.86	**0.81**	
Mean F1		0.70			0.70			**0.71**		0.20
				Granularity 4						
SS (82)	0.72	0.72	**0.72**	0.69	0.72	0.71	0.68	0.68	0.68	
A (107)	0.51	0.85	0.64	0.54	0.85	0.66	0.58	0.85	**0.69**	
ST (26)	0.50	0.13	**0.20**	0.50	0.13	**0.20**	0.59	0.13	**0.20**	
J (54)	0.57	0.21	0.31	0.88	0.37	**0.52**	0.86	0.32	0.46	
MA (86)	0.62	0.46	0.53	0.75	0.54	**0.63**	0.75	0.54	**0.63**	
O (283)	0.74	0.81	0.78	0.76	0.84	**0.79**	0.74	0.86	**0.79**	
Mean F1		0.53			**0.58**			0.57		0.21

7 Conclusions and Future Work

The objective of this study is to show that the Tree Kernels (TKs) can be successfully combined with traditional features such as TFIDF n-grams to create classifiers able to differentiate among different kinds of argumentative stances of opposition. This differentiation can facilitate the detection of those argumentative units that are specifically related to Argument Schemes (e.g., argument from Expert Opinion, "Slippery Slope" argument). Since Tree Kernels can calculate the similarity between tree-structured sentences by comparing their fragments, this kind of classification can be performed without the need of extracting sophisticated features.

All the classifiers were created combining a Partial Tree Kernel (PTK) related to the GCRT representations and a linear kernel related to the TFIDF BoW/n-gram vector representations.

This kind of classification can be applied to premises to facilitate the discrimination among different Argument Schemes, which is a crucial sub-task in the Argument Mining pipeline. In the future, we will compare TKs performances with the performances of traditional textual representation, to assess whether and to what extent TKs outperform traditional features. Another important future improvement involves the modelization of Argument Schemes [21] in LegalRuleML [1] in order to manage, using the above mentioned Tree Kernels methods, the attacks to some parts of the "Slippery Slope arguments" [20] and so to apply defeasible legal reasoning in order to defeat some precedents in this kind of Argument Scheme. A serialization in LegalRuleML of a "Slippery Slope" argument using constitutive and prescriptive rules could develop strategies to attack its premises, or to attack the inferential link between premises and conclusion, or to attack the conclusion directly by posing a counterargument against it.

References

1. Athan, T., Governatori, G., Palmirani, M., Paschke, A., Wyner, A.: LegalRuleML: design principles and foundations. In: Faber, W., Paschke, A. (eds.) Reasoning Web 2015. LNCS, vol. 9203, pp. 151–188. Springer, Cham (2015). https://doi.org/10.1007/978-3-319-21768-0_6
2. Cabrio, E., Villata, S.: Five years of argument mining: a data-driven analysis. In: IJCAI, pp. 5427–5433 (2018)
3. Croce, D., Moschitti, A., Basili, R.: Structured lexical similarity via convolution kernels on dependency trees. In: Proceedings of the Conference on Empirical Methods in Natural Language Processing, pp. 1034–1046. Association for Computational Linguistics (2011)
4. Feng, V.W., Hirst, G.: Classifying arguments by scheme. In: Proceedings of the 49th Annual Meeting of the Association for Computational Linguistics: Human Language Technologies, pp. 987–996 (2011)
5. Filice, S., Castellucci, G., Croce, D., Basili, R.: KeLP: a kernel-based learning platform for natural language processing. In: Proceedings of ACL-IJCNLP 2015 System Demonstrations, pp. 19–24 (2015)
6. Filice, S., Moschitti, A.: Learning pairwise patterns in community question answering. Intelligenza Artificiale 12(2), 49–65 (2018)
7. Hovy, D., et al.: Identifying metaphorical word use with tree kernels. In: Proceedings of the First Workshop on Metaphor in NLP, pp. 52–57 (2013)
8. Lawrence, J., Reed, C.: Argument mining using argumentation scheme structures. In: COMMA, pp. 379–390 (2016)
9. Liga, D.: Argumentative evidences classification and argument scheme detection using tree kernels. In: Proceedings of the 6th Workshop on Argument Mining, pp. 92–97 (2019)
10. Lippi, M., Lagioia, F., Contissa, G., Sartor, G., Torroni, P.: Claim detection in judgments of the EU court of justice. In: Pagallo, U., Palmirani, M., Casanovas, P., Sartor, G., Villata, S. (eds.) AICOL 2015-2017. LNCS (LNAI), vol. 10791, pp. 513–527. Springer, Cham (2018). https://doi.org/10.1007/978-3-030-00178-0_35

11. Lippi, M., et al.: Claudette: an automated detector of potentially unfair clauses in online terms of service. Artif. Intell. Law **1**, 1–23 (2018)
12. Lippi, M., Torroni, P.: Argument mining: a machine learning perspective. In: Black, E., Modgil, S., Oren, N. (eds.) TAFA 2015. LNCS (LNAI), vol. 9524, pp. 163–176. Springer, Cham (2015). https://doi.org/10.1007/978-3-319-28460-6_10
13. Lippi, M., Torroni, P.: Margot: a web server for argumentation mining. Expert Syst. Appl. **65**, 292–303 (2016)
14. Mayer, T., Cabrio, E., Lippi, M., Torroni, P., Villata, S.: Argument mining on clinical trials. In: Computational Models of Argument: Proceedings of COMMA 2018, vol. 305, p. 137 (2018)
15. Moschitti, A.: Efficient convolution kernels for dependency and constituent syntactic trees. In: Fürnkranz, J., Scheffer, T., Spiliopoulou, M. (eds.) ECML 2006. LNCS (LNAI), vol. 4212, pp. 318–329. Springer, Heidelberg (2006). https://doi.org/10.1007/11871842_32
16. Moschitti, A., Pighin, D., Basili, R.: Tree kernels for semantic role labeling. Comput. Linguist. **34**(2), 193–224 (2008)
17. Nguyen, T.V.T., Moschitti, A., Riccardi, G.: Convolution kernels on constituent, dependency and sequential structures for relation extraction. In: Proceedings of the 2009 Conference on Empirical Methods in Natural Language Processing: Volume 3, pp. 1378–1387. Association for Computational Linguistics (2009)
18. Rooney, N., Wang, H., Browne, F.: Applying kernel methods to argumentation mining. In: Twenty-Fifth International FLAIRS Conference (2012)
19. Shawe-Taylor, J., Cristianini, N., et al.: Kernel Methods for Pattern Analysis. Cambridge University Press, Cambridge (2004)
20. Walton, D.: The basic slippery slope argument. Informal Logic **35**(3), 273–311 (2015)
21. Walton, D., Macagno, F.: A classification system for argumentation schemes. Argument Comput. **6**(3), 219–245 (2015)

Fuzzy Logic Programming for Tuning Neural Networks

Ginés Moreno[✉], Jesús Pérez, and José A. Riaza

Department of Computing System, U. Castilla-La Mancha, 02071 Albacete, Spain
{Gines.Moreno,JoseAntonio.Riaza}@uclm.es, j.perezmar@posgrado.uimp.es

Abstract. Wide datasets are usually used for training and validating neural networks, which can be later tuned in order to correct their final behaviors according to a few number of test cases proposed by users. In this paper we show how the FLOPER system developed in our research group is able to perform this last task after coding a neural network with a fuzzy logic language where program rules extend the classical notion of clause by including on their bodies both fuzzy connectives (useful for modeling activation functions of neurons) and truth degrees (associated with weights and biases in neural networks). We present an online tool which helps to select such operators and values in an automatic way, accomplishing with our recent technique for tuning this kind of fuzzy programs. Moreover, our experimental results reveal that our tool generates the choices that better fit user's preferences in a very efficient way and producing relevant improvements on tuned neural networks.

Keywords: Neural networks · Fuzzy logic programming · Tuning

1 Introduction

In order to deal with partial truth and reasoning with uncertainty in a natural way, many fuzzy logic programming languages implement (extended versions of) the resolution principle introduced by Lee [5], such as Elf-Prolog [3], Fril [1], MALP [7] and FASILL [4]. For the last couple of languages, we have developed the FLOPER system devoted to execute, debug, unfold and tune fuzzy programs.

In this paper we focus on the so-called *multi-adjoint logic programming* approach MALP [7] a powerful and promising approach in the area of fuzzy logic programming. In this framework, a program can be seen as a set of rules whose bodies admit the presence of connectives for linking atoms and/or truth degrees and each rule can also be annotated by a truth degree.

When specifying a MALP program, it might sometimes be difficult to assign weights—truth degrees—to program rules, as well as to determine the right

This work has been partially supported by the EU (FEDER), the State Research Agency (AEI) and the Spanish *Ministerio de Economía y Competitividad* under grant TIN2016-76843-C4-2-R (AEI/FEDER, UE).

P. Fodor et al. (Eds.): RuleML+RR 2019, LNCS 11784, pp. 190–197, 2019.
https://doi.org/10.1007/978-3-030-31095-0_14

connectives in their bodies. In order to overcome this drawback, in [8,9] we have recently introduced a symbolic extension of MALP programs called *symbolic multi-adjoint logic programming* (sMALP). Here, we can write rules containing *symbolic* truth degrees and *symbolic* connectives, i.e., connectives which are not defined on its associated lattice. These sMALP programs can tuned w.r.t. a given set of test cases, thus easing what is considered the most difficult part of the process: the specification of the right weights and connectives for each rule.

The main goal of this paper consists in the application of our tuning technique to fuzzy logic programs modeling neural networks (visit the online tool https:// dectau.uclm.es/fasill/sandbox?q=nn). Unlike the traditional training process used for building and validating neural networks and other computational models, here we focus on tuning key components (weights and activation functions) of previously trained models, in order to adjust the behavior of the final neural networks w.r.t. a small set of test cases. It is important to note here that the tuning process is able to guess not only weights, as training techniques also do, but also activation functions, which are not altered at training time.

Keras [2] is a high-level, open source neural network library written in Python. It is designed to enable the fast experimentation of deep neural networks in a modular, extensible and user-friendly way. In this paper we will focus on a basic neural network architecture in order to simplify our explanations: the MLP or multilayer perceptron [6] (see Fig. 3).

The structure of this paper is as follows. Section 2 is devoted to translate Keras models representing fully trained neural networks to MALP programs. Next, before applying our tuning techniques in order to correct some possible structural deficiencies of the MLP, we show in Sect. 3 how to transform the associated MALP program into a symbolic (sMALP) one, thus allowing us to automatically tune its symbolic components (i.e. some random weights/biases and activation functions) according to a set of test cases proposed by users. Finally, in Sect. 4 we conclude by showing some lines for future work.

2   Coding Neural Networks as MALP Rules

In this section, we address the problem of translating neural networks to MALP rules by taking into account the following guidelines:

- Activation functions: since a huge amount of activation functions have been created during the last four decades, it is decisive to select an appropriate subset of them for our initial prototype.
- Weights and biases: there are two essential parts of every neuron, and it is important to choose a good data structure for representing them.
- Structure of the network: we will follow the universal representation of a multilayer perceptron, in which each node in one layer connects with a certain weight to any other node in the following layer (except for nodes in the output layer).
- Lattice of truth degrees and connectives: in order to implement a fuzzy logic program, it is necessary to define a complete lattice where all our activation functions are modeled.

Input

</> Program

```
1  node_1_2(X1, X2, X3, X4)<- #@s1( @add(0.62446076, @add( @prod(0.7310704,X1), @add( @prod(0.7056907,X2), @
2  node_2_2(X1, X2, X3, X4)<- #@s1( @add(-0.50864863, @add( @prod(-0.5254192,X1), @add( @prod(0.32891437,X2)
3  node_3_2(X1, X2, X3, X4)<- #@s1( @add(0.8460652, @add( @prod(0.14646524,X1), @add( @prod(0.9592526,X2), @
4  node_4_2(X1, X2, X3, X4)<- #@s1( @add(0.0, @add( @prod(-0.39125273,X1), @add( @prod(0.4458459,X2), @add( @
5  node_5_2(X1, X2, X3, X4)<- #@s1( @add(-0.47975096, @add( @prod(0.33556423,X1), @add( @prod(-0.73593193,X2)
6  node_6_2(X1, X2, X3, X4)<- #@s1( @add(0.4133083, @add( @prod(0.30373776,X1), @add( @prod(0.763268,X2), @ad
```

Unfold program

● Lattice

```
1  % Elements
2  member(+infty).   member(-infty).   member(X) :- number(X).
3  members([0.95, 0.30, 0.60, -0.30, -0.95, -0.60]).
4
5  % Distance
6  distance(X,Y,Z) :-
```

Fig. 1. Screenshot of the input area of the FLOPER online tool.

As shown in Fig. 1, MALP always associates to each fuzzy program a lattice of truth degrees. It consists on a set of Prolog clauses intended to define three mandatory predicates for identifying the bottom and top elements of the lattice, as well as for testing its members (valid truth degrees). In our particular case, the following set of clauses is required:

```
bot(-infty).                          top(+infty).
member(+infty). member(-infty). member(X) :- number(X).
```

Moreover, it is also compulsory to define the ordering relation among truth degrees (in particular, it suffices by testing whether the first truth degree is less or equal than the second one, for each pair of elements) and computing the distance between each couple of truth degrees, being this last predicate especially important when applying our tuning techniques in the next section:

```
leq(_, +infty). leq(-infty, _). leq(X, Y) :- X =< Y.
distance(X,Y,Z) :- Z is abs(Y-X).
```

Next, the lattice is completed by adding as much definitions as needed for fuzzy connectives (also called aggregators, which in essence are operators acting on truth degrees) modeling activation functions in our case, as described by Prolog clauses such as the following one referring to the *softplus* function:

```
agr_softplus(+infty,+infty).
agr_softplus(-infty,-infty).
agr_softplus(X,Y) :- Y is log(1+exp(X)).
```

The way in which these functions are defined is very important in order to create a consistent lattice. So, it is necessary to define appropriate boundaries for the analytical formula describing each function. In this paper, we will work with ten popular activation functions, whose analytic definitions appear in Fig. 2.

$$@_{sigmoid}(x) \triangleq \frac{1}{1+e^{-x}} \qquad @_{softplus}(x) \triangleq ln(1+e^x)$$

$$@_{relu}(x) \triangleq \begin{cases} 0 \text{ if } x < 0 \\ x \text{ if } x \geq 0 \end{cases} \qquad @_{leaky\_relu}(x) \triangleq \begin{cases} 0.01x \text{ if } x < 0 \\ x \qquad \text{ if } x \geq 0 \end{cases}$$

$$@_{softsign}(x) \triangleq \frac{1}{1+|x|} \qquad @_{sinusoid}(x) \triangleq \begin{cases} 1 \qquad \text{ if } x = 0 \\ \frac{sin(x)}{x} \text{ if } x \neq 0 \end{cases}$$

$$@_{arctan}(x) \triangleq arctan(x) \qquad @_{tanh}(x) \triangleq tanh(x)$$

$$@_{binary}(x) \triangleq \begin{cases} 1 \text{ if } x > 0 \\ 0 \text{ if } x \leq 0 \end{cases} \qquad @_{identity}(x) \triangleq x$$

Fig. 2. Standard activation functions used in neural networks.

The final step of the translation procedure simply transforms all the content given by the intermediate data structure into MALP rules. Due to this fact, it is not necessary to explicitly declare a rule for each neuron in input nodes, so they will always be codified in the first hidden layer. Figure 3 shows the three patters followed by the rules modeling the network, which need to be adapted to the MALP syntax, as we are going to see in the following example.

Example 1. For the MLP trained with the Iris Dataset using Keras, the fourth neuron of the hidden layer can be represented as the following MALP rule, whose head and body are connected by the <- symbol and the names of connectives always start by @ (note, for instance, that $@_{prod}$ and $@_{add}$ obviously refer to the basic product and sum operators):

```
node_2_4(X1, X2, X3, X4) <-
    @relu(@add(0.0, @add(
        @prod(-0.39125273, X1), @add(
            @prod(0.4458459, X2), @add(
                @prod(-0.31539902, X3),
                @prod(-0.4348098, X4)))))).
```

This rule models a node that receives four inputs directly from the input layer. The activation function is *ReLU*, given by the aggregator $@_{relu}$. The first number, that is 0.0, defines the bias of the network, and the remaining ones that come along with each parent node are the weights for the different inputs of the neuron. For other neurons on inner layers, the corresponding MALP rules are quite similar to this one.

Finally, the following MALP rule (which has been shortened for readability reasons) codes a neuron in the output layer and it represents the concrete class under consideration, for instance, iris_setosa:

```
iris_setosa(X1, X2, X3, X4) <-
    @sigmoid(@add(-0.011225972,
        @add(@prod(1.4837127, node_2_3(X1,X2,X3,X4)),
        ...)))))).
```

Here, the activation function is the sigmoid and the bias for this neuron corresponds to the negative value -0.011225972.

3   Tuning Neural Networks Through Symbolic MALP Rules

As detailed in [8,9], it is possible to transform a MALP program into a symbolic one, called sMALP program in brief, by simply replacing truth degrees in program

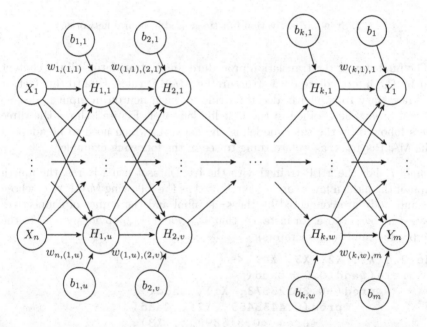

A) For each node $H_{1,i}$ (with activation function $f_{1,i}$) in the first hidden layer, for $1 \leq i \leq u$:
$$node_{1,i}(X_1, \ldots, X_n) \leftarrow f_{1,i}(b_{1,i} + w_{1,(1,i)} * X_1 + \cdots + w_{n,(1,i)} * X_n)$$
B) For each node $H_{j,i}$ (with activation function $f_{j,i}$) in any other hidden layer, for $2 \leq j \leq k$ and $1 \leq i \leq q(j)$, where $q(j)$ refers to the number of nodes in the hidden layer j:
$$node_{j,i}(X_1, \ldots, X_n) \leftarrow f_{j,i}(b_{j,i} + w_{(j-1,1),(j,i)} * node_{j-1,1}(X_1, \ldots, X_n) + \cdots + w_{(j-1,q(j-1)),(j,i)} * node_{j-1,q(j-1)}(X_1, \ldots, X_n))$$
C) For each class Y_i in the output layer (with activation function f_i), for $1 \leq i \leq m$:
$$class_i(X_1, \ldots, X_n) \leftarrow f_i(b_i + w_{(k,1),i} * node_{k,1}(X_1, \ldots, X_n) + \cdots + w_{(k,w),i} * node_{k,w}(X_1, \ldots, X_n))$$

Fig. 3. Rule schemes for coding a neural network.

rules by *symbolic constants* with shape *#identifier*. In order to evaluate these programs, in [8] we have introduced a symbolic operational semantics that delays the evaluation of symbolic expressions. Therefore, a *symbolic computed answer* could now include symbolic (unknown) truth values. We have proved the correctness of the approach, i.e., the fact that using the symbolic semantics and then replacing the unknown values by concrete ones gives the same result than replacing these values in the original sMALP program and, then, applying the concrete semantics on the resulting MALP program. Moreover, we show how symbolic programs can be used to tune a program w.r.t. a given set of test cases.

The precision of our tuning algorithm described in [8] and implemented in [9] depends on the set of truth degrees that can be associated to the symbolic constants in the sMALP program at tuning time (when trying to minimize the deviation between the computed answers and the test cases). Obviously, the larger the domain of values is, the more precise the results are (but the algorithm is more expensive, of course). This set is expressed as a list in the Prolog clause defining predicate *members* inside the lattice of truth degrees associated to the program. In the examples we are going to see in this section, we will try to guess the best symbolic substitutions with the following list of valid truth degrees:

```
members([0.95, 0.50, 0.25, -0.25, -0.50, -0.95]).
```

Our algorithm uses thresholding-based techniques and represents a much more efficient method for tuning the fuzzy parameters of a program than repeatedly executing the program from scratch. The results of an experimental evaluation reported in [8] (using a desktop computer equipped with an i3-2310M CPU @ 2.10 GHz and 4.00 GB RAM) considers some tuning examples dealing with 9000 different symbolic substitutions and 11 symbolic constants, where the problem is solved in just a few seconds.

Before illustrating some tuning examples we have worked with the MALP program obtained after translating the Keras model associated to the neural network seen in Example 1. As expected, we got an error of 103.82526 with the original neural network[1], which motivates us for tuning the original MALP program after introducing symbolic constants on it (thus becoming a sMALP program). In Fig. 4 we can see a screenshot of the online tool we have developed for testing our tuning tool freely available at https://dectau.uclm.es/floper/nn. In order to find the best substitution for the symbolic constants just introduced in the sMALP program, the first step before launching the tuning process consists of introducing a relevant set of test cases. In this paper we focus on the complete Iris Dataset. As shown in Fig. 4, each test case adopts the following shape:

```
1 -> iris_setosa(5.1, 3.5, 1.4, 0.2).
0 -> iris_versicolor(5.1, 3.5, 1.4, 0.2).
0 -> iris_virginica(5.1, 3.5, 1.4, 0.2).
```

---

[1] This error is a non-negative real number (zero in the best case) obtained as the sum of the deviations of all registers in the dataset used at training time w.r.t. the execution of the trained network with the same cases.

196   G. Moreno et al.

Running   Tuning
```

**✎ Test cases**

```
1  1.0 -> iris_setosa(5.1, 3.5, 1.4, 0.2).
2  0.0 -> iris_versicolor(5.1, 3.5, 1.4, 0.2).
3  0.0 -> iris_virginica(5.1, 3.5, 1.4, 0.2).
4  1.0 -> iris_setosa(4.9, 3.0, 1.4, 0.2).
5  0.0 -> iris_versicolor(4.9, 3.0, 1.4, 0.2).
6  0.0 -> iris_virginica(4.9, 3.0, 1.4, 0.2).
```

```
FASILL - Thresholded method ▾                    Tune
```

Output

Q Symbolic substitution

```
1  best symbolic substitution: {#@s1/@identity, #@s2/@sigmoid}
2  deviation: 98.16979776674235
3  execution time: 2768 milliseconds
```

Fig. 4. Screenshot of the FLOPER online tool after tuning a symbolic program.

This case means that an instance with parameters 5.1, 3.5, 1.4 and 0.2 always belongs to the class `iris_setosa`, and never belongs to the groups `iris_versicolor` and `iris_virginica`.

Example 2. In the MALP representation seen in Example 1, we find different activation functions inside the same layer and, after introducing the couple of symbolic aggregators $\#@s1$ (in all nodes of the second layer) and $\#@s2$ (in all nodes of the last layer), we obtain the following sMALP code:

```
node_2_4(X1, X2, X3, X4) <-
    #@s1(@add(0.0, @add(
        @prod(-0.39125273, X1), @add(
            @prod(0.4458459, X2), @add(
                @prod(-0.31539902, X3),
                @prod(-0.4348098, X4)))))).

iris_setosa(X1, X2, X3, X4) <-
    #@s2(@add(-0.011225972, @add(
        @prod(1.4837127, node_3_2(X1,X2,X3,X4)),
        ...)))))).
```

Note that this sMALP program differs from the initial MALP one by the fact that the original $@_{relu}$ and $@_{sigmoid}$ aggregators have been changed now by the symbolic aggregators $\#@s_1$ and $\#@s_2$, respectively. Next, the tuning process needs a few milliseconds to find the symbolic substitution $\{\#@s_1/@_{identity}, \#@s_2/@_{sigmoid}\}$, as seen in Fig. 4. Once applied this substitution to the symbolic program, we obtain a final MALP program whose further execution would decrease the error to 98.16823, which means an improvement of 5.45%.
```

Focusing on the second layer, note that in the tuned MLP we prefer the use of the identity activation function instead of the original $@_{relu}$ aggregator. This is due to the fact that, since some input values of the neuron can be negative, the identity function is able to produce better results than the sigmoid one.

4 Conclusions and Future Work

In this paper we have used the FLOPER system developed in our research group for designing a flexible application coping now with neural networks. After implementing an automatic technique for translating MLP's to the fuzzy logic language MALP, we have seen that some (random) weights/biases and activation functions of such neural networks can be safely tuned in order to satisfy users expectations according their proposed test cases.

With the online tool at https://dectau.uclm.es/fasill/sandbox?q=nn, we have just initiated a new research line in our group, where some pending task are: manipulating neural networks beyond the simpler MLP case, dealing with more activation functions, increasing the efficiency of the tuning method by using SAT/SMT solvers, experimenting with other real problems and so on.

References

1. Baldwin, J.F., Martin, T.P., Pilsworth, B.W.: Fril-Fuzzy and Evidential Reasoning in Artificial Intelligence. Wiley, Hoboken (1995)
2. Chollet, F., et al.: Keras: deep learning library for Theano and TensorFlow. **7**(8), T1 (2015). https://keras.io/k
3. Ishizuka, M., Kanai, N.: Prolog-elf incorporating fuzzy logic. New Gener. Comput. **3**(4), 479 (1985)
4. Julián-Iranzo, P., Moreno, G., Penabad, J.: Thresholded semantic framework for a fully integrated fuzzy logic language. J. Log. Algebraic Methods Program. **93**, 42–67 (2017)
5. Lee, R.C.T.: Fuzzy logic and the resolution principle. In: Readings in Fuzzy Sets for Intelligent systems, pp. 442–452. Elsevier (1993)
6. Van Der Malsburg, C.: Frank Rosenblatt: principles of neurodynamics: perceptrons and the theory of brain mechanisms. In: Palm, G., Aertsen, A. (eds.) Brain Theory, pp. 245–248. Springer, Berlin Heidelberg (1986). https://doi.org/10.1007/978-3-642-70911-1_20
7. Medina, J., Ojeda-Aciego, M., Vojtáš, P.: Similarity-based Unification: a multi-adjoint approach. Fuzzy Sets Syst. **146**, 43–62 (2004)
8. Moreno, G., Penabad, J., Riaza, J.A., Vidal, G.: Symbolic execution and thresholding for efficiently tuning fuzzy logic programs. In: Hermenegildo, M.V., Lopez-Garcia, P. (eds.) LOPSTR 2016. LNCS, vol. 10184, pp. 131–147. Springer, Cham (2017). https://doi.org/10.1007/978-3-319-63139-4_8
9. Moreno, G., Riaza, J.A.: An online tool for tuning fuzzy logic programs. In: Costantini, S., Franconi, E., Van Woensel, W., Kontchakov, R., Sadri, F., Roman, D. (eds.) RuleML+RR 2017. LNCS, vol. 10364, pp. 184–198. Springer, Cham (2017). https://doi.org/10.1007/978-3-319-61252-2_13

Querying Key-Value Stores Under Single-Key Constraints: Rewriting and Parallelization

Olivier Rodriguez$^{(\boxtimes)}$, Reza Akbarinia, and Federico Ulliana

INRIA & LIRMM, Univ. Montpellier, Montpellier, France
{olivier.rodriguez,reza.akbarinia,federico.ulliana}@inria.fr

Abstract. We consider the problem of querying key-value stores in the presence of semantic constraints, expressed as rules on keys, whose purpose is to establish a high-level view over a collection of legacy databases. We focus on the rewriting-based approach for data access, which is the most suitable for the key-value store setting because of the limited expressivity of the data model employed by such systems. Our main contribution is a parallel technique for rewriting and evaluating tree-shaped queries under constraints which is able to speed up query answering. We implemented and evaluated our parallel technique. Results show significant performance gains compared to the baseline sequential approach.

1 Introduction

Semantic constraints are knowledge on the structure and on the domain of data which are used in contexts such as data integration and ontology mediated query answering to establish a unified view of a collection of a database. Constraints allow users to better exploit their data thanks to the possibility of formulating high-level queries, which use a vocabulary richer than that of the single sources. In the last decade, the use of constraints in the form of ontologies has been intensively studied in the knowledge representation domain [3,4,10]. A key factor in the rise of the paradigm has been the reuse of off-the-shelf data management systems as the underlying physical layer for querying data under constraints. This resulted in a successful use of the paradigm especially on top of relational and RDF systems [5]. However, the use of constraints to query NOSQL systems like key-value stores (e.g., MongoDB [1], CouchDB [2]) has just begun to be investigated [6–8]. Key-value stores are designed to support data-intensive tasks on collections of JSON records, this last one being a language which is becoming the new de facto standard for data exchange.

To illustrate the use of semantic constraints for querying key-value records, consider the records in Example 1 which describe university departments. Query Q_1 selects all department records having a professor with some contact details. Query Q_2 selects all computer science departments with a director. It can be easily seen that these two queries do not match any of the records. Indeed, Q_1 asks for the key contact which is not used in both r_1 and r_2, while Q_2 asks, on

© Springer Nature Switzerland AG 2019
P. Fodor et al. (Eds.): RuleML+RR 2019, LNCS 11784, pp. 198–206, 2019.
https://doi.org/10.1007/978-3-030-31095-0_15

the one hand, for the key director, which is not used in r_1 and, on the other hand, for the value *"CS"* for the department name, which does not match that of r_2.

(r_1) { dept : {
        name : *"CS"* ,
        prof : { name : *"Bob"* ,
            mail : *"bob@uni.com"* } } }

(r_2) { dept : {
        name : *"Math"* ,
        director : { name : *"Alice"* ,
            phone : null } } }

(σ_1) phone $\rightarrow$ contact
(σ_2) mail $\rightarrow$ contact
(σ_3) director $\rightarrow$ prof
(σ_4) prof $\rightarrow$ $\exists$director

(Q_1) find({ dept : { prof : { contact : \$exists } } })

(Q_2) find({ dept : { name : *"CS"*, director : \$exists } })

Example 1. Data, queries, and rules.

This is where semantic constraints come into play. Indeed, although the key contact is not used in the records, this can be seen as a *high-level* key generalizing both phone and mail, as captured by rules σ_1 and σ_2. Therefore, by taking into account these semantic constraints, r_1 satisfies the query Q_1. Moreover, since σ_3 says that the director of a department must be a professor, also r_2 satisfies Q_2. Finally, σ_4 says that whenever a professor is present, then a director exists. Again, with this rule in hand, r_1 would also satisfy Q_2. This example outlines how semantic constraints allow users to better exploit their data.

The two main algorithmic approaches usually considered to account for semantic constraints during data access are materialization and query rewriting. Intuitively speaking, for constraints of the form k $\longrightarrow$ k', materialization means creating a *fresh copy* of the value of k and then associating this copy to the key k'. It is important to notice that, being the JSON data model based on trees, materialization can result in exponential blowups of the data. Also, not only it is computationally expensive to repeatedly copy subrecords, but it is also hard to efficiently implement such mechanism on top of key-value stores whose primitives, despite handling bulk record insertions, are not oriented towards the update of a single record. This is exacerbated by the fact that data is voluminous. In contrast, queries are usually small. From this perspective, it is thus interesting to explore query rewriting approaches that can take into account semantic constraints while accessing data without modifying the data sources. The idea of query rewriting is to propagate constraints "into the query". This process yields a set (or a union) of rewritings whose answers over the input database is exactly the same as the initial query on the database where materialization would have been done. Being rewritings independent from the sources, this approach is well suited for accessing legacy databases, in particular with read-only access rights.

The query facilities of key-value stores systems include primarily a language for selecting records matching several conditions based on tree-shaped queries called find() queries [1,2]. The MongoDB store also includes an expressive language for aggregate queries which is equivalent to nested relational algebra [9]. In this work, we focus our attention on the evaluation of find() queries under single-key constraints built on pairs of keys, as those of Example 1. It is worth noting that NOSQL systems still lack the standardization of a common query language and therefore of a standard syntax and semantics for queries. Therefore, in the formal development presented in Sect. 2, we chose to abstract away from the conventions of existing systems and adopt a syntax for queries akin to that of key-value records and a natural semantics based on tree-homomorphisms.

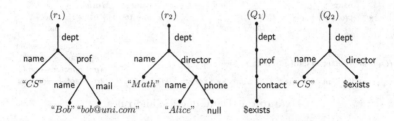

Fig. 1. Tree representation of records and queries of Example 1

In spite of the many advantages we already mentioned it is understood that, depending on the target language chosen for the rewritings, the query rewriting approach can suffer from combinatorial explosions - even for rule languages with a limited expressivity [4]. This happens as well for the key-value store languages. This means that the rewriting set of a query to generate may be large, which has a consequent impact on the slowing down of query answering.

To mitigate this problem, we present a novel technique for parallelizing both the generation and the evaluation of the rewriting set of a query serving as the basis for distributed query evaluation under constraints. Our solution is presented in Sect. 3, and relies on a schema for encoding the possible rewritings of a query on an integer interval. This allows us to generate equi-size partitions of rewritings, and thus to balance the load of the parallel working units that are in charge of generating and evaluating the queries. The experimental evaluation of our technique reported in Sect. 4 shows a significant reduction of query rewriting and execution time by means of parallelization.

2   Single-Key Constraints and Query Rewriting

This section is dedicated to the formalization of the setting we consider, and follows the lines of [8]. For concision, we will focus on a simplified JSON model. However, this is w.l.o.g, as the technique we present applies to the full language.

Data. A *key-value record* is a finite (non-empty) set of key-value pairs of the form $r = \{(k_1, v_1), \ldots, (k_n, v_n)\}$ where all k_i are distinct keys and each v_i is a value. A *value* is defined as *i*) a constant or the null value, *ii*) a record r, or *iii*) a non-empty sequence $v = [v_1 \ldots v_n]$ where each v_i is a constant, null, or a record. A value v in a key-value record can be associated with a *rooted labelled forest* T_v, where each tree of the forest has a root and nodes and edges can be labelled. If v is a constant or null then T_v is a single (root) node labelled by that value. If v is a sequence $v = [v_1 \ldots v_n]$ then T_v is a forest of n rooted trees $T_1, \ldots, T_n$ where T_i is the tree associated to the value v_i. Finally, if v is a record, then T_v is as follows. Let, k be a key of v and T_k be the forest associated to the value of k. Then T_v contains (i) all nodes and edges of each tree T_k, (ii) a root node s, and (iii) an edge from s to s_k labelled by k whenever s_k is the root of a tree in T_k. Clearly, T_v is a rooted tree whenever v is a record. In the remainder, we will see a key-value record as its associated tree, as illustrated in Fig. 1. Note the order of the elements of a sequence is not represented in the associated trees. Also, we will assume a fixed way to associate a tree with a unique (representative) record.

Rules. We focus on semantic constraints we call *single-keys*, also studied in [6,8]. These are expressed as rules σ of the form

$$\text{k} \longrightarrow \text{k}' \text{ (key inclusion)} \qquad \text{k} \longrightarrow \exists \text{k}' \text{ (mandatory key)}$$

enabling the definition of hierarchies of keys as well as the existence of mandatory keys. The semantics of constraints is defined on the tree associated to a record. Next, we denote by (u, w, k) an edge from u to w labelled by k. We say that a tree T satisfies a constraint σ: k $\longrightarrow$ k′ if for each of its edges of the form (u, w, k) there also exists an edge (u, z, k') and an isomorphism φ from the subtree rooted at w to that rooted at z such that $\varphi(w) = z$. Then, T satisfies σ: k $\longrightarrow$ $\exists$k′ if for each of its edges of the form (u, w, k) there also exists an edge (u, z, k'), whatever the subtree rooted at z. Let Σ be a set of constraints. Then, we say that a tree T is a model of r and Σ when (i) T_r and T have the same root, (ii) T_r is a subtree of T, and (iii) T satisfies all constraints of Σ. For single-key constraints, it can be easily shown that every pair (r, Σ) admits a finite model.

Queries. We consider the problem of answering find() queries that are integrated in the facilities of popular key-value stores such as MongoDB and CouchDB [1,2]. These queries select all records satisfying some structural and value conditions, and can be seen as Boolean queries, in that query evaluation on a record yields an answer which is either the empty set or the record itself. A query is thus of the form find(ϕ) where ϕ is a key-value record. Importantly, we assume that (1) queries do not use the null value and (2) the reserved constant \$exists is used to require the existence of any value associated with a key, as illustrated by Q_1 in Example 1. As for records, queries can be associated with labelled trees. Figure 1 illustrates the tree representation of the queries of Example 1.

Then, a query find(ϕ) answers *true* on a record r if there exists a mapping h from the nodes of T_ϕ to that of T_r such that (i) the root of T_ϕ is mapped to the root of T_r, (ii) for every edge (u, w, k) of T_ϕ, $(h(u), h(w), \text{k})$ is an edge of T_r, and (iii) every leaf node u of T_ϕ which is labelled by a constant different from \$exists is mapped to a node $h(u)$ with the same label. Finally, with constraints, a query find(ϕ) answers true on r and Σ if it answers true on all models of r and Σ.

Query Rewriting. Query rewriting is an algorithmic procedure for taking into account a set of semantic constraints Σ that starts from a query Q and produces a set of rewritings $Rew(Q, \Sigma)$ such that, for all records r, Q answers true on r and Σ if and only if there exists a query $Q' \in Rew(Q, \Sigma)$ that answers true on r. As for rules, we define the rewriting of a query find(ϕ) on its associated tree T_ϕ. So, find(ϕ) can be rewritten with σ: k $\rightarrow$ k′, if T_ϕ contains an edge (u, w, k'). Similarly, find(ϕ) can be rewritten with σ: k $\rightarrow$ $\exists$k′, if T_ϕ has an edge (u, w, k')

where w is a leaf node labelled by $exists. In both cases, the rewriting consists at replacing the edge (u, w, k') in T_ϕ with (u, w, k). Let $T_{\phi'}$ be the resulting tree, whose representative record is ϕ'. Then we say that find(ϕ') is a direct rewriting of find(ϕ) with σ. We denote by $Rew(Q, \Sigma)$ the set of Q' for which there exists a (possibly empty) sequence of direct rewritings from Q to Q' using the rules of Σ. This means that Q belongs to $Rew(Q, \Sigma)$. The size of $Rew(Q, \Sigma)$ is bounded by $|\Sigma|^{|Q|}$, where $|Q|$ is the number of edges of Q. The correctness of the rewriting algorithm can be shown by extending the proofs of [8].

	rewriting	*array*	*integer*
(Q_1^1)	find({ dept : { director : { contact : $exists } } })	$\langle 0, 1, 0 \rangle$	1
(Q_1^2)	find({ dept : { prof : { phone : $exists } } })	$\langle 0, 0, 1 \rangle$	2
(Q_1^3)	find({ dept : { director : { phone : $exists } } })	$\langle 0, 1, 1 \rangle$	3
(Q_1^4)	find({ dept : { prof : { mail : $exists } } })	$\langle 0, 0, 2 \rangle$	4
(Q_1^5)	find({ dept : { director : { mail : $exists } } })	$\langle 0, 1, 2 \rangle$	5
(Q_2^1)	find({ dept : { name : "*CS*" , prof : $exists } })	$\langle 0, 0, 1 \rangle$	1

Fig. 2. Rewritings of the queries in Example 1 (left) and their encoding (right)

Figure 2 illustrates the rewritings of queries given in Example 1. Here, $Q_1^1 - Q_1^5$ are rewritings of Q_1 with $\sigma_1, \sigma_2, \sigma_3$, while Q_2^1 is a direct rewriting of Q_2 with σ_4. It holds that Q_1^4 selects r_1, Q_1^3 selects r_2, and Q_2^1 selects r_1. Note that rules for mandatory keys apply only on the leaves of a query that are labelled with $exists. To see why consider the query find({dept : { director : "*Alice*" }}). Here, if σ_4 is used, we get find({dept : { prof : $exists }}) which is not a valid query rewriting.

3   Parallelization

We now present a parallel method that can be used to distribute both the generation and the evaluation of the rewriting set of a query to a set of independent computing units $u_1, \ldots, u_m$, each being a local thread or a machine of a cluster. Our approach relies on an *interval-encoding* of the rewritings. The general idea is to establish a bijection between $Rew(Q, \Sigma)$ and the integers in $[0, \ldots, N-1]$, where $N = |Rew(Q, \Sigma)|$. Then each of the m computing units is communicated an interval $[i, j]$ of size $\lambda \approx N/m$ corresponding to the subset of rewritings it has to generate. This will result in a parallel rewriting method that enjoys the following three properties.

(1) *partitioning*: no rewriting is computed twice by two distinct units
(2) *load balance*: the number of rewritings is equally distributed across all units
(3) *bounded-communication*: units receive a constant size interval information

Encoding Queries. In contrast to the general case [6,8], when considering single-key constraints, one can exploit the fact that, the rewriting process we described in the previous section, yields queries that are structurally similar. This enables a compact representation of (the edges of) a query as fixed size arrays, which we now describe. Let $\mathsf{find}(\phi)$ be a query. By fixing a total order on the edges of T_ϕ, we can see the query as an array $\langle k_1, \ldots, k_n \rangle$, where k_i is the key labelling the i-th edge of T_ϕ. Thus, to reconstruct a rewriting from an array it just suffices to replace the i-th edge of Q with the i-th key of the array. Moreover, given that an edge can be rewritten only in a finite number of ways, we can even use integers to denote the possible labels of the query edges. These ideas are the basis of the definition of an encoding function $\gamma_{Q,\Sigma}$ which is illustrated next.

Consider the query Q_1 and $\sigma_1, \sigma_2, \sigma_3$ of Example 1 yielding rewritings Q_1^1–Q_1^5 as in Fig. 2. For simplicity, assume the edges of Q_1 being ordered by depth. So the edges labelled by dept, prof, and contact are indexed by 1, 2, and 3, respectively. To begin, we represent the query Q_1 with $\langle 0,0,0 \rangle$ where the value 0 at position 1, 2, and 3, of the array denote the fact that no edge is rewritten. Then, the rewritings Q_1^2, Q_1^4 can be represented by the arrays $\langle 0,0,1 \rangle$, $\langle 0,0,2 \rangle$, denoting the fact that the edge labelled by contact has been rewritten either by phone or mail while the rewritings Q_1^1, Q_1^3, Q_1^5 can be represented by the arrays $\langle 0,1,0 \rangle$, $\langle 0,1,1 \rangle$ and $\langle 0,1,2 \rangle$ where prof is replaced by director and the edge labelled with contact is rewritten (or not) as before. The resulting encoding function is $\gamma_{Q,\Sigma} = \{(1,0,\mathsf{dept}), (2,0,\mathsf{prof}), (2,1,\mathsf{director}), (3,0,\mathsf{contact}), (3,1,\mathsf{phone}), (3,2,\mathsf{mail})\}$.

It is important to notice that at this point $\gamma_{Q,\Sigma}$ establishes a bijection from arrays to the rewritings of a query. The next step towards our goal of mapping rewritings to integers is to map the arrays encoding the rewritings to a sequence of successive integers. To do so, we see an array as a number in a multiple base $(b_1, \ldots, b_n)$ where each b_i denotes the number of possible rewritings of the i-th edge of Q. An array $\langle c_1 \ldots c_n \rangle$ in the base $(b_1, \ldots, b_n)$ corresponds to the integer $\mathbf{p} = c_1 + c_2 * B_1 \ldots + c_n * B_{n-1}$ with $B_1 = b_1$ and $B_i = b_i * B_{i-1}$ for $i \geq 2$. In the example, the arrays $\langle 0,0,0 \rangle, \langle 0,1,0 \rangle, \langle 0,0,1 \rangle, \langle 0,1,1 \rangle, \langle 0,0,2 \rangle, \langle 0,1,2 \rangle$ in base $(b_1, b_2, b_3) = (1, 2, 3)$ correspond to the integers in the interval $[0, 5]$, respectively. For instance, $\langle c_1, c_2, c_3 \rangle = \langle 0,1,1 \rangle$ correspond to the integer $\mathbf{3}$ as, given that $B_1 = b_1 = 1$ and $B_2 = 2$, we have $0 + 1 * B_1 + 1 * B_2 = \mathbf{3}$. Conversely, the integer $\mathbf{p}$ in base $(b_1, \ldots, b_n)$ corresponds to the array $\langle c_1, \ldots, c_n \rangle$ where $c_i = (d_i \bmod b_i)$ where $d_1 = \mathbf{p}$ and $d_i = (d_{i-1} \div b_{i-1})$ for all $i \geq 2$. Of course, it must be that $0 \leq \mathbf{p} < B_n$. The correspondence between rewritings and integers is outlined in Fig. 2. Finally note that by using the same formula we can compute the size of the rewriting set of a query, which is B_n, with n the number of edges of Q.

Building the Encoding Function. In the general case not only two rules σ_1 and σ_2 can rewrite the same edge of the query, but also the application of σ_1 can enable that of σ_2. Hence, the number of alternative keys for a single edge has to be inferred by looking at the dependencies between the keys in Σ. In doing so, we have to distinguish between the different types of edges of the query. For every edge of the query labelled by k the set of possible rewritings is made of all k' for which there exists a sequence of rules $\sigma_1, \ldots, \sigma_n$ of the form $\sigma_i = k'_i \longrightarrow k_i$ such that $k_i = k'_{i+1}$ for all $1 \leq i < n$, with $k'_1 = k'$ and $k_n = k$. For every edge of the query labelled by k ending on a node labelled by \$exists the set of possible rewritings is made of all k' for which there exists a sequence of rules this time either of the form $\sigma_i = k'_i \longrightarrow \exists k_i$ or $\sigma_i = k'_i \longrightarrow k_i$ satisfying the same condition as before. Note that it is possible to analyze Σ *independently of any query*, and therefore compute once the possible rewritings of a key depending on the cases described before. Then, the construction of $\gamma_{Q,\Sigma}$ follows by fixing any total order on the edges of T_Q. The size of $\gamma_{Q,\Sigma}$ is bounded by $|Q| \times |\Sigma|$. This avoids to communicate to the units the whole $Rew(Q, \Sigma)$, whose size can be exponential.

In conclusion, the key properties achieved with our interval encoding are that (1) we avoid a "centralized" enumeration of the rewritings (which is parallelized) and (2) minimize communication costs by sending to each unit only a pair of values (i, j) denoting a (possibly exponentially large) query set it has to handle.

4   Performance Evaluation

We implemented our approach in Java and parallelized query rewriting and evaluation by executing concurrent threads and using different cores of a machine. Nevertheless, our approach is suitable for any shared nothing parallel framework. For example, the threads can also be executed in the nodes of a distributed cluster, if such a cluster is available. The three main modules of our tool are dedicated to (i) the interval encoding, (ii) rewriting generation, and (iii) query evaluation. Next, we use the term *query answering* for the combination of the three tasks, which amounts to the whole task of answering queries under constraints.

We performed an experimental evaluation whose goal is to show the benefits of parallelization when querying key-value stores under semantic constraints. We deployed our tool on top of key-value store MongoDB version 3.6.3. Our experiments are based on the XMark benchmark which is a standard testing suite for semi-structured data [11]. XMark provides a document generator whose output was translated to obtain JSON records complying with our setting. Precisely, we performed our experiments on a key-value store instance created by shredding XMark generated data in JSON records. The results reported here concern an instance created from 100 MB XMark and split in $\sim$60 K records of size $\sim$1KB. XMark also provides a set of queries that were translated to our setting. To test query evaluation in the presence of constraints, we then extended the benchmark by manually adding a set of 68 rules on top of the data. These are "specialization" rules of the form $k_{new} \rightarrow k_{xmark}$ where k_{xmark} is a key of the XMark data vocabulary and k_{new} is a fresh key that does not appear in XMark. The benchmark data employs a vocabulary made of 91 keys and the rules define the specialization of 40 among them. More precisely, 20 keys have 1 specialization, 14 keys have two specializations, 5 have three specializations and 1 key has 5. Accordingly, the generated XMark data has been modified by randomly replacing one of such keys by one of its specializations thereby mimicking the fact that datasets use more specific keys while the user asks a high-level query.

Tests were performed on an Ubuntu 18.10 64-bit system, running on a machine that provides an Intel Core^{TM} i7-8650U CPU 4 cores, 16 GB of RAM, and an Intel SSD Pro 7600p Series. Figure 3 summarises (i) the query answering time under constraints and (ii) the speed up of our parallel technique for

10 XMark queries, by varying the number of threads. The speed-up is defined as the ratio between the case of 1 thread (*i.e.*, without parallelization) and the case with n threads. As expected, our results show that the query answering time depends on the size of the rewriting set, and the queries are thus sorted according to this criterion. Query answering time with *one thread* takes up to 1.3 s for queries with less 150 rewritings (i.e., q_4, q_{10}, q_1, q_2) and increases to 2.8 s for q_3, which has 324 rewritings. However, by using four threads, answering time for q_3 drops to $1.3s$ (55% time reduction). Answering q_7, which has 1296 rewritings, takes 11 s. This falls to 4.7 s by using four threads (58% time reduction). The same can be observed for q_8 and q_9. More generally, our results show that already by only using *two* threads, there is a 1.5x speedup (33% reduction) of query answering time for almost all queries. This increases to a 2/2.3x speedup (50–58% time reduction) when *four* threads are used. Interestingly, this is the maximum number of concurrent physical threads of our test machine, and we observe that when using eight virtual threads essentially no improvement can be further remarked. Naturally, when the number of rewritings of a query is too small, the impact of parallelization is less important. For example, as illustrated by q_4, which has only 18 rewritings, only a 1.2x speedup is achieved with four threads. Summing up, this shows the interest of parallelization in querying key value stores under semantic constraints.

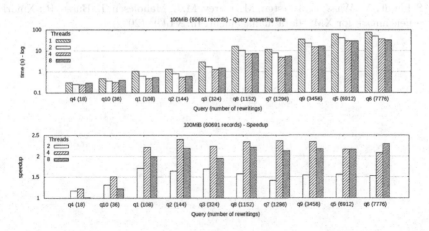

Fig. 3. Evaluation time and speedup of our method for XMark queries on MongoDB

Conclusion. In this paper, we proposed a parallel technique for the efficient rewriting and evaluation of tree-shaped queries under constraints based on an interval encoding of the rewriting set of a query. We implemented our solution and measured its performance using the XMark benchmark. The results show significant performance gains compared to the baseline sequential approach.

Acknowledgements. This work has been partially supported by the ANR CQFD Project (ANR-18-CE23-0003).

References

1. MongoDB. www.mongodb.com
2. CouchDB. couchdb.apache.org
3. Mugnier, M.-L., Thomazo, M.: An introduction to ontology-based query answering with existential rules. In: Koubarakis, M., et al. (eds.) Reasoning Web 2014. LNCS, vol. 8714, pp. 245–278. Springer, Cham (2014). https://doi.org/10.1007/978-3-319-10587-1_6
4. Calvanese, D., Giacomo, G.D., Lembo, D., Lenzerini, M., Rosati, R.: Tractable reasoning and efficient query answering in description logics: the DL-Lite family. J. Autom. Reason. **39**, 385–429 (2007)
5. Poggi, A., Lembo, D., Calvanese, D., De Giacomo, G., Lenzerini, M., Rosati, R.: Linking data to ontologies. J. Data Semant. (2008)
6. Mugnier, M., Rousset, M., Ulliana, F.: Ontology-mediated queries for NOSQL databases. In: AAAI (2016)
7. Botoeva, E., Calvanese, D., Cogrel, B., Rezk, M., Xiao, G.: OBDA beyond relational DBs: a study for MongoDB (2016)
8. Bienvenu, M., Bourhis, P., Mugnier, M., Tison, S., Ulliana, F.: Ontology-mediated query answering for key-value stores. In: IJCAI (2017)
9. Botoeva, E., Calvanese, D., Cogrel, B., Xiao, G.: Expressivity and complexity of MongoDB queries. In: ICDT (2018)
10. Xiao, G., et al.: Ontology-based data access: a survey. In: IJCAI (2018)
11. Schmidt, A., Waas, F., Kersten, M., Carey, M.J., Manolescu, I., Busse, R.: Xmark: a benchmark for XML data management. In: VLDB (2002)

Author Index